STP

NEW NÁTIONAL CURRICULUM
MATHEMATICS

11A

L. BOSTOCK B.Sc.

S. CHANDLER B.Sc.

A. SHEPHERD B.Sc.

E. SMITH M.Sc.

First published in 1998 by:
Stanley Thornes (Publishers) Ltd

Revised edition printed in 2002 by:
Nelson Thornes Ltd
Delta Place
27 Bath Road
CHELTENHAM
GL53 7TH
United Kingdom

02 03 04 05 06/ 10 9 8 7 6 5 4 3 2

A catalogue record of this book is available from the British Library.

ISBN 0-7487-6551-4

Illustrations by Linda Jeffrey and Peters & Zabransky (UK) Ltd
Page make-up by Tech-Set Ltd.

Printed and bound in Spain by GraphyCems.

Acknowledgements
The publishers are grateful to the following for permission to reproduce copyright
material:

Her Majesty's Stationery Office: extract from *The Highway Code* – p. 21; extract
from *Social Trends 22* – p. 144

The authors and publishers are grateful to the following examination boards for
permission to reproduce questions from their past examination papers. (Any answers
included have not been provided by the examining boards, they are the responsibility
of the authors and may not necessarily constitute the only possible solutions.)

    London Examinations, A division of Edexcel Foundation (Edexcel)
    Northern Examinations and Assessment Board (AQA)
    University of Cambridge Local Examinations Syndicate (OCR)
    The Associated Examining Board (AQA)
    Welsh Joint Education Committee (WJEC)

Front cover image produced using material kindly supplied by I LOVE LOVE CO,
makers of The Happy Cube © Laureyssens/Creative City Ltd 1986/91.
Distributed in UK by: RIGHTRAC, 119 Sandycombe Road, Richmond Surrey TW9
2ER
Tel: 0208 940 3322

CONTENTS

INTRODUCTION

This book continues to help you to learn, enjoy and progress through Mathematics in the National Curriculum. As well as a clear and concise text covering the remaining topics necessary to complete the syllabus, the book offers help in developing strategies for problem solving and investigational work, and a thorough revision programme of all the work necessary for the Higher Tier GCSE of the Revised National Curriculum.

The new work is followed by a substantial revision section in four parts that revises all the important topics you need to be familiar with. Each part concludes with an extensive exercise of past examination questions.

Several of the chapters dealing with the new work have a 'mixed exercise' after the main work of the chapter has been completed. This will help you to revise what you have done, either when you have finished the chapter or at a later date.

Everyone needs success and satisfaction in getting things right. With this in mind we have divided many of the exercises into three types of question.

The first type, identified by plain numbers, e.g. **15**, helps you to see if you understand the work. These questions are considered necessary for every chapter.

The second type, identified by an underline, e.g. **15**, are extra, but not harder, questions for quicker workers, for extra practice or for later revision.

The third type, identified by a coloured square, e.g. **15** , are for those of you who like a greater challenge.

You are probably now using a calculator for many of your calculations but it is still unwise to rely on a calculator for number work that can be done in your head or easily on paper. Remember, you will have a non-calculator paper as part of your GCSE examinations. Whether you use a calculator or do the working yourself, always estimate your answer and always ask yourself the question, 'Is my answer a sensible one?'

Mathematics is an exciting and enjoyable subject when you understand what is going on. Never forget, if you don't understand something, ask someone who can explain it to you. If you still don't understand ask again. Good luck with your studies.

To the teacher
This is the final book of the STP National Curriculum Mathematics series. It is based on the ST(P) Mathematics series but has been extensively rewritten and is firmly based on the latest Programme of Study for Key Stage 4. A significant part of the book is devoted to a thorough revision of the whole course.

Most scientific calculators now on sale use direct keying sequences for entering functions such as sin 47° and this is the order used in this book. Many calculators have the facility to solve quadratic equations; one way to discourage their use as a method of 'first resort' is to ask for answers in surd form.

The A series of books aims to prepare pupils for the higher tier at GCSE.

THE LANGUAGE OF MATHEMATICS

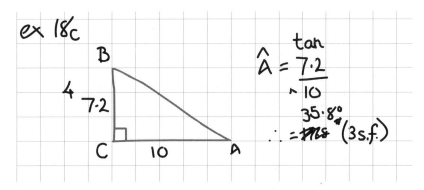

Look carefully at this extract from a student's book: angle A is *not* equal to tan $\frac{7.2}{10}$ and tan $\frac{7.2}{10}$ is *not* equal to $35.8°$.

It shows either muddled thinking or the attitude 'I know what I mean and the answer is right so what is the problem?' The problem is that it does not communicate with others so that *they* know *exactly* what you mean and can follow your reasoning; at the very least you need to be able to do this for GCSE examiners.

The ability to use clear, correct, unambiguous mathematical language and reasoning cannot be acquired overnight. It depends on a frame of mind which sees the need to do so and on knowing the meaning of the symbols used.

We begin by looking at examples that seem almost too trivial to bother with but they give an introduction to the rigour needed so that you can express yourself clearly and unambiguously.

The language of mathematics is a combination of words and symbols where each symbol is the shorthand form for a word or phrase. When the words and symbols are correctly used a piece of mathematical reasoning can be read, as prose can, in properly constructed sentences.

You have already used a fair number of symbols but not, perhaps, always with enough care for their precise meaning. As we now take a look at some familiar symbols we find that some can be translated correctly in more than one way.

THE USE AND MISUSE OF SYMBOLS AND WORDS

First consider the symbol '+'.
This can be read as 'plus' or 'and' or 'together with' or 'positive'.

For example, $3 + 2$ means 3 plus 2 or 3 and 2
and $3x + 2y$ means $3x$ together with $2y$
and $+(+2)$ means plus positive 2.

The symbol '−' has a similar variety of translations.

e.g. $5 - 4$ means 5 minus 4
     $5 - (-4)$ means 5 minus 'negative 4'.

Now consider '×' which can be read as 'multiplied by' or 'times' or 'of'.

e.g. 7×5 means 7 multiplied by 5 or 7 times 5 or 7 (lots) of 5

     $\frac{1}{100} \times x$ means one hundredth of x.

Note that '7×5' can be read as 'multiply 7 by 5',

The next sign we consider is '=' which means 'is equal to'. This symbol should be used *only* to link two quantities that are equal in value. Used in this way a short complete sentence is formed,

e.g. $x = 3$ is read as 'x is equal to 3'.

It is bad practice to use '=' in place of the word 'is'. For example, when defining a symbol such as the radius of a circle we should say 'the radius of the circle is r cm' and *not* 'the radius of the circle $= r$ cm', because r cm and the radius are not two separate quantities of equal value; r cm *stands for* the radius. If the radius is later found to be 4 cm, it is *then* correct to say '$r = 4$'.

EXERCISE 1A

In questions **1** to **6** we give some problems followed by solutions which, although ending with the correct answer, contain nonsense on the way. These solutions have been taken from actual students' work and are examples of very common misuses of language.
In each case criticise the solution, and write a correct version.

1 Simplify $2\frac{1}{2} + 1\frac{1}{4} - 2\frac{1}{3}$.

$$2\frac{1}{2} + 1\frac{1}{4} = 3\frac{3}{4} - 2\frac{1}{3}$$
$$= 1\frac{5}{12}$$

2 Solve the equation $6x + 5 = 3x + 11$.

$$6x + 5 = 3x + 11$$
$$= 3x + 5 = 11$$
$$= 3x = 6$$
$$= x = 2$$

3 Find $\widehat{A}$ when $\sin \widehat{A} = 0.5$.

$$\sin \widehat{A} = 0.5$$
$$= 30°$$

4 Write down the formula for the circumference of a circle.

The formula for the circumference of a circle is $2\pi r$.

5 Two angles of an isosceles triangle each measure $70°$. Find the size of the third angle.

$$\text{Third angle} = 70° + 70°$$
$$= 180° - 140°$$
$$= 40°$$

6 Three buns and two cakes cost $54\,\text{p}$ and five buns and one cake cost $62\,\text{p}$. Find the cost of one bun and of one cake.

$$\text{Buns} = x\,\text{p} \qquad \text{Cakes} = y\,\text{p}$$
$$3x + 2y = 54\,\text{p}$$
$$5x + y = 62\,\text{p}$$
$$10x + 2y = 124\,\text{p}$$
$$7x = 70\text{p}$$
$$x = 10\,\text{p} \quad \text{and} \quad y = 12\,\text{p}$$

Questions **7** to **11** are examples based on actual items culled from radio, television or newspapers. The numbers and names involved may have been changed. Comment on the use or misuse of mathematical language and, where possible, write a better version.

7 An article from the sport section of a newspaper.

'... shows he can give 110% of himself...'

8 From an advertisement.

'Pay five times less for your MOBILE PHONE!'

9 From a test for seven-year-olds.

'Add $4 + 3$, $5 + 5$ and $6 + 2$.'

10 From a national newspaper giving tables of results from Key Stage 2 tests.

The percentage of pupils achieving Level 3 or above in English, Mathematics and Science are added to give the overall percentage score for each school.

Brookfield	83%
Hindwood	126%
Abbeyfield	203%

11 From a national newspaper.

Dominco, a domestic appliance insurer has issued the following statement in an attempt to diffuse damaging publicity following Washco's admission that it pays commission to its service engineers who sell new washing machines. 'From the 100 000 policies currently insured only 0.1 percent have required a replacement machine in the first year.'

REASONING AND SYMBOLS THAT CONNECT STATEMENTS

The symbol '$\therefore$', meaning 'therefore', introduces a fact, complete in itself, which follows from a previous complete fact. It is correct to write

$$x^2 = 9$$
$$\therefore \quad x = \pm 3$$

We can also use the symbol '$\Rightarrow$' which means 'giving' or 'which gives', so it is correct to write

$$x^2 = 9$$
$$\Rightarrow \quad x = \pm 3$$

An alternative to the symbols '$\therefore$' and '$\Rightarrow$' is the abbreviation 'i.e.'. This is often used when statements give the same information but in a different form, as in

$$3 = x$$
$$\text{i.e.} \quad x = 3$$

It is not correct, however, to use $\therefore$, $\Rightarrow$ or i.e. to link $3x + 2x$ and $5x$. Each of these is simply an expression, not a complete fact or statement, and in this case we can link them by using '$=$',

that is, $\qquad\qquad\qquad 3x + 2x = 5x.$

As $3x + 2x$ and $5x$ are different ways of writing the same expression, we can also link them using the symbol '$\equiv$' which means 'is identical to',

that is, $$3x + 2x \equiv 5x.$$

The words and symbols mentioned so far are sufficient to write and read most mathematics at this level but variety can be added by using other words with the same meaning. The word 'therefore' can be replaced with any of the following: so, then, hence, thus.

EXERCISE 1B

In questions **1** to **4**, a problem is given together with a solution without any linking symbols. Rewrite the solutions using appropriate linking symbols.

1 Solve the equation $x^2 - 3x - 4 = 0$

$$x^2 - 3x - 4 = 0$$
$$(x - 4)(x + 1) = 0$$
$$x = 4$$
$$x = -1$$

2 Given that $a = b + c$, find a when $b = 2$ and $c = -1$.

When $b = 2$ and $c = -1$, $a = 2 + (-1)$
$$a = 2 - 1$$
$$a = 1$$

3 In $\triangle ABC$, $\widehat{A} = 90°$, $\widehat{B} = 30°$. Find $\widehat{C}$.

$$\widehat{A} + \widehat{B} = 90° + 30° = 120°$$
$$120° + \widehat{C} = 180°$$
$$\widehat{C} = 180° - 120° = 60°$$

4 In $\triangle ABC$, $\widehat{C} = 90°$, $AB = 3\,\text{cm}$ and $CB = 2\,\text{cm}$. Find $\widehat{A}$.

$$\sin \widehat{A} = \tfrac{2}{3}$$
$$\widehat{A} = 41.81\ldots°$$

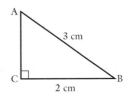

Note that 'angle A' can be abbreviated to either $\angle A$ **or** $\widehat{A}$.

5 Write the solution to this problem in words, not symbols, using just one sentence. Check your solution by reading it to make sure that it makes sense.

Apples cost 59 p per pound.
Find the cost of 3 pounds of these apples.

Giving reasons for statements that you make is another very important aspect of communication. Reasons should show the facts and processes involved and work that does not include them can be difficult if not impossible to follow.

Consider, for example, this incorrect solution to the problem in question **3** in the last exercise.

$$\widehat{A} + \widehat{B} = 90° + 30° = 120°$$
$$\widehat{C} = 100°$$

There is no reason given for why $\widehat{C} = 100°$ (which is wrong) so it is unclear whether the mistake comes from getting the arithmetic wrong or not knowing that the angles of a triangle add up to 180°. (In particular, GCSE examiners can give little credit for this solution so, if you are not already in the habit of giving reasons in mathematics, it is in your own interest to start now.)

Reasons for mathematical statements can usually be expressed very concisely provided that they are clear and unambiguous. They do not always need to be explicit; in the solution to question **3** given in the exercise, line 2 of the solution makes it clear that the fact that the angles of a triangle add up to 180° is being used.

Some people, however, try to be too brief with their reasons for steps in a solution.

In the solution of a pair of simultaneous equations, for example, we sometimes see

$$3x - 4y = 5 \qquad \times 2$$
$$2x + 7y = 13 \qquad \times 3$$
$$6x - 8y = 10$$
$$6x + 21y = 39$$

These four lines are disjointed, do not really explain what is happening and cannot be read sensibly in words. A better way to present this piece of work is as follows:

a
$$3x - 4y = 5 \qquad\qquad [1]$$
$$2x + 7y = 13 \qquad\qquad [2]$$
$$2 \times [1] \Rightarrow \qquad 6x - 8y = 10$$
$$3 \times [2] \Rightarrow \qquad 6x + 21y = 39$$

After the initial naming of the equations as [1] and [2] this now reads '2 times equation [1] gives $6x - 8y = 10$' and '3 times equation [2] gives $6x + 21y = 39$'.

This presentation is also clear:

b
$$3x - 4y = 5 \qquad [1]$$
$$2x + 7y = 13 \qquad [2]$$
$$6x - 8y = 10 \qquad (\text{ multiplying } [1] \text{ by } 2)$$
$$6x + 21y = 39 \qquad (\text{ multiplying } [2] \text{ by } 3)$$

This version too can be read:

$6x - 8y = 10$ comes from multiplying equation $[1]$ by 2, etc.

Notice that in **a** an instruction was given first, describing the operation to be carried out and leading to an equation, whereas in **b** the operation was carried out and then explained.

NECESSARY AND SUFFICIENT CONDITIONS

As well as giving reasons for deductions in a solution, it is obviously vital that those deductions are correct. A common mistake is to give insufficient conditions to make a deduction true.

For example, it is true that

$$\triangle ABC \text{ is equilateral} \Rightarrow AB = BC$$

but if we swap the statements round then

$$\text{'In } \triangle ABC, AB = BC\text{'} \Rightarrow \text{'}\triangle ABC \text{ is equilateral'}$$

is *not* true because, although $AB = BC$ is a necessary condition for $\triangle ABC$ to be equilateral, it is not sufficient. We also need the third side, AC, to be equal to the other two sides otherwise $\triangle ABC$ could be isosceles without being equilateral.

Alternatively, the statement 'In $\triangle ABC$, $\widehat{A} = \widehat{B} = 60°$' is another set of necessary and sufficient conditions for $\triangle ABC$ to be equilateral. (They are sufficient because $\widehat{A} = \widehat{B} = 60° \Rightarrow \widehat{C} = 60°$ as the sum of the interior angles in a triangle is $180°$.)

So either 'In $\triangle ABC$, $AB = BC = AC$' or 'In $\triangle ABC$, $\widehat{A} = \widehat{B} = 60°$' give necessary and sufficient conditions for $\triangle ABC$ to be equilateral.

Note that swapping the statements in a deduction, gives the *converse*,

i.e. the converse of '$\triangle ABC$ is equilateral' $\Rightarrow$ '$AB = BC$' is 'In $\triangle ABC$, $AB = BC$' $\Rightarrow$ '$\triangle ABC$ is equilateral' and, as we have seen, if a statement is true, its converse is not always true.

Note also that to negate the meaning of a connecting symbol we use a forward slash, e.g. $\not\Rightarrow$ means 'does not give' and $\neq$ means 'is not equal to' so we can correctly write

$$\text{'In } \triangle ABC, AB = BC\text{'} \not\Rightarrow \text{'}\triangle ABC \text{ is equilateral'}$$

Write out solutions to the problems in questions **1** to **5** using linking symbols and reasons where appropriate.

1 Solve the equations
$$2x + 3y = 7$$
$$8x - 5y = 11$$

2 Three of the angles in a quadrilateral are each of size $100°$. Find the size of the fourth angle.

3 In $\triangle ABC$, $\hat{B} = 90°$, $AB = 4\,cm$ and $BC = 3\,cm$. Find AC.

4 Use a sketch graph to explain why the equation $x^2 + 4 = 0$ has no solutions.

5 The price of a compact disk is £12.80 including VAT at $17\frac{1}{2}$%. Find the price excluding VAT.

In questions **6** to **10** say whether the deductions are correct. If they are not, give reasons why.

6 The four sides of quadrilateral ABCD are equal in length $\Rightarrow$ ABCD is a square.

7 Each angle of quadrilateral ABCD is $90° \Rightarrow$ ABCD is a square.

8 The diagonals of quadrilateral ABCD bisect each other $\Rightarrow$ ABCD is a parallelogram.

9 $x = 3 \Rightarrow x^2 = 9$

10 $x^2 = 9 \Rightarrow x = 3$

11 Write down a set of necessary and sufficient conditions for a quadrilateral ABCD to be a square.

12 Write down a set of necessary and sufficient conditions for the equation $ax^2 + bx + c = 0$ to have equal roots.

PROOF Problems which require a proof fall into two categories: proving that a statement is false or proving that a statement is true.

To prove that a statement is false
Proving that a statement is false is straightforward; all we have to do is to find an example which demonstrates that the statement is not true (this is called a counter example). Finding that example, however, is not always so easy – the main problem is that we often have a fixed idea of the type of example we are looking for.

Consider this example.

Show that the statement 'The sum of two irrational numbers is always irrational' is false.

If we start with the idea that we are looking for positive numbers and single terms, we will never find a counter example. Remember that an irrational number can be positive or negative and that $1 + \sqrt{2}$, for example, is irrational.

Now $(1 + \sqrt{2}) + (3 - \sqrt{2}) = 4$ which shows that the sum of two irrational numbers can be rational, i.e. the sum of two irrational numbers is *not* always irrational.

To prove that a statement is true

Proving that a statement is true requires a logical argument that convinces other people, so it is essential to give reasons for all the steps. It must also exclude the possibility that a counter example exists; this usually means that the argument must be framed in general terms. In the case of numbers, letters need to be used to represent those numbers, that is, algebra must be used. Remember that, unless you are lucky or know the proof, your first line of reasoning can often lead to a dead end. When this happens you clearly have to start again; but first identify why your reasoning went wrong and see what lessons you can learn from it.

Consider this example.

Prove that the sum of the squares of any two positive numbers is greater than their product.

If a and b are two numbers, we have to prove that

$$\text{for all positive values of } a \text{ and } b, \ a^2 + b^2 > ab$$

It is easier to prove that something is greater than zero than any other number, so we will start by rearranging the statement,

i.e. $\quad\quad\quad$ if $a^2 + b^2 > ab$, then $a^2 + b^2 - ab > 0$

Now we can consider the left-hand side: to prove that any given expression is greater than zero, it often helps to start with a perfect square.

We will try $(a + b)^2$ which we know is positive, i.e. greater than zero,

$$(a + b)^2 > 0 \quad \Rightarrow \quad a^2 + 2ab + b^2 > 0$$

Adding or subtracting the same quantity on both sides of an inequality does not change the inequality,

$\therefore \quad\quad\quad\quad\quad\quad a^2 + 2ab + b^2 - 3ab > -3ab$

i.e. $\quad\quad\quad\quad\quad\quad a^2 + b^2 - ab > -3ab$

but this is a dead end because it does not prove that $a^2 + b^2 - ab > 0$, only that it is greater than a negative number.

We got this negative number by starting with a square from which we had to subtract a quantity to give the expression required. We will now try a square to which we have to add an expression to give the form we want.

$(a-b)^2$ is positive, i.e. $(a-b)^2 > 0$

$\Rightarrow$ $a^2 - 2ab + b^2 > 0$

Adding ab to both sides gives $a^2 - 2ab + b^2 + ab > ab$

i.e. $a^2 + b^2 - ab > ab$

$\therefore$ $a^2 + b^2 - ab$ is greater than a positive number, so it is definitely greater than zero, i.e. $a^2 + b^2 - ab > 0 \Rightarrow a^2 + b^2 > ab.$

EXERCISE 1D

1 Sam writes the numbers 1 to 40 in the following rectangular array.

$$\begin{array}{cccccccc} 1 & 2 & 3 & 4 & 5 & 6 & 7 & 8 \\ 9 & 10 & 11 & 12 & 13 & 14 & 15 & 16 \\ 17 & 18 & 19 & 20 & 21 & 22 & 23 & 24 \\ 25 & 26 & 27 & 28 & 29 & 30 & 31 & 32 \\ 33 & 34 & 35 & 36 & 37 & 38 & 39 & 40 \end{array}$$

He selects the block

21	22
29	30

from the array and finds that the difference between the products of the numbers in the two diagonals is $22 \times 29 - 21 \times 30 = 8$.

a If the first entry in a block of four is x, copy and complete the block in terms of x.

b Hence prove that the difference in the products of the diagonals is always 8.

c Terri writes the numbers 1 to 40 in a rectangular array so that she has 1 to 5 in the first row, 6 to 10 in the second, and so on. If a block of four numbers is taken from this array write down the difference you would expect between the products of the diagonals. Prove your result.

2

	M	T	W	Th	F	S	S
	1	2	3	4	5	6	7
	8	9	10	11	12	13	14
	15	16	17	18	19	20	21
	22	23	24	25	26	27	28
	29	30	31				

This is a month from a calendar on which a block of nine numbers has been marked.

The difference between the products of the numbers in opposite corners is

$$24 \times 12 - 10 \times 26 = 28.$$

Prove that for every block of nine numbers that can be selected from this array the difference in the products of the numbers in opposite corners is always 28.

3 Consider the sequence $1, 5, 9, 13, 17, 21, \ldots$ Can we find the sum of the first 20 terms, or even the sum of the first 100 terms, without writing them all down first?

a Write down the sum of the first 10 terms and call this sum S_{10},

i.e. $S_{10} = 1 + 5 + 9 + 13 + 17 + 21 + 25 + 29 + 33 + 37$ [1]

Next write the same line again but with the terms on the right-hand side in the reverse order.

i.e. $S_{10} = 37 + 33 + 29 + 25 + 21 + 17 + 13 + 9 + 5 + 1$ [2]

  i What is the sum of each number in [1] and the number below it in [2]?

  ii How many pairs of numbers do you get?

  iii Hence show that the sum of the first 10 numbers in the sequence is 190.

b Use this method to find the sum of the first 20 terms without writing them all down.

c i Find an expression, in terms of n, for the nth term of the sequence.

  ii Hence find an expression, in terms of n, for the sum S_n of the first n terms of the given sequence.

4 Consider the sequence $1, 3, 9, 27, \ldots$

   a How do you find any term from the preceding term?

   b Find, as a power of 3,

      i the 8th term

      ii the 10th term.

   c If the sum of the first 6 terms is S_6

     i.e. $$S_6 = 1 + 3 + 3^2 + 3^3 + 3^4 + 3^5 \qquad [1]$$

     then $[1] \times 3$ gives $$3S_6 = 3 + 3^2 + 3^3 + 3^4 + 3^5 + 3^6 \qquad [2]$$

     Now $[2] - [1]$ gives $2S_6 = 3^6 - 1$

     i.e. $$S_6 = \frac{3^6 - 1}{2}$$

     Find similar expressions for **i** S_8, the sum of the first 8 terms

                                      ii S_{10}, the sum of the first 10 terms.

   d Use your calculator to find the exact value of **i** S_8 **ii** S_{10}.

   e **i** Find an expression, in terms of n, for the nth term of the sequence.

     ii Hence find an expression for S_n, the sum of the first n terms.

5 **a** Prove that the following statements are false.

     i The difference between two irrational numbers is irrational.

     ii If a and b are any two positive numbers, then $a^2 + b^2 > a + b$.

   b Prove that the following statements are true.

     i If n is an odd number, n^2 is also an odd number.

     ii The sum of any five consecutive numbers is a multiple of 5.

6 Determine which of the following statements are true and which are false. If a statement is true, prove it; if it is false give a counter example.

   a If n is a prime number, $n^2 + 1$ is an even number.

   b If n is an even number, n^2 is a multiple of 4.

   c The positive difference between a number n and the cube of that number is never odd.

   d If an irrational number is divided by a different irrational number the result is a rational number.

Footnote

You can now amuse yourself by finding examples of 'bad practice' in the use of language and symbols in this book!

MORE EQUATIONS

THE CIRCLE

If P is any point on this circle with coordinates (x, y) then, using Pythagoras' Theorem

$$x^2 + y^2 = OP^2$$
$$= 4^2$$

i.e. $x^2 + y^2 = 16$

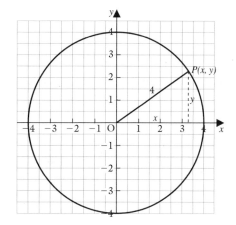

From this we can see that the equation of a circle, centre the origin, radius r units is

$$x^2 + y^2 = r^2$$

and that the equation $x^2 + y^2 = a^2$ represents a circle centre the origin, radius a.

For example, the equation $x^2 + y^2 = 100$ represents a circle whose centre is the origin and whose radius is $\sqrt{100}$, i.e. 10.

Similarly, to find the equation of the circle in the diagram below, we first need to find its radius.

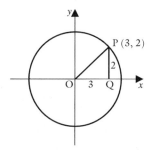

OP is a radius of the circle and we can use Pythagoras' Theorem in triangle OPQ to find its length.

$$OP^2 = 2^2 + 3^2 \quad \text{so} \quad OP = \sqrt{13}$$

Therefore the equation of the circle is

$$x^2 + y^2 = (\sqrt{13})^2$$

i.e. $x^2 + y^2 = 13$

When you draw the graph to represent a circle, it is important that the same scale is used on both axes, otherwise the circle will appear distorted.

1 Draw the circle, centre the origin, which has a radius of 2 units. What is the equation of this circle?

2 Repeat question 1 when the radius is 8 units.

Write each equation in the form $x^2 + y^2 = r^2$ then draw, on a set of axes, the circle whose equation is

3 $x^2 + y^2 = 9$ **5** $x^2 + y^2 = \frac{16}{9}$ **7** $4x^2 + 4y^2 = 49$

4 $x^2 + y^2 = 25$ **6** $x^2 + y^2 = \frac{25}{4}$ **8** $16x^2 + 16y^2 = 81$

GRAPHS OF CIRCLES AND STRAIGHT LINES

When two graphs intersect, the coordinates of the points of intersection give the solutions when the equations of the graphs are solved simultaneously.

a Draw the graph of $x^2 + y^2 = 25$.
b Draw, on the same axes, the graph of $2x - y + 3 = 0$.
c Write down the coordinates of the points where the graph intersect.
d Write down the simultaneous equations for which these values are the solutions.

a

> The graph of $x^2 + y^2 = 25$, i.e. $x^2 + y^2 = 5^2$, is a circle centre at the origin, radius 5.

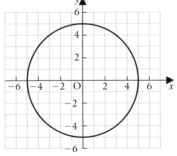

b $2x - y + 3 = 0$

so $2x + 3 = y$

i.e. $y = 2x + 3$

> Rearrange the equation $2x - y + 3 = 0$ in the form $y = mx + c$.

> The graph of $y = 2x + 3$ is a straight line so we need take only three values of x and find the corresponding values of y.

x		-4	-1	0
$y(= 2x + 3)$		-5	1	3

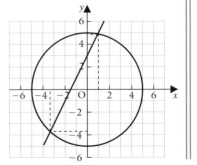

c The straight line intersects the circle at the points whose coordinates are $(0.9, 4.9)$ and $(-3.3, -3.7)$.

d The values given in part **c** are the solutions of the simultaneous equations

$$x^2 + y^2 = 25$$
$$2x - y + 3 = 0$$

For each graph write down

a the equation of the circle

b the equation of the straight line

c the coordinates of the points where the straight line and the circle intersect

d the pair of equations for which the coordinates you have given in part **c** are the solutions.

1

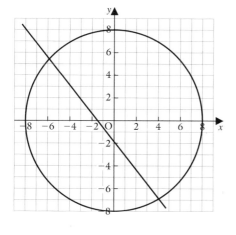

2

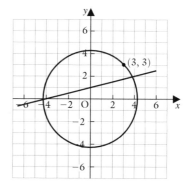

3

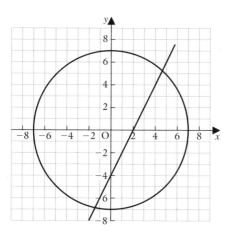

4

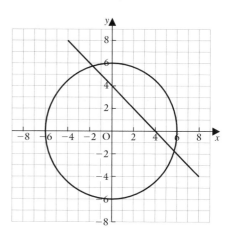

(3, 3)

5 a Draw the graph of $x^2 + y^2 = 4$.

b Draw, on the same axes, the graph of $y = x + 1$.

c Write down the coordinates of the points where the graphs intersect.

d Write down the simultaneous equations for which these values are the solutions.

6 a Draw the graph of $x^2 + y^2 = 36$.

b Draw, on the same axes, the graph of $y = 2x + 2$.

c Write down the coordinates of the points where the graphs intersect.

d Write down the simultaneous equations for which these values are the solutions.

7 a Draw the graph of $4x^2 + 4y^2 = 25$.

b Draw, on the same axes, the graph of $y = 1 - x$.

c Write down the coordinates of the points where the graphs intersect.

d Write down the simultaneous equations for which these values are the solutions.

SIMULTANEOUS EQUATIONS, ONE LINEAR, ONE QUADRATIC

We saw in the previous section that if we draw the graph of an equation of the form $x^2 + y^2 = r^2$ we get a circle, centre at the origin and radius r. If, further, the graph of the line $y = mx + c$ intersects this circle, the points of intersection give the solutions to the simultaneous equations

$$x^2 + y^2 = r^2$$
$$y = mx + c$$

Next we see how we can solve such pairs of equations algebraically.

For example, to solve the equations
$$x^2 + y^2 = 17$$
$$x + y = 3$$

it is clear that we cannot eliminate either x or y by addition or subtraction of the equations whatever we multiply them by.

However we can reduce these two equations to one equation containing one unknown by using the method of *substitution*.

We do this by using the linear equation to find one letter in terms of the other and then make a substitution in the more complicated equation.

EXERCISE 2C

Solve the pair of equations $x^2 + y^2 = 17, x + y = 3$

$$x^2 + y^2 = 17 \qquad\qquad [1]$$
$$x + y = 3 \qquad\qquad [2]$$

Using equation $[2]$ we can express either x or y in terms of the other variable and substitute this value in equation $[1]$. Here x and y are equally convenient to use.

From $[2]$ $\qquad\qquad x = 3 - y \qquad\qquad [3]$
so $\qquad\qquad x^2 = (3 - y)(3 - y)$
$\qquad\qquad\qquad\quad = 9 - 6y + y^2$

Substitute for x^2 in $[1]$

$$9 - 6y + y^2 + y^2 = 17$$
$$2y^2 - 6y + 9 = 17$$
$$2y^2 - 6y - 8 = 0$$
$$y^2 - 3y - 4 = 0$$
$$(y - 4)(y + 1) = 0$$

Either $y = 4$ or $y = -1$.

From $[3]$, when $y = 4$, $x = -1$
$\qquad\qquad$ when $y = -1$, $x = 4$

Check in [1] When $x = 4$ and $y = -1$, $x^2 + y^2 = 16 + 1 = 17$
$\qquad\qquad\qquad$ When $x = -1$ and $y = 4$, $x^2 + y^2 = 1 + 16 = 17$

The solution is $x = 4, y = -1$ or $x = -1, y = 4$.

Solve the following pairs of equations.

1 $x^2 + y^2 = 20, y = x + 2$ $\qquad$ **5** $x^2 + y^2 = 10, x - y = 4$

2 $x^2 + y^2 = 25, y = x + 1$ $\qquad$ **6** $x^2 + y^2 = 34, y - x = 2$

3 $x^2 + y^2 = 13, x + y = 5$ $\qquad$ **7** $x^2 + y^2 - 25 = 0, x - y = 1$

4 $x^2 + y^2 = 5, x + y = 3$ $\qquad$ **8** $x^2 + y^2 - 29 = 0, y = 7 - x$

Solve the following pairs of equations and illustrate your solutions with sketches.

9 $9 - x^2 = y^2, x + y = 3$ $\qquad$ **10** $16 - x^2 = y^2, x + y + 4 = 0$

Sometimes the equation of the straight line is more difficult to handle.

Solve the equations $x^2 + y^2 = 13, 2x + 3y = 13$

$$x^2 + y^2 = 13 \qquad [1]$$
$$2x + 3y = 13 \qquad [2]$$

From [2]
$$2x = 13 - 3y$$

$$x = \frac{13 - 3y}{2} \qquad [3]$$

$$x^2 = \frac{(13 - 3y)(13 - 3y)}{4}$$

$$= \frac{169 - 78y + 9y^2}{4}$$

Substitute in [1]
$$\frac{169 - 78y + 9y^2}{4} + y^2 = 13$$

Multiply both sides by 4 $\quad 169 - 78y + 9y^2 + 4y^2 = 52$
$$\Rightarrow \qquad 13y^2 - 78y + 117 = 0$$

Divide both sides by 13
$$y^2 - 6y + 9 = 0$$
$$\Rightarrow \qquad (y - 3)(y - 3) = 0$$

$$\therefore y = 3 \text{ (twice)}$$

From [3], when $y = 3$, $x = 2$

To interpret this graphically, we see that the line $2x + 3y = 13$ cuts the circle $x^2 + y^2 = 13$ in two coincident points, i.e. the line is a tangent to the circle.

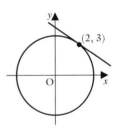

Solve the following pairs of equations.

11 $x^2 + y^2 = 10$
$\quad y = 2x + 1$

12 $x^2 + y^2 = 29$
$\quad y = 3x - 1$

13 $x^2 + y^2 = 10$
$\quad x + 2y = 1$

14 $x^2 + y^2 = 2$
$\quad 3x - 2y = 5$

15 $x^2 + y^2 = 4$
$\quad 2x + 3y = 4$

16 $x^2 + y^2 = 10$
$\quad x - y = 4$

Some quadratic equations do not give circles when we draw their graphs. Pairs of simultaneous equations such as $\quad 3xy + y^2 = 13$
$$y = x - 3$$
may be difficult to solve graphically but can be solved algebraically using the method given above. They can also be solved graphically but plotting some of the curves involved is not straightforward even with the help of a graphics calculator.

Solve the pair of simultaneous equations $\quad 3xy + y^2 = 13 \qquad$ [1]
$$y = x - 3 \qquad [2]$$

(It is simpler to substitute for x in [1], to avoid squaring.)

From [2] $\qquad\qquad\qquad\qquad x = y + 3 \qquad\qquad\qquad\qquad$ [3]

In [1] $\qquad\qquad\qquad\qquad 3y(y + 3) + y^2 = 13$
$$3y^2 + 9y + y^2 = 13$$
$$4y^2 + 9y - 13 = 0$$
$$(4y + 13)(y - 1) = 0$$

$\therefore\quad$ either $\quad 4y + 13 = 0$ or $y = 1$

i.e. $\qquad y = \dfrac{-13}{4}$ or $y = 1$

From [3], when $y = -3\frac{1}{4}$, $x = -3\frac{1}{4} + 3 = -\frac{1}{4}$
and $\qquad$ when $y = 1$, $x = 4$.

Check $\quad$ When $x = 4$ and $y = 1$, $3xy + y^2 = 12 + 1 = 13$

When $x = -\frac{1}{4}$ and $y = -3\frac{1}{4}$, $3xy + y^2 = \frac{39}{16} + \frac{169}{16} = 13$

The solution is $x = 4$, $y = 1$ or $x = -\frac{1}{4}$, $y = -3\frac{1}{4}$

Solve the following pairs of simultaneous equations.

17 $\quad 2xy + y = 10$
$\qquad x + y = 4$

18 $\quad x^2 - 2xy = 32$
$\qquad y = 2 - x$

19 $\quad 9 - x^2 = y^2$
$\qquad x + y = 3$

20 $\quad xy = 12$
$\qquad x + y = 7$

21 $\quad x^2 - y^2 = 0$
$\qquad 3x + 2y = 5$

22 $\quad xy = 6$
$\qquad 4x - 3y = 6$

23 $\quad x(x - y) = 6$
$\qquad x + y = 4$

24 $\quad x^2 + xy + y^2 = 7$
$\qquad 2x + y = 4$

25 $\quad xy + x = -3$
$\qquad 2x + 5y = 8$

26 $\quad x^2 = y + 3$
$\qquad 2x - 3y = 8$

USING A
GRAPHICS
CALCULATOR

It is possible to plot the graph of a circle on a graphics calculator, but as graphs to be drawn must be entered in the form $y = f(x)$, it has to be done with two equations.

For example, to plot the circle $x^2 + y^2 = 25$

first rearrange the equation as $y^2 = 25 - x^2$

which gives $\qquad\qquad\qquad\qquad y = \pm\sqrt{25 - x^2}$

so you need to enter two equations,

i.e. $y = \sqrt{25 - x^2}$ and $y = -\sqrt{25 - x^2}$

You then need to make the scales on both axes as near equal as possible to make the shape look like a circle.

To solve the equations $x^2 + y^2 = 25$ and $y = x - 1$, you will also need to add the graph of the line. Then you can use the trace function together with the zoom function to find the coordinates of the points of intersection.

If you have a graphics calculator, try this and compare the solution with that for question **7** in **Exercise 2C**.

VARIATION

Lee Ann is looking forward to learning to drive. To be ready to start she has been studying *The Highway Code*. One thing that strikes her immediately is that she needs to understand the significance of 'Shortest Stopping Distances'. These must be very important as they are displayed prominently on the back cover.

Shortest Stopping Distances

20mph   = 12m (40ft) or 3 car lengths
6m(20ft) 6m(20ft)

30mph   = 23m (75ft) or 5 car lengths
9m(30ft)   14m(45ft)

40mph   = 36m (120ft) or 9 car lengths
12m(40ft)   24m(80ft)

50mph   = 53m (175ft) or 13 car lengths
15m(50ft)   38m(125ft)

60mph   = 73m (240ft) or 18 car lengths
18m(60ft)   55m(180ft)

70mph   = 96m (315ft) or 24 car lengths
21m(70ft)   75m(245ft)
Thinking distance   Braking distance

It is very obvious that the faster a vehicle travels the greater the distance needed for it to brake to a halt. A good driver should always be aware of his speed and avoid being too close to the vehicle in front. Such a driver needs to be able to estimate the distance between his vehicle and the vehicle immediately ahead and needs to have an idea whether or not this distance is sufficient to stop in, should there be an emergency.

To be a safe driver Lee Ann needs to know

- how 'thinking distance' is related to the speed
- how 'braking distance' is related to the speed
- how 'shortest stopping distance' is related to the speed.

The work that follows will show any driver how to work out the way in which these quantities are related.

RELATIONSHIPS

Frequently, in everyday life, we come across two quantities that appear to be related to each other in some way. The amount I spend on potatoes, when they cost 60 p a bag, depends on the number of bags I buy; the distance I travel in a car, at a constant speed, depends on the length of time that I am travelling; the number of blank tapes I can buy for £30, depends on the price of one such tape. These are examples of quantities that are related by a simple algebraic equation.

On the other hand, there is no simple algebraic relationship between the amount a person earns and how much that person spends on food; between our weight and our height; or between how far we travel to school and the time we get up in the morning.

The first exercise in this chapter helps us to recognise some of the simple relationships that connect sets of varying quantities.

EXERCISE 3A

Write down the equation connecting the two variables given in the table.

x	2	3	5	10	12
y	6	9	15	30	36

We observe that in each case the value of y is three times the value of x.

$$y = 3x$$

In each of the following questions write down the equation connecting the variables.

1

x	1	2	4	7	10
y	3	6	12	21	30

2

p	0	1	2	3	4
q	0	1	4	9	16

3

x	1	2	3	4	5
V	1	8	27	64	125

4

A	0	4	9	16	25
r	0	2	3	4	5

5

x	2	4	6	24
y	12	6	4	1

6

r	0	2	4	6	10
s	0	0.2	0.4	0.6	1

7

x	-3	-1	0	2	4
y	36	4	0	16	64

8

p	-9	-6	-3	2	4
q	4	6	12	-18	-9

9 A set of rectangles is such that, in each one, the length is twice the breadth. Use the lengths given in the table and complete the table to give the area of each rectangle. What is the relationship between the area A and the length L?

Length of rectangle, L cm	2	4	5	6	8	10
Area of rectangle, A cm^2						

10 Copy the triangle given below on squared paper. Its base is 3 cm and its height is 2 cm.

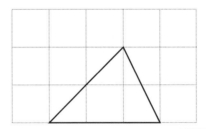

Draw three similar triangles with bases 6 cm, 9 cm and 12 cm whose heights will be 4 cm, 6 cm and 8 cm respectively. Find the area of each triangle. Use these values to complete the following table.

Base, b cm	3	6	9	12
Area, A cm^2				

What relationship connects A and b?

11 For this question imagine that you have a quantity of identical cubes. Cubes of sugar or Oxo cubes would be suitable.

Use these cubes to build bigger cubes whose edges are larger than the basic cube by factors of 2, 3, 4 and 5. It will help if you draw diagrams. For each of these cubes find how many times larger its volume is than the volume of the basic cube, i.e. how many of the smallest cubes are required to make each of the larger cubes.

Use your results to complete the following table.

Factor, x, by which the edge of the basic cube is multiplied	2	3	4	5
Factor, y, by which the volume of the basic cube is multiplied				

What is the relationship between x and y?

The questions in this exercise have illustrated several different ways in which varying quantities may be related. An increase in one quantity may lead to an increase or a decrease in the other.

DIRECT VARIATION

We have already seen in Book 9A, Chapter 4, that if two varying quantities are always in the same ratio, they are said to be directly proportional to one another.

Consider the total cost for a group of people to attend a concert. The varying costs, depending on the size of the group, are given in the table.

Number of people, N	5	10	15	25	35	50
Total cost in £, C	20	40	60	100	140	200

The two quantities C and N are connected by the equation $C = 4N$ i.e. the value of C is always four times the value of N. Therefore C is directly proportional to N.

This relation is called *direct proportion*.

In general, if two variables Y and X are in direct proportion then $Y = kX$ where k is the constant of proportion.

Alternatively we can say that Y *varies directly with* X.
This gives exactly the same equation,
i.e. $Y = kX$ where k is the constant of variation.

If we plotted the data given in the table on a graph, with N along the horizontal axis and C along the vertical axis the points would all lie on the straight line with equation $C = 4N$. Comparing the equation $C = 4N$ with the general equation $y = mx + c$, we see that the gradient of the line $C = 4N$ is 4, and this gradient represents the cost for one person to attend the concert. The fact that the straight line passes through the origin confirms that the cost if no people attend is £0.

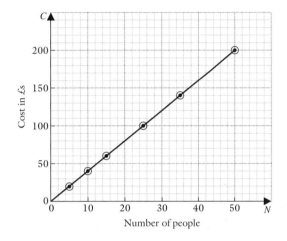

Number of people

The information in the table can also be written

Number of people attending : Cost in £s $= 1 : 4$

It is important to realise that we now have three different ways in which we can illustrate the relationship between the number of people attending the concert and the total cost. The notation we have introduced on page 24 is important for anyone going on to study mathematics further.

In mathematics we are always looking for shorter ways of writing things. Instead of writing 'y is directly proportional to x' or 'y varies directly as x' we sometimes write $y \propto x$, where '$\propto$' means 'varies as', from which we can write the equation $y = kx$ where k is some constant.

EXERCISE 3B

1 Copy and complete the table so that $y \propto x$.
What is the equation connecting x and y?

x	2		7	8	
y	20	40			95

2 Copy and complete the table so that $C \propto r$.
What is the equation connecting C and r?

r		3	5		8
C	6	18		36	48

3 Copy and complete the table so that $C \propto n$.

Number of units of electricity used, n	100	120	142	260	312	460
Total cost in pence, C	600		852	1560		2760

What meaning can you give to the constant of proportion?

4 Copy and complete the table so that $Y \propto X$.

Number of oranges bought, X	2	4	7	9	11	15
Total cost in pence, Y	20	40		90		150

What meaning can you give to the constant of variation?

If y varies directly as x and $y = 2$ when $x = 3$, find

a y when x is 9   **b** x when y is 18

$$y \propto x$$

i.e. $\quad\quad\quad\quad\quad\quad y = kx \quad$ where k is a constant

> We know that $y = 2$ when $x = 3$ so we can substitute 2 for y and 3 for x into $y = kx$

$\therefore \quad\quad\quad\quad\quad\quad 2 = k \times 3$

i.e. $\quad\quad\quad\quad\quad\quad 3k = 2 \quad\Rightarrow\quad k = \frac{2}{3}$

Hence $\quad\quad\quad\quad\quad y = \frac{2}{3}x$

a When $x = 9,\quad y = \frac{2}{3} \times 9$

$\quad\quad\quad\quad\quad\quad\quad = 6$

b When $y = 18,\quad 18 = \frac{2}{3}x$

i.e. $\quad\quad\quad\quad\quad\quad 54 = 2x$

$\Rightarrow \quad\quad\quad\quad\quad\quad 27 = x$

i.e. $\quad\quad\quad\quad\quad\quad x = 27$

---

5 y varies directly as x and $y = 21$ when $x = 7$.
Find   **a** y when $x = 3$ **b** x when $y = 48$.

6 $y \propto x$ and $y = 6$ when $x = 24$.
Find **a** y when $x = 6$ **b** x when $y = 5$.

7 s varies directly as t and $s = 35$ when $t = 5$.
Find **a** s when $t = 3$ **b** t when $s = 49$.

8 P is directly proportional to Q and $P = 15$ when $Q = 50$.
Find **a** P when Q is 70 **b** Q when P is 12.

9 W is directly proportional to S and $W = 8$ when S is 10.
Find **a** W when S is 30 **b** S when W is 12.

10 Y is directly proportional to X and $Y = 45$ when $X = 18$.
Find **a** Y when X is 6 **b** X when Y is 20.

11 y varies directly as $3x - 4$ and $y = 33$ when $x = 5$.
Find **a** y when $x = 2$ **b** x when $y = 15$.

DEPENDENT AND INDEPENDENT VARIABLES

Usually one of the quantities varies because of a change in the other. In question **3** opposite the total cost goes up because the number of units of electricity used goes up, that is, the variation in the first quantity, C, *depends* on the change in the other, n. C is called a *dependent variable* while n is referred to as an *independent variable*.

Similarly, when the radius of a circle increases, the area of the circle increases. Therefore the radius is the independent variable and the area is the dependent variable.

The dependent variable is sometimes proportional to a power of the independent variable. For example, the area of a circle, A, is directly proportional to the *square* of its radius, R, and we can write

$$A \propto R^2 \quad \text{or} \quad A = kR^2$$

(k has a special value in this case; can you say what it stands for?)

Similarly, if the safe speed, V, at which a car can round a bend varies as the square root of the radius of the bend, R, then

$$V \propto \sqrt{R} \quad \text{or} \quad V = k\sqrt{R}.$$

Copy and complete the table for positive values of x so that $y \propto x^2$.

x	1	3	5		11
y	2		50	98	

What is the equation connecting x and y?

Since $y \propto x^2$, $\qquad\qquad y = kx^2$

When $x = 1$, $y = 2$ so $2 = k \times 1^2$

i.e. $\qquad\qquad\qquad k = 2$

$\therefore \qquad\qquad\qquad y = 2x^2$

When $x = 3$, $\qquad y = 2 \times 9 = 18$

When $x = 11$, $\qquad y = 2 \times 121 = 242$

When $y = 98$, $\quad 98 = 2 \times x^2$

$\Rightarrow \qquad\qquad\qquad x^2 = 49$

$\therefore \qquad\qquad\qquad x = 7$

> We are only asked for the positive value so we discard $x = -7$

The completed table is

x	1	3	5	7	11
y	2	18	50	98	242

and the equation connecting x and y is $y = 2x^2$.

1 Copy and complete the table for positive values of x so that $y \propto x^2$.

x	0		3	4	5	
y		12	27		75	192

What is the equation connecting x and y?

2 Copy and complete the table for positive values of t so that $s \propto t^2$.

t	2	4		6	10
s		80	125	180	

What is the equation connecting s and t?

3 Copy and complete the table so that $y \propto x^2$.

x	-3	-1	0	2	4	7
y				16		196

What is the equation connecting x and y?

If y is directly proportional to the square of x and $y = 3$ when $x = 1$, find

a y when x is 4   **b** x when y is $\frac{3}{4}$.

$$y \propto x^2$$

i.e. $y = kx^2$ where k is a constant

But $y = 3$ when $x = 1$

∴ $3 = k \times 1^2$

i.e. $k = 3$

so $y = 3x^2$

a If $x = 4$, then $y = 3 \times 4^2$
$$= 3 \times 16$$
$$= 48$$

b If $y = \frac{3}{4}$, then $\frac{3}{4} = 3x^2$

$$x^2 = \frac{1}{4}$$ dividing both sides by 3

$$x = \pm \frac{1}{2}$$

---

4 y is directly proportional to the square of x and $y = 18$ when $x = 3$. Find

    a y when $x = 4$ **b** x when $y = 2$

5 y varies as the square of x and $y = 48$ when $x = 4$. Show that $y = 3x^2$ and find

    a y when $x = \frac{1}{2}$ **b** x when $y = \frac{1}{3}$

6 $P \propto Q^2$ and $P = 12$ when $Q = 4$.

    Find   **a** P when $Q = 12$

              b the positive value of Q when $P = 48$

If y is directly proportional to the cube of x and $y = 4$ when $x = 2$, find

a y when $x = 4$ **b** x when $y = \frac{1}{2}$.

$$y \propto x^3$$

i.e. $y = kx^3$

But $y = 4$ when $x = 2$

$\therefore$ $4 = k \times 2^3$

i.e. $8k = 4$

$\Rightarrow$ $k = \frac{1}{2}$

so $y = \frac{1}{2}x^3$

a If $x = 4$, $y = \frac{1}{2} \times 4^3$

$\qquad\qquad\qquad = 32$

b If $y = \frac{1}{2}$, $\frac{1}{2} = \frac{1}{2}x^3$

i.e. $x^3 = 1$

$\Rightarrow$ $x = 1$

7 Copy and complete the table so that $V \propto H^3$.

H	2		6	8	10
V	2	16	54		

What is the equation connecting V and H?

8 Copy and complete the table so that $y \propto x^3$.

x	3	6	9	12	15
y		72		576	

What is the equation connecting x and y?

9 $y \propto x^3$ and $y = 3$ when $x = 2$.

Find   **a** y when $x = 4$ **b** x when $y = 81$.

10 If y varies directly as the cube of x and $y = 64$ when $x = 2$,

find   **a** y when $x = 3$ **b** x when $y = 8$.

11 W is proportional to the cube of H and $W = 32$ when $H = 4$.

Find   **a** W when $H = 6$ **b** H when $W = 4$.

12 Copy and complete the table so that $V \propto \sqrt{R}$.

R	0	1	4		25
V			8	12	

What is the equation connecting V and R?

13 Y varies directly as the square root of X and $Y = 1$ when $X = 100$. Find **a** Y when $X = 400$ **b** X when $Y = 3$.

14 Plot the graph of y against x for the following data.

x	1	4	9	16	25
y	1	2	3	4	5

Is the graph a straight line? If it is not, complete the following table and plot the graph of y against $\sqrt{x}$.

$\sqrt{x}$					
y	1	2	3	4	5

Is this graph a straight line? What is the equation connecting x and y?

15 Plot the graph of y against x for the following data.

x	1	2	3	4	5
y	0.5	4	13.5	32	62.5

Is the graph a straight line? If it is not, plot the graphs of y against x^2 and y against x^3. Use your results to find the equation connecting x and y.

INVERSE VARIATION

In Book 9A, Chapter 4 we saw that when two quantities are inversely proportional their product remains constant.

Consider two varying quantities whose product is constant, for example the number of stamps of a particular value that Len can buy for 480 p. At 10 p each he can buy 48 while at 12 p each he can buy 40. Some of the varying numbers of stamps he can buy are listed in the table.

Cost of each stamp, c pence	60	48	40	20	12	10
Number, n, of stamps Len can buy for 480 p	8	10	12	24	40	48

The two quantities c and n are connected by the equation $n = \dfrac{480}{c}$ i.e. $nc = 480$.

We say that n is inversely proportional to c or n varies inversely as c.

Plotting the data given in the table on a graph gives the curve shown opposite. The shape of this curve is characteristic for any two variables that are inversely proportional.

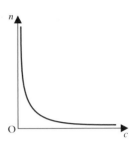

In general,

> if X and Y are inversely proportional,
>
> then $Y = \dfrac{k}{X}$ where k is the constant of proportion.
>
> Alternatively, we say that if Y varies inversely as X
>
> then $Y = \dfrac{k}{X}$ where k is the constant of variation.

Note that if we compare the equation $Y = \dfrac{k}{X}$ with the equation $y = mx + c$ then plotting Y against $\dfrac{1}{X}$ will give a straight line passing through the origin. This means that we can test whether or not two quantities are inversely proportional by plotting one quantity against the reciprocal of the other. If the points lie on a straight line that passes through the origin then one quantity varies inversely as the other, and the gradient of the line gives the constant of variation.

EXERCISE 3D

In question **1** to **3** complete the table given that the product of the varying quantities is constant. Write down the equation connecting these varying quantities.

1

Cost of a birthday card in pence, C	25	50	100	125
Number of cards that can be bought for £5, N	20		5	

2

Number of similar magazines a boy could buy with his pocket money, N	12	9	8	6
Cost of one magazine in pence, C	60		90	

3

Pressure, P pascals	4	5	6	8	12
Volume, V m^3	30		20		10

In questions **4** and **5** write down the equations connecting x and y.

4

x	36	24	18	12	8
y	2	3	4	6	9

5

x	0.8	0.9	1.2	1.8	2.7
y	2.7	2.4	1.8	1.2	0.8

x	1	2	3	4	6	12
y	12	6	4	3	2	1

a Write down the equation connecting x and y.

b Construct a table showing the values of $\dfrac{1}{x}$ and y.

c Plot the values of y against the corresponding values of $\dfrac{1}{x}$. Do these points lie on a straight line that passes through the origin? If they do, what does this confirm about the relationship between x and y? What is the gradient of this line? Interpret the gradient.

a $y = \dfrac{12}{x}$

b

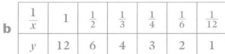

$\dfrac{1}{x}$	1	$\dfrac{1}{2}$	$\dfrac{1}{3}$	$\dfrac{1}{4}$	$\dfrac{1}{6}$	$\dfrac{1}{12}$
y	12	6	4	3	2	1

c
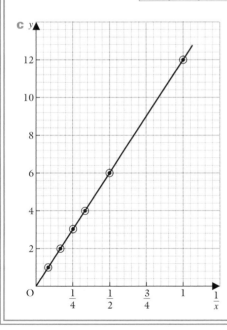

The points lie on a straight line passing through the origin. This confirms that y varies inversely as x.

Gradient of line $= \dfrac{\text{increase in } y}{\text{increase in } \dfrac{1}{x}}$

$= \dfrac{12}{1} = 12$

The gradient of the line is the constant of variation.

In questions **6** and **7** repeat the worked example for the data given in the table.

6

x	10	5	1	0.5	0.25
y	0.1	0.2	1	2	4

7

x	2.4	1.8	1.2	0.9	0.8
y	3	4	6	8	9

To summarise this work so far, if two quantities p and q are inversely proportional, we say that p varies inversely as q.

This relationship can be written as $p \propto \dfrac{1}{q}$, or as $p = \dfrac{k}{q}$ which implies that $pq = k$.

The idea can then be extended so that if two quantities x and y are such that y varies inversely as the square of x

then $\qquad\qquad\qquad\qquad y \propto \dfrac{1}{x^2} \quad \Rightarrow \quad y = \dfrac{k}{x^2}$

Similarly, y varies inversely as $\sqrt{x} \quad \Rightarrow \quad y \propto \dfrac{1}{\sqrt{x}} \quad \Rightarrow \quad y = \dfrac{k}{\sqrt{x}}$.

Copy and complete the table for positive values of x so that $y \propto \dfrac{1}{x^2}$

x	3	5		10	15
y	100	36	25		

If $y \propto \dfrac{1}{x^2}$ then $y = \dfrac{k}{x^2}$

But $y = 100$ when $x = 3$

$\therefore \ 100 = \dfrac{k}{9}$, i.e. $k = 900$ so $y = \dfrac{900}{x^2}$

Check when $x = 5$, $y = \dfrac{900}{25} = 36$

If $x = 10$, $y = \dfrac{900}{100} = 9$ and if $x = 15$, $y = \dfrac{900}{225} = 4$

If $y = 25$, $25 = \dfrac{900}{x^2}$

i.e. $25x^2 = 900 \quad \Rightarrow \quad x^2 = \dfrac{900}{25} = 36$

i.e. $x = 6$

$\therefore$ the completed table is

x	3	5	6	10	15
y	100	36	25	9	4

1 Copy and complete the table so that $y \propto \dfrac{1}{x}$.

x	2	4	6	9	12	
y	18	9			3	2

What is the equation connecting x and y?

2 Copy and complete the table for positive values of x so that $y \propto \dfrac{1}{x^2}$.

x	0.5		2	3	6	10
y		36	9			

What is the equation connecting x and y?

3 Copy and complete the table so that $q \propto \dfrac{1}{\sqrt{p}}$.

p	0.25		4	9		25
q	120	60	30		15	12

What is the equation connecting p and q?

4 If y is inversely proportional to x, and $y = 8$ when $x = 5$,

find   **a** y when x is 10   **b** x when y is 2   **c** y when x is -4.

5 $y \propto \dfrac{1}{\sqrt{x}}$ and $y = 2$ when $x = 4$.

Find   **a** y when $x = 9$ **b** x when $y = 1$.

6 If p is inversely proportional to v, and $p = 15$ when $v = 20$,

find   **a** p when $v = 30$ **b** v when $p = 7.5$.

7 If P varies inversely as $Q + 2$ and $P = 5$ when $Q = 4$

find   **a** P when $Q = 3$ **b** Q when $P = 15$.

8 If y is inversely proportional to the square of x, and $y = 4$ when $x = 5$,

find   **a** y when $x = 2$ **b** x when $y = 1$.

9 If y varies inversely as x and $y = 6$ when $x = 8$,

find   **a** y when $x = 12$ **b** x when $y = 4$.

10 If y varies inversely as the cube of x and $y = 7$ when $x = 6$,

find   **a** y when $x = 3$ **b** x when $y = 189$.

If y is the constant speed of a train and x is the time it takes to travel a fixed distance k, find the value of n if x and y are related by a law of the form $y \propto x^n$.

Since   distance travelled = constant speed × time

$$k = y \times x$$

i.e. $xy = k$

and $y = \dfrac{k}{x}$

or $y = kx^{-1}$

∴ $y \propto x^n$ where $n = -1$

11 In each of the following cases, x and y are related by a law of the form $y \propto x^n$. Find the value of n.

a y is the area of a square and x is the length of one side.

b y is the area of a circle and x is its radius.

c y is the volume of a sphere and x is its radius.

d y is the length of a rectangle of constant area and x is its breadth.

e y is the radius of a circle and x is its area.

f y is the length of a line in centimetres and x is its length in millimetres.

g y is the speed and x is the time taken to travel round one lap of a racing track.

**MIXED
EXAMPLES**

EXERCISE 3F

1 p varies directly as the square of q and $p = 9$ when $q = 6$.

Find p when q is   **a** 2   **b** −2   **c** 5.

2 A is directly proportional to L and $A = 28$ when $L = 4$.

Find   **a** A when $L = 3$ **b** L when $A = 42$.

3 $y \propto x^3$ and $y = 48$ when $x = 4$.

Find   **a** the formula for y in terms of x

b y when $x = 2$

c x when $y = 6$.

4 y varies inversely as x and $y = 7$ when $x = 6$.

Find   **a** y when $x = 3$ **b** x when $y = 14$.

5 y is inversely proportional to x^2 and $y = 4.5$ when $x = 4$.

Find   **a** y when $x = 3$ **b** x when $y = 8$.

6 R is directly proportional to the positive square root of S and $R = 4$ when $S = 64$.

a Calculate R when S is   **i** 16   **ii** 6.25

b Calculate S when R is   **i** 2   **ii** 3.5

7 Copy and complete the table so that $y \propto x^2$.

x	0	1		4	8
y		0.25	1		16

8 Copy and complete the table so that $t \propto \sqrt{s}$.

s		4	9		
t	0	0.5		1	2

9 Given that $y \propto \sqrt{x}$

a what is the effect on y if

   i x is multiplied by 4   **ii** x is divided by 4?

b what is the effect on x if

   i y is multiplied by 3   **ii** y is divided by 3?

10 Given that y varies inversely as x^2

a what is the effect on y if

   i x is multiplied by 2   **ii** x is divided by 2?

b what is the effect on x if

   i y is multiplied by 9   **ii** y is divided by 9?

11 Given that y varies as x^n, write down the value of n in each of the following cases.

a y is the area of a square of side x

b y is the volume of a cube of edge x

c y is the volume of a cylinder with constant base area A and height x

d y and x are the sides of a rectangle with a given area.

A stone falls from rest down a mine shaft. It falls D metres in T seconds where D varies as the square of T. If it falls 20 m in the first 2 seconds and takes 5 seconds to reach the bottom, how deep is the shaft?

$$D \propto T^2$$

i.e. $$D = kT^2$$

But $D = 20$ when $T = 2$

$\therefore$ $$20 = k \times 2^2$$

i.e. $$4k = 20$$

$$k = 5$$

$\therefore$ $$D = 5T^2$$

If $T = 5$, $D = 5 \times 5^2$

$$= 125$$

Therefore the shaft is 125 metres deep.

12 The mass, M kg, of a circular disc of constant thickness varies as the square of its radius, R cm. If a disc of radius 5 cm has a mass of 1 kilogram find

a the mass of a disc of radius 10 cm

b the radius of a disc of mass 25 kg.

13 The safe speed, V km/h, at which a car can round a bend of radius R metres varies as $\sqrt{R}$. If the safe speed on a curve of radius 25 m is 40 km/h, find the radius of the curve for which the safe speed is 64 km/h.

14 The time of swing, T seconds, of a simple pendulum is directly proportional to the square root of its length, L cm.
If $T = 2$ when $L = 100$ find

a T when $L = 64$ **b** L when $T = 1\frac{1}{2}$.

15 The extension, x cm, of an elastic string varies as the force, F newtons, used to extend it. If a force of 4 newtons gives an extension of 10 cm find

a the extension given by a force of 10 newtons

b the force required to give an extension of 12 cm.

16 The cost of buying a rectangular carpet is directly proportional to the square of its longer side. If a carpet whose longer side is 3 m costs £180 find

  a the cost of a carpet with a longer side of 4 m

  b the length of the longer side of a carpet costing £405.

17 The radius of a circle, r cm, varies as the square root of its area, A cm^2. How does the radius change if the area is increased by

  a a factor of 4   **b** a factor of 25   **c** 44% ?

18 Mathematically similar jugs have capacities that vary as the cubes of their heights. If a jug 10 cm high holds $\frac{1}{8}$ litre find

  a the capacity of a jug that is 12 cm high

  b the height of a jug that will hold 1 litre.

19 For a given mass of gas at a given temperature the pressure p varies inversely as the volume, v. If $p = 100$ when $v = 2.4$ find

  a v when $p = 80$ **b** p when $v = 2$.

20 For a vehicle travelling between two motorway service stations the time taken is inversely proportional to its speed. If it takes $2\frac{1}{2}$ hours when its speed is 48 mph find

  a its average speed if it takes 3 hours

  b by how much its average speed must increase if the journey time is to be reduced to 2 hours.

21 The number, n, of plastic squares of edge x cm, and with a fixed thickness, that can be made from a given volume of plastic can be found using the formula $n = \dfrac{k}{x^2}$ where k is a constant.

  a Given that 1000 squares of side 2 cm can be made from a given volume of the material, calculate how many squares can be made from an equal volume of material if the edge of the square is

   i 4 cm   **ii** 10 cm   **iii** 3 cm?

  b Rearrange $n = \dfrac{k}{x^2}$ to make x the subject.

  c 800 squares are to be made using the same volume of plastic. Calculate, correct to two decimal places, the length of an edge of the largest square. State, giving reasons, whether you have rounded your answer up or down.

22 The energy generated by a solar panel varies directly as the area of the panel. At midday on a Saturday in December a rectangular panel measuring 1.2 m by 0.8 m produced 0.84 units of energy.

  a How many units of energy would be produced by a solar panel measuring   **i** 1.8 m by 0.8 m   **ii** 2 m by 1.4 m?

  b The solar panels supplied by one manufacturer are all 1 m wide but can be made to any length. What length of panel is needed if the amount of energy produced is   **i** 1.4 units   **ii** 1.05 units?

23 The volume of perfume, $V\,\text{cm}^3$, contained in a perfume bottle varies as the cube of the height of the bottle, $h\,\text{mm}$. The graph representing the way that V varies with h for a collection of these perfume bottles could be

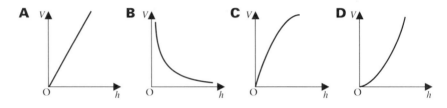

24 By what factor does y change when x is doubled if

  a $y \propto x$ **b** $y \propto \dfrac{1}{x}$ **c** $y \propto x^2$ **d** $y \propto x^3$

25 State the percentage change in y when x is increased by 25% if

  a $y \propto x$ **b** $y \propto \dfrac{1}{x}$ **c** $y \propto x^2$

26 Consider now the problem stated at the beginning of the chapter.

  a How is 'Thinking distance' in feet related to speed in miles per hour?

  b How is 'Braking distance' in feet related to speed in miles per hour?

  c Use your relationships to work out the total stopping distance when the speed is   **i** 45 mph   **ii** 100 mph

  d Find the speed, in mph, when the total stopping distance is 280 ft.

  e What is the rough length, in metres, of a car?

  f When the Highway Code was first written all distances were in feet and speeds were in mph. In recent years it has been modified so that distances are also given in metres. Do the figures in metres relate accurately to the figures in feet? Justify your answer. Why do you think both metric and imperial units are given for distance but only miles per hour is given for speed?

  g Make a new table giving the speeds in kilometres per hour and the distances in metres. (Start at 20 km/h and go up 30 km/h at a time until you get to 140 km/h.)

This is an experiment to find out how the time of swing of a simple pendulum is related to the length of the pendulum.

You will need

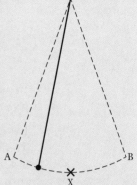

a length of fine string about 1 metre long
a small heavy mass to fix to one end of the string
a means of suspending the other end of the string from a fixed point (A ruler and some Blu Tack should help you to do this.)
a stopwatch which measures time in seconds.

Attach the pendulum to a fixed point and measure its length, L centimetres.

Start the pendulum swinging and measure the time it takes to pass a fixed point X marked behind the pendulum, say 50 times. From this you can calculate the time it takes to swing from one extreme position A to the other extreme position B and back again. Suppose it takes T seconds.

Repeat the experiment for different lengths of string and collect your results in the following table.

Length of pendulum, L cm						
Time of swing, T seconds						

Draw four graphs by plotting

a T against L **b** T against $\dfrac{1}{L}$ **c** T against L^2 **d** T against $\sqrt{L}$

Use your graphs to decide, with reasons, which of the following statements is true.

The time of swing of a simple pendulum varies directly as its length.
The time of swing of a simple pendulum varies inversely as its length.
The time of swing of a simple pendulum varies directly as the square of its length.
The time of swing of a simple pendulum varies directly as the square root of its length.

Repeat using different masses. Does it make any difference to the time of swing? Use an appropriate graph to find the time of swing for a given length or the length for a given time of swing.

FUNCTIONS

In mathematics we often use a rule that can be applied to one number to give another number.

For example, consider the line whose equation is $y = 2x - 6$.
To find the y-coordinate of the point on this line whose x-coordinate is 4, the right-hand side of the equation tells us we have to double 4 and then subtract 6; i.e. when $x = 4$, $y = 2(4) - 6 = 2$.

We can think of '$2x - 6$' as the rule that operates on the x-coordinate of a point on this line to give the y-coordinate of the point.

A similar situation arises with sequences,
e.g. when the nth term of a sequence is $n^2 + 2$, we can find the sixth term by squaring 6 and adding 2, i.e. 6th term $= 6^2 + 2 = 38$.

In this case, '$n^2 + 2$' is the rule that uses the position number of a term to give the value of that term.

Because we use such rules in a variety of different contexts, we need language and notation to describe the rules.

DEFINITIONS AND NOTATION

A rule that operates on one number to give another number is called a *function*.

The function 'halve' can be illustrated by this flow chart.

The letter f is used to represent a function.
The result of applying a function to x is written $f(x)$, read as 'f of x'.

Hence, when f is the function 'halve' we write $f(x) = \frac{1}{2}x$.
Similarly, if $f(x) = x^2$, then the function is 'square'.

Also, if $f(x) = 2x - 5$, then $f(3)$ means the result of applying f to the number 3, i.e. $f(3) = 2(3) - 5 = 1$

The set of numbers that we put into a function is called the *domain* of the function.

The set of numbers that result from using the function is called the *range* of the function.

1 Describe the given functions in words

a $f(x) = 2x$

c $f(x) = \dfrac{1}{x}$

b $f(x) = x + 2$

d $f(x) = x^3$

If $f(x) = \frac{1}{2}x^2 - 4$ find $f(-2)$.

$f(x) = \frac{1}{2}x^2 - 4$

$f(-2) = \frac{1}{2}(-2)^2 - 4$

$f(-2)$ means the value of $\frac{1}{2}x^2 - 4$ when $x = -2$

$= \frac{1}{2}(4) - 4 = -2$

---

2 Given that $f(x) = 2x - 1$ find

a $f(0)$ **b** $f(4)$ **c** $f(-1)$ **d** $f(-5)$

3 If $f(x) = 3 - 4x$ find

a $f(0)$ **b** $f(1)$ **c** $f(-1)$ **d** $f(-\frac{1}{2})$

4 If $f(x) = (1 - x)^2$ find

a $f(1)$ **b** $f(0)$ **c** $f(-1)$ **d** $f(-2)$

5 If $f(x) = x^2 - 2x + 3$ find

a $f(4)$ **b** $f(0)$ **c** $f(3)$ **d** $f(-1)$

6 The function f is defined by $f(x) = \dfrac{3}{1 + x}$, $x \neq -1$. Find

a $f(2)$ **b** $f(-2)$ **c** $f(0)$ **d** $f(-4)$

7 The function f is defined by $f(x) = 2x - \dfrac{4}{x}$, $x \neq 0$. Find

a $f(2)$ **b** $f(4)$ **c** $f(-1)$ **d** $f(-8)$

Explain why this function cannot be applied to zero.

8 For $f(x) = 3x^3 - 4x$ find

a $f(3)$ **b** $f(1)$ **c** $f(-1)$ **d** $f(-2)$

> If $f(x) = x^2 - 4$ find the values of x for which $f(x) = 5$.
>
> $$f(x) = x^2 - 4$$
>
> If $f(x) = 5$ then $\qquad x^2 - 4 = 5$
>
> $$x^2 = 9$$
>
> $$x = \pm\sqrt{9}$$
>
> i.e. $\qquad\qquad\qquad x = 3 \ \text{ or } \ x = -3.$

In each question from **9** to **17** find the value(s) of x for which the given function has the given value.

9 $f(x) = 5x - 4, f(x) = 2$

10 $f(x) = \dfrac{1}{x}, f(x) = 5$

11 $f(x) = 3 - x, f(x) = -4$

12 $f(x) = 2x + 1, f(x) = -9$

13 $f(x) = x^2, f(x) = 9$

14 $f(x) = \dfrac{1}{x^2}, f(x) = 1$

15 $f(x) = x^2 - 2x, f(x) = 3$

16 $f(x) = x + \dfrac{1}{x}, f(x) = 2$

17 $f(x) = (x + 1)(x - 2), \quad f(x) = 0$

18 If $f(x) = x^2 + x - 6,$ find

   a $f(1)$

   b $f(-1)$

   c the values of x for which $f(x) = 0$.

19 If $f(x) = x^2 - 2x - 15,$ find

   a $f(2)$

   b $f(-2)$

   c the values of x for which   **i** $f(x) = 0$ **ii** $f(x) = 9$

20 Given that $f(x) = x^3 - 8$ find

   a $f(0)$

   b $f(1)$

   c the value of x for which $f(x) = 0$

   d the value of x for which $f(x) = 1$.

21 Find the value of k if $f(3) = 3$ where

    a $f(x) = kx - 1$ **b** $f(x) = x^2 - k$.

22 The function f is defined by $f(x) = kx^2 - 5$

    a If $k = 2$, find   **i** $f(-3)$ **ii** $f(1)$

    b If $f(4) = 11$, find k.

23 The function f is defined by $f(x) = 6 - x$.

    a Find $f(-2)$, $f(0)$ and $f(4)$.

    b Is $f(x)$ increasing or decreasing in value as x increases in value?

    c Give the value of x for which $f(x) = 0$.

    d Give the range of values of x for which $f(x)$ is negative.

24 Given that $f(x) = x^2 + 2$

    a Find $f(0)$, $f(2)$, $f(4)$, $f(10)$

    b Is $f(x)$ increasing or decreasing in value as x increases in value from zero?

    c Find $f(0)$, $f(-1)$, $f(-3)$, $f(-10)$

    d What is the least possible value for $f(x)$? Justify your answer.

GRAPHICAL REPRESENTATION

Consider the function $f(x) = 4x - 2$.

If we write $y = f(x)$ then the equation $y = 4x - 2$ can be used to give a graphical representation of $f(x)$.

Now $y = 4x - 2$ is the equation of a straight line whose gradient is 4 and which cuts the y-axis where $y = -2$.

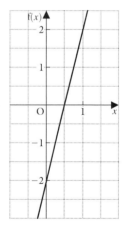

The vertical axis can be labelled either $f(x)$ or y, since $y = f(x)$.
A graphical representation like this gives a picture of how $f(x)$ changes in value as the value of x varies.
In this case the graph shows that, as x increases in value the corresponding values of $f(x)$ also increase.

For questions **1** to **10**, sketch a graph to represent the function and give the range of values of x for which

a $f(x) < 0$ **b** $f(x)$ is increasing as x increases.

1 $f(x) = 2x$

2 $f(x) = x + 1$

3 $f(x) = x^2$

4 $f(x) = \dfrac{1}{x}, \quad x \neq 0$

5 $f(x) = x^3$

6 $f(x) = 1 - x$

7 $f(x) = 1 - x^2$

8 $f(x) = \dfrac{2}{x^2}, \quad x \neq 0$

9 $f(x) = (x - 1)(x - 2)$

10 $f(x) = x^3 + 2$

11 On the same set of axes draw a sketch graph of

a $y = x$

b $y = x + 1$

c $y = x - 2$

Describe the transformation that maps the graph of part **a** to the graph of part **b**.

12 On the same set of axes draw a sketch of the graph of

a $y = x^2$

b $y = x^2 + 1$

c $y = x^2 - 2$

Describe the transformation that maps the graph of part **a** to the graph of part **c**.

13 Repeat question **12** for the following curves for values of $x > 0$.

a $y = \dfrac{1}{x}$ **b** $y = \dfrac{1}{x} + 1$ **c** $y = \dfrac{1}{x} - 1$

Describe the transformation that maps the graph of part **a** to the graph of part **c**.

14 On the same set of axes sketch the graph of

a $y = x^2$

b $y = -x^2$

Describe the transformation that maps the curve for part **a** to the curve for part **b**.

15 Repeat question **14** for the graphs of

a $y = x - 1$

b $y = 1 - x$

16 Repeat question **14** for the graphs of

    a $y = x^2 + 2$ **b** $y = -x^2 - 2$

17 On the same set of axes sketch the graphs of

    a $y = x^3 + 3$ **b** $y = (-x)^3 + 3$

18 On the same set of axes sketch the graphs of

    a $y = x^2$ **b** $y = (x-2)^2$ **c** $y = (x+3)^2$

    Explain how you can use the graph of $y = x^2$ to sketch the graph
of $y = (x-a)^2$.

TRANSFORMATIONS OF GRAPHS

The previous exercise has shown that if the graph of $y = f(x)$ is
known we can obtain sketches of the graphs of several related
equations. These are now considered in greater detail.

THE GRAPH OF $y = f(x) + k$

Consider the curves whose equations are

$$y = f(x) \text{ and } y = f(x) + k \text{ where } k \text{ is a number.}$$

Questions **11** to **13** of the last exercise demonstrate that

> the curve $y = f(x) + k$ is the translation parallel to the y-axis
> of the curve $y = f(x)$.
>
> If k is positive the translation is k units upwards
> and if k is negative the translation is k units downwards.

This fact can be very useful when sketching curves.

For example, to sketch the curve
$y = x^3 - 4$, we can start with the
known shape and position of
$y = x^3$ and then move it 4 units
downwards.

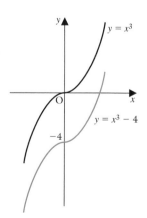

THE GRAPH OF
$y = f(x - a)$

Question **18** of the last exercise demonstrates that

> the curve $y = f(x - a)$ is given by translating the curve $y = f(x)$ a distance a in the positive direction of the x-axis. For $y = f(x + a)$, the graph moves to the left.

Hence to sketch the graph of $y = (x - 2)^3$ we start with a sketch of $y = x^3$, then move it 2 units to the right.

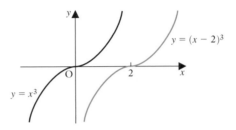

Similarly, if $y = (x + 3)^2$ i.e. $y = (x - \{-3\})^2$, we start by sketching $y = x^2$ and then slide it 3 units to the left as shown.

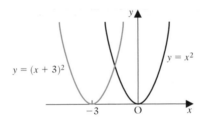

THE GRAPH OF
$y = -f(x)$

Questions **14** to **16** of the last exercise demonstrate that

> the curve $y = -f(x)$ is the reflection in the x-axis of the curve $y = f(x)$.

Hence to sketch the curve
$y = -\dfrac{1}{x}$, we can start with
the known curve $y = \dfrac{1}{x}$
and reflect it in the x-axis.

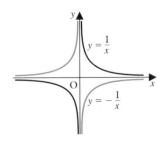

THE GRAPH OF
$y = f(-x)$

If $\quad f(x) = (x - 2)(x - 1)$

then $\quad f(-x) = (-x - 2)(-x - 1) = -(x + 2) \times -(x + 1)$

$\qquad\qquad = (x + 2)(x + 1)$

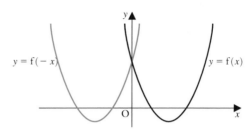

The sketch, together with question **17** of the last exercise, demonstrate that

$y = f(-x)$ is the reflection of $y = f(x)$ in the y-axis.

THE GRAPH OF
$y = af(x)$

We sketch the graph of $y = af(x)$ by stretching the graph $y = f(x)$ by a factor of a in the y direction.

To sketch the graph of $y = 2x^3$,
first sketch the graph of $y = x^3$
and then, for each value of x,
double the y value.

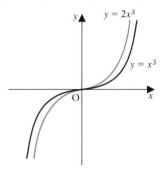

THE GRAPH OF
$y = f(ax)$

For $a > 1$, we sketch the graph of $y = f(ax)$ by reducing the width of the graph of $y = f(x)$ to $\dfrac{1}{a}$ of its original width.

To sketch the graph of $y = (3x)^2$,
first sketch the graph of $y = x^2$
and reduce the width of it to $\frac{1}{3}$ of
its original width.

If $a < 1$, we stretch the graph of
$y = f(x)$ by a factor $\dfrac{1}{a}$ parallel to
the x-axis, e.g. $y = \left(\frac{1}{2}x\right)^2$ is twice
as wide as $y = x^2$

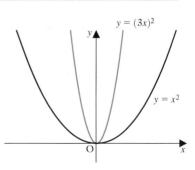

To draw a sketch graph, start by using a ruler to draw x- and y-axes but do not scale them. Show clearly all the significant features of a graph, such as where it crosses the axes.

1 This is the graph of $y = x^2$. Make four sketches of this graph. Use these, one at a time, to sketch the graphs of

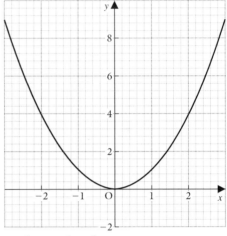

   a $y = x^2 - 3$

   b $y = (x-1)^2$

   c $y = 2 - x^2$

   d $y = 2x^2$

2 Sketch, on separate diagrams, the graphs of

   a $y = 2x$ **c** $y = 1 - 2x$

   b $y = 2x - 1$ **d** $y = 3(1 - 2x)$

3 On separate axes, sketch the graphs of

   a $y = x^2 + 5$ **c** $y = 4 - x^2$

   b $y = x^2 - 4$ **d** $y = 2(x^2 + 5)$

4 On separate axes, sketch the graphs of

   a $y = x^2 - 9$ **c** $y = (x - 3)^2$

   b $y = 9 - x^2$ **d** $y = (x - 3)^2 + 3$

5 **a** Write the expression $x^2 + 4x + 6$ in the form $(x + a)^2 + b$.

   b i Use the result from part **a** to sketch the graph of
      $y = x^2 + 4x + 6$.

   ii What is the minimum value of y?

6 Sketch, on separate axes, the graphs of

   a $y = x^3$ **d** $y = (x + 1)^3$

   b $y = -x^3$ **e** $y = x^3 + 3$

   c $y = 8 - x^3$ **f** $y = \frac{1}{2}x^3$

7 This is a sketch of the graph of $f(x) = 2^x$. Copy this sketch and, on the same diagram, sketch the graph of

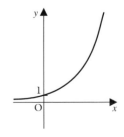

   a $f(x) = -2^x$ **b** $f(x) = 1 - 2^x$

8 This is a sketch of the graph
of $f(x) = 4x - x^2$.
Copy this sketch and extend both axes.
On the same diagram, sketch the
graph of

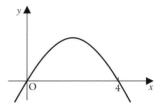

a $f(x) = x^2 - 4x$

b $f(x) = x^2 - 4x - 4$

c $f(x) = 4(2x) - (2x)^2$

9 Complete the square for $f(x) = x^2 - 6x + 3$ (i.e. express
$x^2 - 6x + 3$ in the form $(x + a)^2 + b$. Hence sketch the graph of
$y = f(x)$ and write down the minimum value of y.

10 The sketch shows the graph of $y = f(x)$.

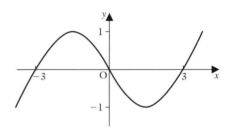

On separate diagrams draw sketches of

a $y = -f(x)$ **d** $y = f(x - 3)$

b $y = f(-x)$ **e** $y = f(3x)$

c $y = f(x + 3)$ **f** $y = 2f(x)$

11 Each of the curves given below is a transformation of the curve
$y = x^3$. Write down the letter corresponding to the equation of the
curve.

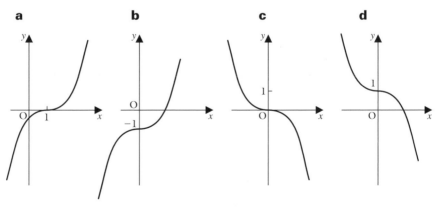

a   **b**   **c**   **d**

A $y = 1 - x^3$ **B** $y = (x - 1)^3$ **C** $y = -x^3$ **D** $y = x^3 - 1$

12 Each curve is a transformation of the curve $y = x^2$. Describe the transformation and give the equation of the transformed curve.

a   b   c   d

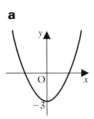 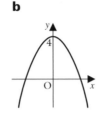 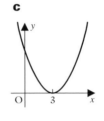 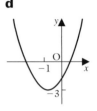

13 This is a sketch of the graph $y = f(x)$ where $f(x) = (1 - x)(x - 2)(x - 3)$.

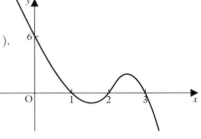

a Copy the diagram and, on the same set of axes, draw the graph of $y = f(-x)$.

b Write down the value of x for which $f(x) = f(-x)$.

c Draw the reflection of the curve in the x-axis and write down its equation.

14 This is a sketch of the function $f(x)$ for $-2 < x < 2$. Copy the diagram and, on separate sets of axes, sketch the graph of
a $f(x + 1)$ **b** $f(2x)$.

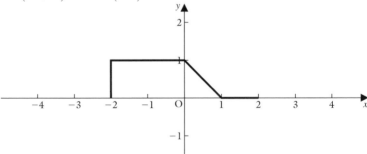

15 a Sketch the curve whose equation is $y = f(x)$ where
$$f(x) = (x + 2)(x + 1)(x - 2)$$

b On the same set of axes, sketch the curve whose equation is $y = f(-x)$.

c How many solutions are there to the equation $f(x) = f(-x)$? Show that one of these solutions is between -1 and 1 and write down the others.

CATEGORIES OF FUNCTION

We know about several categories of function:

any function whose graph is a straight line is a *linear function*, e.g. the function f, where $f(x) = 2x - 3$ is a linear function;

a function f, where $f(x) = ax^2 + bx + c$ is called a *quadratic function*, and the curve representing the function $f(x) = ax^2 + bx + c$ has a distinctive shape (a parabola) with a line of symmetry,

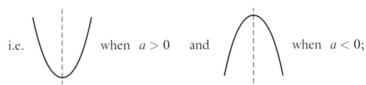

i.e.   when $a > 0$ and   when $a < 0$;

a function of the form $f(x) = ax^3 + bx^2 + cx + d$ is a *cubic function*,

e.g. if $f(x) = x(x - 1)(x - 2)$
then its graph looks like this:

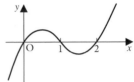

We now look at some functions we have not previously met.

TRIGONOMETRIC FUNCTIONS

We first met sines, cosines and tangents, when dealing with angles in right-angled triangles and so tend to think of these trigonometric ratios in relation to acute angles only. However, $\sin x°$ has a value for any value of x and so has $\cos x°$. Use your calculator to find, say, $\sin 240°, \cos 515°$.

EXERCISE 4D

1 Copy and complete the following table, using a calculator to find $\sin x°$ correct to 2 decimal places for each value of x from 0 to 360 at intervals of 15.

x	0	15	30	45	60	75	90	105	120	135	150	165	180	195	210	...	345	360
$\sin x°$																		

Use the values in your table to draw the graph of $y = \sin x°$ for $0 \leqslant x \leqslant 360$ and $-1 \leqslant y \leqslant 1$. Use a scale of 2 cm for one unit on the y-axis and 1 cm for 60 units on the x-axis.

2 Construct another table, using the same values of x as given in question **1**, for values of $\cos x°$ correct to 2 decimal places, i.e.

x	0	15	30	...	345	360
$\cos x°$	1	0.97				

Use the values in your table to draw the graph of $y = \cos x°$. Use the same ranges and scales as in question **1**.

THE GRAPH OF
$f(x) = \sin x°$

The graph drawn for question **1** in the last exercise should look like this. It is the graph of the function $f(x) = \sin x°$ for values of x between 0 and 360.

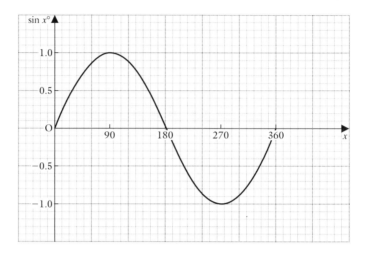

The curve has a distinctive shape: it is called a sine wave.

THE GRAPH OF
$f(x) = \cos x°$

The graph drawn for question **2** in the last exercise should look like this. It is the graph of $f(x) = \cos x°$ for values of x between 0 and 360.

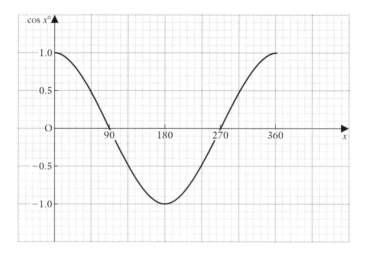

Notice that the cosine curve looks quite different from the sine curve for values of x in the range $0 \leqslant x \leqslant 360$.

However, if both curves are drawn for a larger range of values of x, we find that there is relationship between the curves. Questions **6** and **7** in the next exercise explore this relationship further.

1 a Sketch the curve $y = \sin x°$ for values of x from 0 to 360.

b On the same diagram sketch the curve $y = -\sin x°$.

c On the same diagram sketch the curve $y = 1 + \sin x°$.

2 a Sketch the curve $y = \cos x°$ for values of x from 0 to 360.

b On the same diagram sketch the curves

i $y = 1 + \cos x°$

ii $y = -\cos x°$.

3 a Draw a sketch of the curve $f(x) = \sin x°$ for $0 \leqslant x \leqslant 360$.

b From your sketch, find the values of x for which $f(x) = 0$.

c On the same axes draw a line to show how the value(s) of x can be found for which $f(x) = 0.4$.

4 a Draw a sketch of the curve $f(x) = \cos x°$ for $0 \leqslant x \leqslant 360$.

b For what values of x is $f(x) = 0$?

c On the same axes draw lines to show how the values of x can be found for which

i $f(x) = 0.5$

ii $f(x) = -0.8$.

5 For $0 \leqslant x \leqslant 360$, give the range of values of

a $\sin x°$ **b** $\cos x°$

6 Draw the graph of $y = \sin x°$ for $0 \leqslant x \leqslant 720$ using the following steps.

a Make a table of values of $\sin x°$ for values of x from 0 to 720 at intervals of 30 units. Use a calculator and give values of $\sin x°$ correct to 2 decimal places.

b Draw the y-axis on the left-hand side of a sheet of graph paper and scale it from -1 to 1 using 2 cm to 0.5 units. Draw the x-axis and scale it from 0 to 720 using 1 cm to 60 units.

c Plot the points given in the table produced in part **a** and draw a smooth curve through them.

7 Draw the graph of $y = \cos x°$ for $0 \leqslant x \leqslant 720$ using the same sequence of steps as in question **6**.
Compare the graphs of $y = \sin x°$ and $y = \cos x°$. What do you notice?

8 For values of x between 0 and 360, sketch on the same diagram the graphs of

 a $y = \sin x°$

 b $y = \sin (x + 30)°$

 c $y = \sin (x + 90)°$

9 For values of x between 0 and 360, sketch on the same diagram the graphs of

 a $y = \cos x°$

 b $y = \cos (x + 45)°$

 c $y = \cos (x + 90)°$

You will need a graphics calculator or a computer with graph drawing software for questions **10** to **12**. If you can print out the results, do so and keep them.

10 Set the range to $-360 \leqslant x \leqslant 360$ and $-2 \leqslant y \leqslant 2$.

 a Draw the graph of $y = \sin x°$ and then superimpose the graph of $y = \sin 2x°$.

 b Describe the transformation that transforms the first curve to the second.

 c Clear the screen and repeat parts **a** and **b** for $y = \sin x°$ and $y = \sin 3x°$.

 d *Without drawing the graphs*, describe the transformation that transforms the curve $y = \sin x°$ to the curve $y = \sin \frac{1}{2}x°$.

11 a Set the range to $0 \leqslant x \leqslant 1080$ and draw the graph of $y = \sin x°$.

 b What is the range of values of $\sin x°$?

 c *Sketch* the graph of $y = \sin x°$ for $-360 \leqslant x \leqslant 4 \times 360$.

12 a Set the range to $0 \leqslant x \leqslant 1080$ and draw the graph of $y = \cos x°$.

 b What is
   i the largest value of $\cos x°$
   ii the smallest value of $\cos x°$?

 c Sketch the graph of $y = \cos x°$ for $-360 \leqslant x \leqslant 720$.

THE SINE AND COSINE FUNCTIONS

The graphs drawn for the last exercise demonstrate that the sine of an angle is never greater than 1 and never less than -1.

i.e.   for all values of x, $-1 \leqslant \sin x° \leqslant 1$.

Note also that the curve repeats at intervals of 360 units.

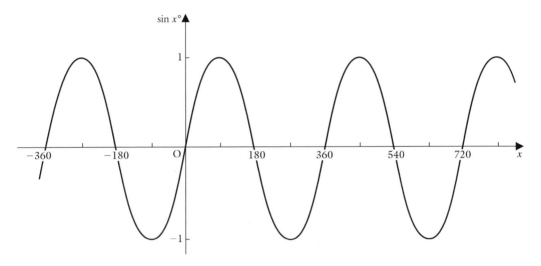

The function $f(x) = \cos x°$ has the same properties. In fact, if the curve $y = \sin x°$ is translated 90 units to the left, we get the curve $y = \cos x°$.

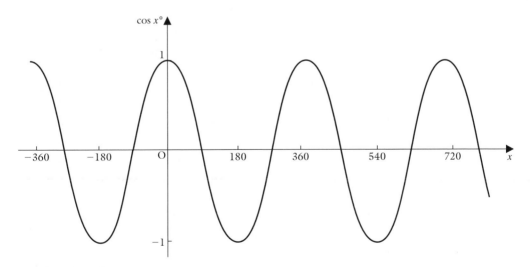

From this we deduce that $\sin(x + 90)° = \cos x°$.

THE TANGENT
FUNCTION

The graph that results from plotting values of tan $x°$ against x in the range $-180 \leqslant x \leqslant 720$ is given below.

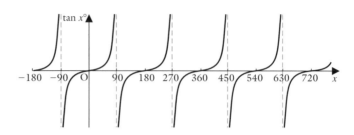

This shows that the function given by $y = \tan x°$ has properties that are quite different from those of the sine and cosine functions. The main differences are:

- tan $x°$ has no greatest or least value
- the graph of $y = \tan x°$ has a repeating pattern but repeats at intervals of 180 units
- the graph has 'breaks' in it which occur when $x = 90$ and after every interval of 180 units from there along the x-axis.

(If you find tan 89°, tan 89.9°, tan 89.99°, etc. on your calculator you get larger and larger values. When you find tan 90° the calculator display reads 'error'. Why do you think this is so?)

EXERCISE 4F

1 Sketch the graph of $y = \tan x°$ for $0 \leqslant x \leqslant 180$ and use it to sketch, on the same axes, the graph of

a $y = \tan(x + 90)°$ **b** $y = \tan(x - 90°)$.

What does the graph of $y = \tan(x + 180)°$ look like?

2 Use the graph of $y = \tan x°$ for $0 \leqslant x \leqslant 90$ to sketch the graph of

a $y = \tan x° + 2$ **b** $y = \tan x° - 2$.

3 Sketch each graph for $-180 \leqslant x \leqslant 180$

a $y = \sin 2x°$ **b** $y = \cos(x - 90)°$

4 Sketch the curves $y = 1 + \sin x°$ and $y = \tan x°$ for $0 < x < 360$. Hence give the number of values of x from 0 to 360 inclusive for which $1 + \sin x° = \tan x°$.
If you have access to a computer or graphics calculator, find these values, each correct to the nearest degree.

5 Repeat question **4**, using $y = \cos x°$ instead of $y = 1 + \sin x°$.

6 The sketch shows part of the graph of
$f(x) = 1 + \cos x°$.
Write down the coordinates of the points marked on the curve.

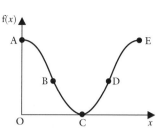

7 Each curve was found by transforming the graph of $f(x) = \sin x°$.
In each case suggest a possible transformation and give the equation of the transformed curve.

a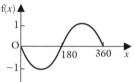

b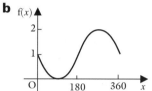

c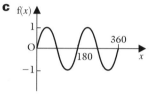

8 a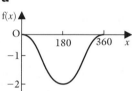

b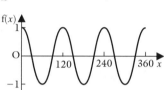

c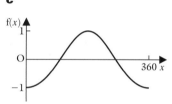

Each of these curves was found by transforming the graph of $f(x) = \cos x°$. In each case suggest a possible transformation and give the equation of the transformed curve.

9 The depth of water, d metres, in a harbour, t hours after noon, is given by $d = 4 + \cos 30t°$ and is illustrated in the sketch below.

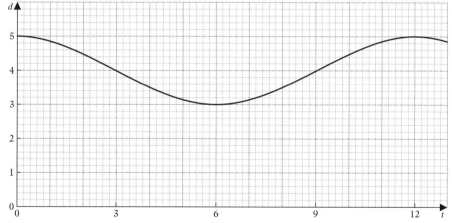

a Estimate the times when the depth of water in the harbour is less than 3.5 m.

b A ship, which has a draught of 3.2 m (that is, it requires water to a depth of 3.2 m to be able to float), ties up in the harbour at noon. What is the longest time it can remain in the harbour without becoming stranded and having to wait for the next tide ?

10 Use sketch graphs for values of x from -360 to 360 to find the *exact* values of x for which

a $\sin x° = -\sin x°$

b $\cos x° = -\cos x°$

11 Use sketch graphs for values of x from -360 to 360 to determine how many values of x there are for which

a $\sin x° = \sin 2x°$

b $\sin x° = \sin 3x°$

12 Use sketch graphs for $0 \leqslant x \leqslant 180$ to find approximate values of x for which

a $\sin x° = \sin(x - 90)°$

b $\cos x° = \cos(x - 90)°$

Use your calculator, together with 'trial and improvement' methods to give your answers to parts **a** and **b** correct to the nearest integer.

13 For values of x between 0 and 180 sketch, on the same axes, graphs for $y = \sin x°$ and $y = \tan x° - 1$. Hence show that there is one solution of the equation $1 + \sin x° = \tan x°$ between $x = 60$ and $x = 70$.

14 Sketch the graphs of $y = \sin x°$ and $y = 2 - \dfrac{x}{30}$ for $0 \leqslant x \leqslant 90$. Hence show that there is one solution of the equation $60 - x = 30 \sin x°$ between $x = 30$ and $x = 50$.

15 Sketch the graphs of $y = \cos x°$ and $y = \dfrac{x}{6}$ for $0 \leqslant x \leqslant 90$.

Hence show that there is one solution of the equation $x = 6 \cos x°$ between 0 and 20. If the values of x are not restricted to those between 0 and 90, how many solutions does the equation have?

16 a What graphs need to be drawn to solve graphically the equation $50 \cos x° = x - 20$?

b What graphs could you draw to solve graphically the equation $30 \sin x° = 20 - 2x$?

c The x-coordinates of the points of intersection of the graphs $y = 14 \tan x°$ and $y = (x - 1)^2$ are found. Write down the equation that has these values of x as roots.

Sketch the graph of $y = \cos x°$ for $-360 < x < 360$ and use the sketch to find the values of x for which $\cos x° = -0.25$, giving values of x to the nearest whole number.

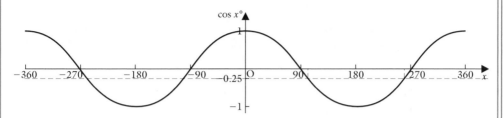

> From the graph we can see that there are four values of x for which $\cos x° = -0.25$.
> Using a calculator, ($\text{SHIFT } \cos^{-1} - 0.25$), gives $x = 104.4\ldots$; this is the smaller
> positive value of x. As the graph is symmetrical about $x = 180$, the other positive value
> of x is $360 - 104.4\ldots$
> As the graph is also symmetrical about the y-axis, there are two negative values.

$\cos x° = -0.25$

so $x = 104.4\ldots$ or $360 - 104.4\ldots$ or $-104.4\ldots$ or $-360 + 104.4\ldots$

i.e. $x = -256, -104, 104$ or 256 to the nearest whole number.

17 Use the sketch in the worked example to help find the values of x between -360 and 360 for which

  a $\cos x° = 0.4$ **b** $\cos x° = -0.7$ **c** $\cos x° = -0.1$

  Give your answers correct to the nearest whole number.

18 Sketch the graph of $y = \sin x°$ for $-360 \leqslant x \leqslant 360$ and use it to help find the values of x in this range for which

  a $\sin x° = 0.2$ **b** $\sin x° = -0.2$ **c** $\sin x° = -0.8$

19 Copy and complete the table to give values of $\sin x° + \cos x°$ for values of x from 0 to 360. Give each value correct to 2 decimal places.

x	0	15	30	45	…	330	345	360
$\sin x°$	0		0.5				-0.26	0
$\cos x°$	1		0.87				0.97	1
$\sin x° + \cos x°$	1		1.37				0.71	1

Hence draw the graph of $f(x) = \sin x° + \cos x°$ for values of x from 0 to 360.

  a Use your graph to find the least value and the greatest value of $f(x)$ within the given range.

  b Describe a compound transformation which maps $y = \sin x°$ to $y = \sin x° + \cos x°$. Hence express $\sin x° + \cos x°$ in the form $a \sin (x + b)°$, giving the values of a and b.

GROWTH
CURVES

Some quantities grow by a constant factor at regular intervals. For example, when cells divide every day,

    1 cell becomes 2 cells after 1 day,
      those 2 cells become 4 cells after another day,
        the 4 cells become 8 cells after a further day, and so on,

i.e. the number of cells doubles every day.

EXERCISE 4G

1 Under ideal conditions, bacteria reproduce by dividing themselves every hour (i.e. each hour one bacterium becomes 2 bacteria). One bacterium is placed on a petri dish and then incubated.

a Copy and complete the table which shows the number of bacteria present t hours after the dish is placed in the incubator.

Number of hours, t	0	1	2	3	4	5	6
Number of bacteria, n	1	2	4	8			

b Plot the points representing corresponding values of n and t on graph paper using a scale of 2 cm for one hour on the horizontal axis and 1 cm for 5 bacteria on the vertical axis. Draw a smooth curve through them.

c Add another row to the table and complete it to give the values of n as powers of 2.

Number of hours, t	0	1	2	3	4	5	6
Number of bacteria, n	1	2	4	8			
				2^3			

By comparing the corresponding values of t and n, deduce a formula for n in terms of t.

d If 10 bacteria were placed on the petri dish at the start, what would the relationship between n and t become?

2 Peter invests £1000 in a savings bond which grows at 10% per annum (i.e. the compound interest is 10% p.a.).

a Copy and complete the table to show the value of the bond n years later. Give the value of P in the first row exactly in index form and in the second row give the value to the nearest £100.

n years after investment	0	1	2	3	10	20	30	40
Value of the bond, £P	1000	1000×1.1^1	1000×1.1^2					

Use the first two rows of the table to deduce the relationship between n and P.

b Plot the points representing these values using a scale of 1 cm for 5 years on the horizontal axis and 1 cm for £2000 on the vertical axis. Draw a smooth curve through them.

c Use your graph to find after how many years the initial value of the bond has doubled. After how many further years does the value double again?

d If Peter invests £5000 in this bond, what is the relationship between P and n?

e Use the relationship found in part **d** to calculate the value of the bond after 25 years.

DECAY CURVES   In other situations quantities decay (shrink in size) by a constant factor at regular intervals.

For example if, in one year, the value of a car depreciates by $\frac{1}{3}$,

then a £9000 car is worth $\frac{2}{3} \times £9000$ $(= £6000)$ a year later

and another year on it is worth $\frac{2}{3} \times £6000$ $(= £4000)$, and so on.

The relationship between the value of the car and its age is investigated in the following exercise.

EXERCISE 4H   **1** A car is bought for £9000 and then depreciates each year to $\frac{2}{3}$ of its value at the start of that year.

   a Copy and complete the table showing the value, £P, of the car t years after it is bought. In the first row for values of P, give the answers in index form and in the second row give the answers correct to the nearest £10.

Number of years, t	0	1	2	3	4	5	6
Value of car, £P	9000	$9000 \times \frac{2}{3}$	$9000 \times (\frac{2}{3})^2$				
	9000	6000	4000				

   b Plot the points representing corresponding values of t and P on a graph using scales of 2 cm for 1 year on the horizontal axis and 1 cm for £500 on the vertical axis. Draw a smooth curve through the points.

   c How long after buying the car does its value halve ? After how much longer does it halve again ? Does the value of the car ever drop to zero ? Explain your answer.

2 A substance loses half its mass through radioactive decay every hour.

   a Starting with 5 grams of this substance, copy and complete the table to give the mass, m grams of the substance t hours later, giving values of m in the third row correct to 3 significant figures.

Number of hours, t	0	1	2	3	4	5	6
Mass of substance, m grams	5	$5 \times \frac{1}{2}$	$5 \times (\frac{1}{2})^2$	$5 \times (\frac{1}{2})^3$			
	5	2.5	1.25	0.625			

   b Plot the points representing corresponding values of t and m on a graph using scales of 1 cm for 1 hour on the horizontal axis and 1 cm for 0.2 grams on the vertical axis. Draw a smooth curve through the points.

   c By writing $\frac{1}{2}$ as 2^{-1}, show that $m = 5 \times 2^{-t}$.

Question **1** in **Exercise 4G** demonstrates that when a quantity grows by a factor of 2 each unit of time, then the relationship between the size of the quantity and time involves a power of 2. In this case, $n = 2^t$. *Exponent* is another word for power (so t is the exponent of 2) and we describe the way in which n increases as *exponential growth*.

The function that changes values of t into values of n is called an *exponential function*.

> In general, a function, f, that gives exponential growth is such that
> $$f(x) = a^x$$
> where a is a number greater than 1.

Question **2** in **Exercise 4G** is another example of exponential growth; in this case $P = 1000 \times 1.1^n$ when the initial investment is £1000. For both questions, the shapes of the curves are similar. It is called an *exponential growth curve*.

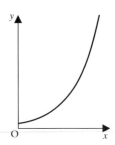

In **Exercise 4H**, the questions involve quantities decaying by the same factor in each unit of time.
In question **1**, the relationship is $P = 9000 \times (\frac{2}{3})^t$.
As $\frac{2}{3} = (\frac{3}{2})^{-1} = 1.5^{-1}$, we can write

$$P = 9000 \times 1.5^{-t} \qquad [1]$$

In question **2**, the relationship between the quantities is $m = 5 \times \frac{1}{2}^t$ which, since $\frac{1}{2} = 2^{-1}$, can be written as

$$m = 5 \times 2^{-t} \qquad [2]$$

Again, the rule governing the decreasing size of each quantity involves a power; we describe the way in which the quantities decrease as *exponential decay*.

There is clearly a connection between exponential growth and exponential decay. The relationships in [1] and [2] show that when we have the power of a number greater than 1, the exponent is negative for exponential decay.

> In general, a function, f, that gives exponential decay is such that
> $$f(x) = a^{-x}$$
> where a is a number greater than 1.

The shape of an exponential decay curve is
typically like the one shown in the diagram.

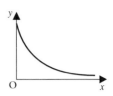

The relationship between the functions for exponential growth and
exponential decay is explored further in the next exercise.

EXERCISE 4I

1 Copy and complete the following table.

x	-3	-2	-1	0	1	2	3	4	5	6
2^x	0.125			1		4				

Use the values in your table to draw the graph of $y = 2^x$ using a
scale of 1 cm for 1 unit on the x-axis and 1 cm for 5 units on the
y-axis.

2 Copy and complete the following table.

x	-3	-2	-1	0	1	2	3
2^{-x}	8	4					0.125

Use the values in your table to draw the graph of $y = 2^{-x}$ using
the same scales as in question **1**.

3 What is the relationship between the curves drawn for questions **1**
and **2**? Is this the relationship you would expect by considering the
equations of the two curves? Explain your answer.

4 The diagram shows
a sketch of the curve $y = 3^x$.

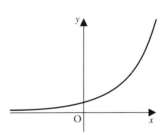

a Give the coordinates of the point where the curve cuts the y-axis.

b Does the curve cross the x-axis? Justify your answer.

c Copy the diagram and add sketches of the curves

    i $y = 2 \times 3^x$ **ii** $y = 5 \times 3^x$

5 Below is a sketch of the curve $y = 1.5^x$.

a

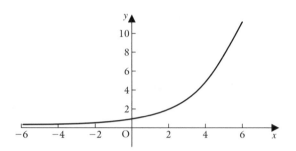

Explain why this curve cuts the y-axis at the same point as the curve $y = 3^x$. Where does the curve $y = a^x$ cut the y-axis? Explain your answer.

b Copy this sketch and add a sketch of the curve $y = 5 \times 1.1^x$.

c Where does the curve $y = 500 \times 1.1^x$ cut the y-axis?

d Repeat part **c** for the curves

    i $y = 10 \times 5^x$ **ii** $y = 10 \times 5^{-x}$

6 The equation of this curve is $y = a \times b^x$.

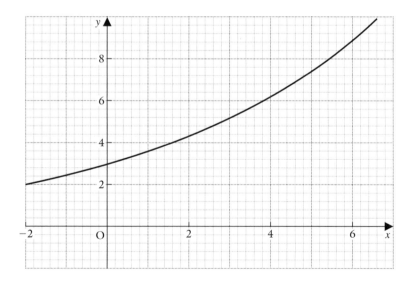

a Write down the value of a.

b From the graph estimate the value of y when $x = 1$. Hence estimate the value of b.

7 The equation of this curve is $y = ab^{-x}$. Estimate the values of a and b and hence write down the equation of the curve.

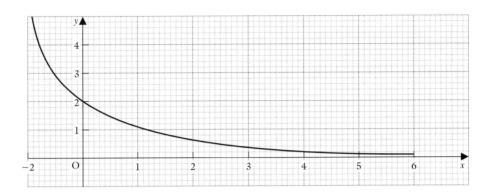

8 The diagram shows a graph plotted from experimental results in a science lesson.

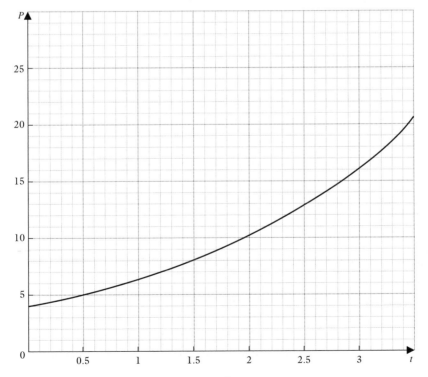

P and t are related by the equation $P = ab^t$.

Use the graph to estimate values of a and b.

9 This graph illustrates the temperature, $T\,°C$ of a metal ingot t minutes after it is removed from a furnace.

The relationship between T and t is $T = ka^{-t}$ where k and a are constants.
Use the graph to estimate the values of k and a.

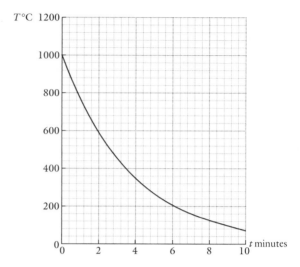

10 Each curve was found by transforming the curve $y = 2^x$.

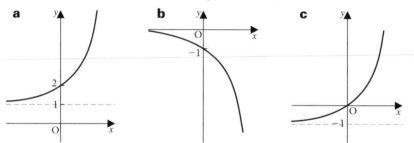

a   **b**   **c**

Suggest a transformation and give the equation of each transformed curve.

11 A sum of money, $£P$ is invested at compound interest of $r\%$ p.a. After T years, the value of the investment is $£A$.

a Copy and complete the table.

T years	0	1	2	3	4
$£A$	P	$P \times \left(1 + \dfrac{r}{100}\right)$	$P \times \left(1 + \dfrac{r}{100}\right)^2$		

b Find a formula for A in terms of P, r and T.

c Sketch a graph showing how A changes as T increases.

d When $P = 3000$ and $r = 4$, write down an expression for the value of the sum of money after 20 years.

e Use your expression to calculate the value after 20 years of a bond initially worth £3000.

f What annual rate of interest is required for a bond whose initial value is £3000 to double its value after 10 years?

12 A radioactive substance decays (i.e. loses mass) at the rate of 0.5% each day. The initial mass of a lump of this substance is 80 kg.

 a What is the mass of the lump one day later?

 b Write down an expression for the mass of the lump n days later.

 c Find the mass of the lump one year (365 days) later.

 d After how many days will the lump be half its original mass?

13 When a particular performance-enhancing drug is taken, its concentration in the blood stream reduces each day by 20% of its concentration at the beginning of that day.

 a If 100 mg is taken into the blood stream at noon on Monday,

   i how many milligrams will be in the blood stream at noon on Tuesday?
   ii on which day will the concentration be half that at noon of Monday?

 b Sketch a graph showing the quantity in the blood stream from noon on Monday to noon on the following Saturday.

14 A population of fruit flies grows to 8 times the initial number in 18 days. Assuming that the growth is exponential, how many days does it take for the population to double?

MIXED EXERCISE

EXERCISE 4J

1 The function f is given by $f(x) = x^2 - 4$.

 a Find $f(0)$, $f(-2)$, $f(5)$.

 b On the same axes, sketch the graphs of $y = f(x)$ and $y = -f(x)$.

2 The diagram shows the graph of $y = f(x)$.

 Draw sketches showing the graphs of

 a $y = f(-x)$
 b $y = 2f(x)$
 c $y = f(2x)$
 d $y = f\left(\dfrac{x}{2}\right)$
 e $y = f(x-1)$
 f $y = f(x) - 1$

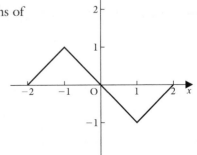

3 Use the graph of $y = \sin x°$ on page 54 to find the values of x between 0 and 360 for which $\sin x° = \frac{1}{2}$.

4 Use the graph of $y = \cos x°$ on page 54 to find the values of x between 0 and 360 for which $\cos x° = -\frac{1}{2}$.

5 Each of the graphs given below is a transformation of the graph of $y = \sin x°$.
Write down the equation of each curve.

a

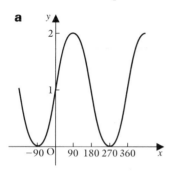

b

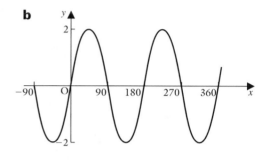

6 This is the graph of $y = ka^{-x}$.
Use the curve to estimate the values
of k and a.

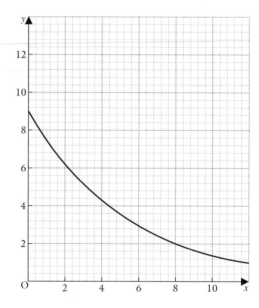

7 Which of the following curves could represent the function f where

a $f(x) = \cos x°$ **b** $f(x) = x^2 - 1$ **c** $f(x) = \frac{1}{x}$ **d** $f(x) = 3^x$

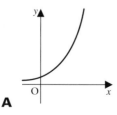

A

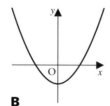

B

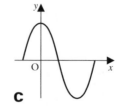

C

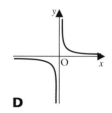

D

INVESTIGATION

Cheryl took some readings from an experiment and then plotted the results on a graph. She hoped that she would be able to discover the relationship between T and s from the shape of the curve through her points.

This is Cheryl's graph.

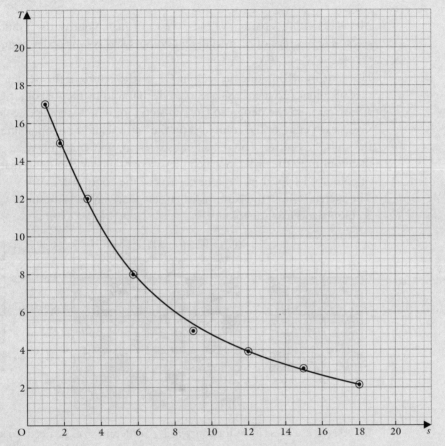

When she looked at it, she realised that her graph could be a section from several curves, e.g.

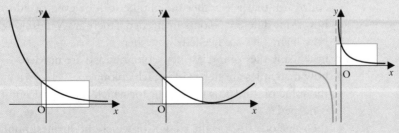

Can you help Cheryl?

SEQUENCES

Mel is designing a pattern of tiles to cover square courtyards of different sizes. He has tiles of many different colours and colour codes them C1, C2, C3, ..., Cx, ..., Cn.

He starts by setting one tile, colour C1, at the centre of the design, then tiles of colour C2 at each side of tile C1, followed by tiles of colour C3 in a 'ring' around the design so far, and so on until he has a tile of colour Cx at each edge of the courtyard on the vertical and horizontal line through the centre. The remaining spaces are bordered in reverse order of colour from the order in which they were originally chosen so that tiles of colour C1 appear in the four corners of the courtyard.

The sketches show the patterns when the courtyard is laid using 3 colours and using 5 colours.

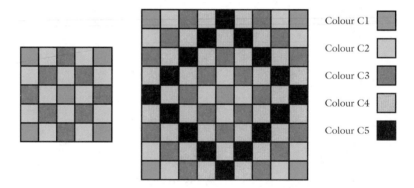

Colour C1
Colour C2
Colour C3
Colour C4
Colour C5

If Mel is to use these ideas for courtyards of different sizes he needs to know

- whether there can be any number of tiles along one edge of the courtyard, that is whether the number can be even or odd
- how many different colours are needed if there are N tiles along each edge
- how many tiles are needed altogether
- how many tiles of the last new colour added are needed
- how many tiles are needed for each colour
- the value of x if Cx is the colour for which the largest number of tiles is needed
- the value of x if Cx is the colour for which the smallest number of tiles is needed.

To solve these problems Mel needs to extend his knowledge of sequences.

THE *n*th TERM OF
A SEQUENCE

In a sequence the terms occur in a *particular order*, that is there is a first term, a second term, a third term and so on. The value of each term depends upon its position in that order.

We use the notation $u_1, u_2, u_3, \ldots$ for the first, second, third, $\ldots$ term in a sequence.

Thus u_{10} means the 10th term

u_{100} means the 100th term

u_n means the *n*th term

If we are given a formula for u_n in terms of n, e.g. $u_n = n(n+1)$, then we can find any term of the sequence.

In this case, the first term, u_1 is found by substituting 1 for n in $n(n+1)$,

i.e. $\qquad\qquad\qquad\qquad u_1 = 1(2) = 2$

Similarly $\qquad\qquad\qquad u_2 = 2(2+1) = 6,$ (substituting 2 for n)

$\qquad\qquad\qquad\qquad u_3 = 3(3+1) = 12,$

and $\qquad\qquad\qquad\qquad u_4 = 4(4+1) = 20$

$$\ldots$$

$$u_{10} = 10(10+1) = 110$$

and so on.

Therefore the sequence defined by $u_n = n(n+1)$ is 2, 6, 12, 20, $\ldots$

The simplest sequence is the sequence of natural numbers,

i.e. $\qquad\qquad\qquad\qquad$ 1, 2, 3, 4, 5, 6, $\ldots$

Other simple sequences are

$\qquad\qquad\qquad$ 1, 3, 5, 7, 9, $\ldots$ $\qquad\qquad$ odd numbers

$\qquad\qquad\qquad$ 2, 4, 6, 8, 10, $\ldots$ $\qquad\qquad$ even numbers

$\qquad\qquad\qquad$ 2, 3, 5, 7, 11, 13, $\ldots$ $\qquad$ prime numbers

To continue a sequence we need to recognise a pattern. Consider, for example,

$$2, 5, 8, 11, \ldots$$

Each number in this sequence is three units greater than the preceding number. Hence the sequence continues 14, 17, 20, $\ldots$

The pattern is not always so obvious. If you cannot immediately see a pattern, try looking for sums, products, multiples, plus or minus a number, squares, etc.

To find an expression for the *n*th term of a sequence in terms of n we need first to see the pattern. For the sequence of natural numbers 1, 2, 3, 4, $\ldots$ it is obvious that the value of a term is the same as the position number of that term i.e. $u_n = n$.

For the sequence $1, 3, 5, 7, \ldots$ the value of any term is 1 less than twice its position number

i.e. $$u_n = 2n - 1.$$

Note that it is sensible to check that a formula obtained for u_n gives the correct values for $u_1, u_2, u_3, u_4, \ldots$

DIFFERENCES

Sometimes the rule or formula to find the next few terms in a sequence is too difficult to spot.

In this case a *difference table* may be helpful.

The first difference sequence is found by subtracting u_1 from u_2, u_2 from u_3, u_3 from u_4, and so on.

As a simple example consider the sequence $5, 6, 9, 14, 21, \ldots$

The terms are	5		6		9		14		21		30		41
1st difference		1		3		5		7	→ 9	→11		

We can see that the first difference row will continue with 9 and 11 and hence the sequence will continue 30 (i.e. $21 + 9$) and 41 (i.e. $30 + 11$). The sequence is now $5, 6, 9, 14, 21, 30, 41, \ldots$

A less obvious example is $2, 5, 13, 31, 64, \ldots$ and we find that we need a second difference row before the pattern becomes clear.

The terms are	2		5		13		31		64		117		195
1st difference		3		8		18		33		53		78	
2nd difference			5		10		15	→20	→25			

In some cases it may be necessary to add a third difference line before being able to see the pattern.

If $u_n = an + b$ where a and b are numbers then we can form the difference table

$$a + b \qquad 2a + b \qquad 3a + b \qquad 4a + b \ldots$$
$$a \qquad\qquad a \qquad\qquad a$$

i.e. the sequence in the first row of differences is constant.

If $u_n = an^2 + bn + c$ where a, b and c are numbers then we can form the difference table

$$a + b + c \qquad 4a + 2b + c \qquad 9a + 3b + c \qquad 16a + 4b + c$$
$$3a + b \qquad\qquad 5a + b \qquad\qquad 7a + b$$
$$2a \qquad\qquad\qquad 2a$$

i.e. the sequence in the second row of differences is constant.

These facts can be used to find the nth term in terms of n when the pattern is difficult to spot.

For the sequence $2, 7, 12, 17, \ldots$ for example, we get the difference table

$$2 \qquad 7 \qquad 12 \qquad 17$$
$$5 \qquad 5 \qquad 5$$

As the first differences are constant we know that $u_n = an + b$ and that $a = 5$

i.e. $u_n = 5n + b$

When $n = 1$, this gives $\qquad 2 = 5 + b$, i.e. $b = -3$

so $\qquad\qquad\qquad\qquad u_n = 5n - 3$

Note that a difference table does not always help because there are some sequences whose first differences, second differences, etc., do not show an obvious pattern.

Sometimes too few terms in a sequence are given for us to be certain that the intended rule has been found.

For example, suppose that we want to define a sequence in the following way: start with 1; multiply each term by 2 to give the next term.
The first three terms in this sequence are $1, 2, 4$ and the next three terms are $8, 16, 32$.
Someone else looking at the first three terms might think that the rule was: start with 1, add 1 to get the second term, then add 2 to get the third term, then add 3 to get the fourth term, and so on.
The next three terms using this rule would be $7, 11, 16$.
To reduce the possibility of ambiguity we normally give at least four terms in a sequence.

FIBONACCI SEQUENCES

In the sequence $2, 4, 6, 10, 16, 26, \ldots$ the terms are generated by starting with 2 and 4, and then adding the previous two terms to get the next term. This is called a Fibonacci sequence.

The simplest Fibonacci sequence is $1, 1, 2, 3, 5, 8, 13, 21, 34, 55, \ldots$
These numbers crop up frequently in nature. For example, if you count the number of spirals in the seed-head of a sunflower you will find that it is one of the numbers in this sequence, say 34 or 55.
Other Fibonacci sequences can be found by starting with different pairs of numbers.

EXERCISE 5A   Find the next two terms in the following sequences.

1 3, 7, 11, 15, ...   **4** $1 + 3, 1 + 3 + 5, 1 + 3 + 5 + 7, ...$

2 3, −6, 12, −24, ...   **5** $1, -\frac{1}{3}, \frac{1}{9}, -\frac{1}{27}, ...$

3 4, 9, 16, 25, ...   **6** 2, 4, 6, 10, 16, 26, ...

7 Give the first four terms of the sequence for which

    a $u_n = 2n + 1$ **d** $u_n = n^2 - 1$

    b $u_n = 2^n$ **e** $u_n = (n+1)(2n-1)$

    c $u_n = (n-1)(n+1)$ **f** $u_n = \dfrac{n}{(n+1)}$

Find a formula for the nth term of the sequence

a $2 \times 3, 3 \times 4, 4 \times 5, ...$ **b** 5, 8, 11, 14, ...

a

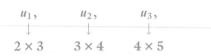

$$
\begin{array}{ccc}
u_1, & u_2, & u_3, \\
\downarrow & \downarrow & \downarrow \\
2 \times 3 & 3 \times 4 & 4 \times 5
\end{array}
$$

Each term is the product of the two integers that follow the position numbers.

Hence $u_n = (n+1)(n+2)$

Check $u_1 = (1+1) \times (1+2) = 2 \times 3$
                 $u_2 = (2+1) \times (2+2) = 3 \times 4$

b

$$
\begin{array}{cccc}
u_1, & u_2, & u_3, & u_4, \\
\downarrow & \downarrow & \downarrow & \downarrow \\
5 & 8 & 11 & 14
\end{array}
$$

Each term is 3 greater than the term before, so these terms involve multiples of 3.

Rearranging each term gives

$$
\begin{array}{cccc}
u_1, & u_2, & u_3, & u_4, \\
\downarrow & \downarrow & \downarrow & \downarrow \\
5 & 5 + 3 & 5 + 2(3) & 5 + 3(3)
\end{array}
$$

From the pattern above $u_n = 5 + (n-1)3$
                                   $= 5 + 3n - 3$
                                   $= 2 + 3n$

Check $u_1 = 2 + 3(1) = 5,$ $u_2 = 2 + 3(2) = 8$

Find a formula for the nth term of each of the following sequences.

8 $5, 7, 9, 11, \ldots$ <u>**11**</u> $1 \times 3, 2 \times 4, 3 \times 5, \ldots$

9 $0, 3, 6, 9, 12, \ldots$ <u>**12**</u> $1, 8, 27, 64, 125, \ldots$

10 $4, 8, 16, 32, 64, \ldots$ <u>**13**</u> $2, 5, 10, 17, 26, \ldots$

14 This is a sequence of pairs of numbers:

$$(1, 2), (2, 5), (3, 10), (4, 17) \ldots$$

Find   **a** the next pair in the sequence

        b the 10th pair in the sequence

        c the nth pair in the sequence.

15 John made these patterns with matchsticks.

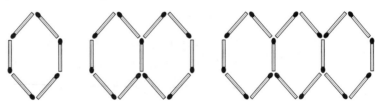

   a Write down the number of matchsticks required for each of the first five patterns in the sequence.

   b How many matchsticks are needed for

      i the nth term in the sequence

     ii the 30th term in the sequence?

   c John has a box of 200 matchsticks. How many patterns can he make in this sequence? How many matchsticks are left over?

16

```
                        1
                    1       1
                1       2       1
            1       3       3       1
        1       4       6       4       1
    1       5       10      10      5       1
```

   a Copy this pattern and write down the next three rows.

   b Add the numbers in each row to form a sequence,

     i.e. $u_1 = 1,$ $u_2 = 1 + 1 = 2, \ldots$

   c Find an expression for the sum of the numbers in the nth row of Pascal's triangle, which is the name of the array of numbers above.

17 These three triples are known as Pythagorean triads.

$$(3, 4, 5), \quad (5, 12, 13), \quad (7, 24, 25)$$

a What is the relationship between the three numbers?
(Their name should give you a clue!)

b Find one more Pythagorean triple (not a multiple of those given above).

18 A boy is given a large bar of chocolate and decides to make it last by eating half of what is left each day. Thus he eats half of the bar on the first day; he eats half of half the bar, i.e. quarter of the bar, on the second day; he eats half of quarter of the bar on the third day, and so on.
Write down the sequence giving the fraction of the bar left at the end of the first, second, third, fourth and fifth days.
In theory, how long will the bar of chocolate last?

19 A sequence is defined by $u_n = \frac{1}{2}n(n + 1)$. Find

a the first five terms of the sequence

b the 20th term of the sequence

c an expression for the term before u_n (i.e. u_{n-1})

d an expression for $u_n - u_{n-1}$

e values of $u_2 - u_1$, $u_3 - u_2$, $u_4 - u_3$, $u_5 - u_4$

f the 20th term of the sequence in part **e**.

20 A sequence is defined by $u_n = \frac{1}{6}(n + 1)(2n + 1)$. Find

a the first five terms of the sequence

b the 20th term of the sequence

c an expression for the term before u_n (i.e. u_{n-1})

d an expression for $u_n - u_{n-1}$

e values of $u_2 - u_1$, $u_3 - u_2$, $u_4 - u_3$, $u_5 - u_4$

f the 20th term of the sequence in part **e**.

For the following sequences make difference tables and use them to find the next two terms in each sequence.

21 $4, 5, 10, 19, 32, \ldots$

24 $1, 2, 5, 10, 17, \ldots$

22 $2, 3, 7, 15, 28, \ldots$

25 $1, 0, 5, 28, 81, 176, \ldots$

23 $11, 17, 33, 71, 143, 261, \ldots$

26 $1, 2, -1, -5, -7, -4, \ldots$

27 For each of the following sequences, there are at least two possible rules or formulas for generating it. Find two possibilities and in each case give three more terms.

   a $3, 9, 27, \ldots$ **b** $3, 6, 12, \ldots$ **c** $1, 2, 5, \ldots$

28 The numbers $87, 7, 2, 18, 58, 121, 35, 210, 163$ when written in order of size, smallest first, would form a sequence if the value of one of them was changed. Find which one should be changed and gives its correct value.

29 The nth term of a sequence is given by $u_n = 3n - 4$.

   a Find the 50th term.

   b Find the value of n when $u_n = 293$.

   c What is the smallest value of n such that $u_n > 1000$?

30 Consider the sequence $2, 7, 12, 17, \ldots$

   a Write down the next two terms and give the formula for the nth term in terms of n.

   b Give the first five terms of the sequence formed by multiplying each term of the given sequence by the term following it. (The first term is 2×7, i.e. 14.)

   c Use a difference table on the five terms of the new sequence formed in part **b** to find the sixth and seventh terms.
   Check that you are correct by using the rule given in part **b**.

   d Give the first five terms of the sequence formed by adding each term of the original sequence to the term following it.

   e Give the formula for the nth term of the sequence in part **c**.

   f   **i** The nth term of a sequence is equal to the sum of the first n terms of the given sequence (e.g. the third term is $2 + 7 + 12$, i.e. 21).
   Write down the first five terms.

      ii A new sequence is formed when the nth term of the sequence in part **i** is divided by n. Write down the first five terms of this sequence using fractions where necessary.

      iii Give a formula for the nth term of the sequence in part **ii**.

      iv Hence give a formula for the nth term of the sequence in part **i**.

31 Form a Fibonacci sequence from each of the following pairs of numbers. Give the first six terms.

    a 1, 3   **b** 2, 3   **c** 3, 4

32 Form a difference table for the sequence you obtained in question **31** part **c**, What do you notice?
Does the same thing happen if you use the sequence obtained for part **b** of the same question?

Find, in terms of n, a formula for the nth term of the sequence
$$1, 6, 15, 28, 45,$$
Use your formula to find the 10th term.

> First find the difference table.

1		6		15		28		45
	5		9		13		17	
		4		4		4		

Since the second row of differences are constant, the nth term of the sequence is given by $u_n = an^2 + bn + c$ where a, b and c are numbers, and, from page 74, we know that $2a = 4$.

$\therefore a = 2$, i.e. $\qquad u_n = 2n^2 + bn + c$

When $n = 1$ this gives $1 = 2 + b + c$

$\Rightarrow \qquad\qquad\qquad b + c = -1 \qquad\qquad\qquad\qquad [1]$

When $n = 2$ this gives $6 = 8 + 2b + c$

$\Rightarrow \qquad\qquad\qquad 2b + c = -2 \qquad\qquad\qquad\qquad [2]$

$[2] - [1]$ gives $b = -1$

$\therefore$ from $[1]$ $c = 0$

$\therefore \qquad\qquad\qquad u_n = 2n^2 - n$

If $n = 10$, the 10th term is $2 \times 10^2 - 10$ i.e. 190

For each of the following sequences, find a formula for the nth term in terms of n. Use your formula to find the 10th term.

33 3, 7, 13, 21, 31, ...   **36** 0, 11, 30, 57, 92, ...

34 5, 16, 33, 56, 85, ...   **37** −2, −2, 0, 4, 10, ...

35 7, 19, 31, 43, 55, ...   **38** 8, 17, 32, 53, 80, ...

Use any method to continue each of the following sequences for three more terms.

39 100, 99, 95, 79, 15, ...

44 1.3, 2.4, 3.5, 4.6, ...

40 $\frac{1}{2}, \frac{2}{3}, \frac{3}{4}, \frac{4}{5}, \ldots$

45 1, −2, 4, −8, ...

41 5, 11, 21, 35, 53, ...

46 $\frac{1}{2}, \frac{3}{5}, \frac{2}{3}, \frac{5}{7}, \frac{3}{4}, \ldots$

42 2, 9, 28, 65, ...

47 9, 20, 37, 60, 89, ...

43 0, 6, 18, 36, 60, ...

48 −4, −6, −6, −4, 0, ...

49 a Give the first eight terms of the Fibonacci sequence that begins 1, 1, ...

  b Form a sequence by expressing each term of the sequence in part **a** as a fraction of its following term. Give the first ten terms.

  c Give the fractions in part **b** as decimals correct to 4 decimal places. What do you notice?

50 a Solve Mel's problems as stated at the beginning of the chapter.

  b Investigate a similar problem if he starts with a block of four squares of the same colour at the centre.

INVESTIGATIONS

1 a

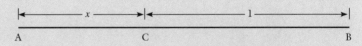

A   C   B

The Greek geometers were very interested in the division of a line AB in such a way that $\frac{AC}{CB} = \frac{CB}{AB}$.

If $AC = x$ units and $CB = 1$ unit find expressions for the ratios $\frac{AC}{CB}$ and $\frac{CB}{AB}$ in terms of x.

If $\frac{AC}{CB} = \frac{CB}{AB}$ form an equation in x and solve it to show that $x = \dfrac{\sqrt{5} - 1}{2}$.

Evaluate x as accurately as your calculator allows.

b Consider the Fibonacci sequence starting with the numbers 1, 1. The next numbers are 2, 3, 5, 8, 13, 21.

If we now find the fractions using the numbers from the sequence two at a time in order we have $\frac{1}{1}, \frac{2}{3}, \frac{5}{8}, \frac{13}{21}, \ldots$

Express these fractions as decimals as accurately as your calculator allows and find the decimal values of the next three fractions in this sequence.

c Repeat part **b** but start with 1, 2, as the first two numbers in the Fibonacci sequence.

d Draw a conclusion that relates your answers to parts **a**, **b** and **c**.

e Use reference books to find out what you can about 'The Golden Ratio'. Is there any connection with your answer to part **d**?

VECTOR METHODS

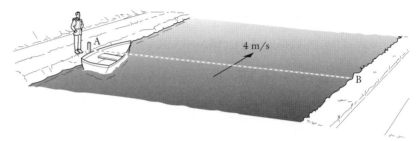

A traveller hires a boat to cross a river from a point A on one bank to, hopefully, a point B on the opposite bank. The river is 100 m wide; the boat can travel in still water at 8 m/s and the current in the river is flowing at 4 m/s.

The traveller is anxious to know whether it is possible to cross the river directly from A to B, the point immediately opposite A. He would like to cross in the shortest possible time. (He is not a very good sailor!)

Before he hires the boat he needs to know

- whether or not the shortest distance necessarily takes the shortest time
- the direction in which he must point the boat if he is to cross directly to B
- the effect of the river's current combining with the velocity of the boat
- whether steering the boat directly across the river is the quickest way of crossing the river
- how long the different possible journeys will take.

To answer these questions the traveller needs to know how to combine quantities that are in different directions, that is he needs to know more about vectors.

DEFINITION OF A VECTOR

Although we usually assume that two and two make four, this is not always the case.

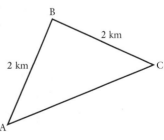

Amberley (A), Beckford (B) and Croxton (C) are three villages. It is 2 km from Amberley to Beckford and 2 km from Beckford to Croxton, but it is easy to see from the diagram that Croxton is not 4 km from Amberley.

8 3

A to B, B to C and A to C are displacements. Each of them has magnitude (that is, size), and each of them has a definite direction in space. They are all examples of a *vector*.

> A vector is a quantity that has both magnitude and a specific direction in space.

A vector can be represented by a directed line segment, i.e. a line of given length in a given direction.

Velocity (e.g. 16 mph west) is a vector but mass (e.g. 4 kg) is not. We can represent 16 mph west by a line.

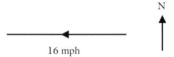

16 mph

N

The length of the line represents the magnitude of the vector. The direction of the line represents the direction of the vector; notice that an arrow is needed on the line.

A *scalar* quantity is one that is fully defined by magnitude alone. For example, length is a scalar quantity since the length of a piece of string does not depend on its direction when it is being measured. Temperature is another quantity that does not have direction, so temperature also is a scalar.

REPRESENTATION OF A VECTOR

We now know that a vector can be represented by a line segment, whose length represents the magnitude of the vector and whose direction, indicated by an arrow, represents the direction of the vector. In general, such vectors can be denoted by a letter in bold type, e.g. **a** or **b**.

It is difficult to hand-write letters in bold type so we indicate that a letter represents a vector by putting a line underneath it. Hence a vector shown in a book as **a** will be handwritten as $\underline{a}$.

Alternatively we can represent a vector by the magnitude and direction of a line joining A to B. We denote the vector from A to B by $\overrightarrow{AB}$ or **AB**, (but by $\overrightarrow{AB}$ when it is handwritten). The vector in the opposite direction, i.e. from B to A, is written $\overrightarrow{BA}$ or **BA**.

THE MAGNITUDE
OF A VECTOR

The magnitude of a vector **a**, which is written as $|\mathbf{a}|$ or a, is equal to the length of the line representing **a**.

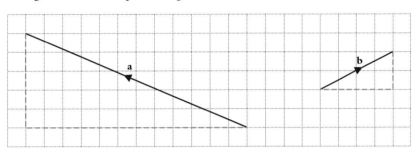

In the diagram, each square is of side 1 unit, so

$|\mathbf{a}| = \sqrt{5^2 + 12^2}$ (Pythagoras' theorem) and $|\mathbf{b}| = \sqrt{4^2 + 2^2}$

$\qquad = \sqrt{25 + 144} \qquad\qquad\qquad\qquad\qquad = \sqrt{20}$

$\qquad = \sqrt{169} \qquad\qquad\qquad\qquad\qquad\quad = 4.47$ (3 s.f.)

$\qquad = 13$

EXERCISE 6A

1 Which of the following quantities are vectors?

   a The length of a piece of wire.

   b The force needed to move a lift up its shaft.

   c A move from the door to your chair.

   d The speed of a galloping horse.

   e The distance between Bristol and Edinburgh.

2 Represent each of the following vector quantities by a suitable directed line.

   a A force of 6 newtons acting vertically downwards.

   b A velocity of 3 m/s on a bearing of 035°.

   c An acceleration of 2 m/s² northwest.

   d A displacement of 7 km due south.

In questions **3** and **4** the vectors are drawn on a unit grid.

3

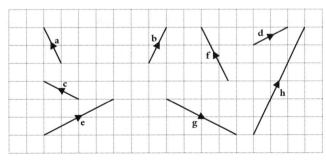

Find the magnitude of each of the vectors given in the diagram.

4

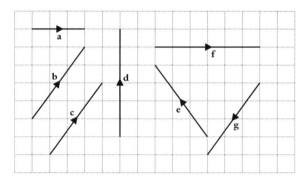

    a Find the magnitude of each of the vectors given in the diagram.

    b Which two vectors are equal in magnitude and in the same direction ?

    c Which two vectors are equal in magnitude but opposite in direction ?

EQUAL VECTORS

Two vectors which have the same magnitude and are in the same direction are equal,

i.e. $\mathbf{a} = \mathbf{b}$ implies that $|\mathbf{a}| = |\mathbf{b}|$ *and* that the directions of $\mathbf{a}$ and $\mathbf{b}$ are the same.

It follows that a vector can be represented by *any* line of the right length and direction, irrespective of its position, so each of the lines in the diagram below represents the vector $\mathbf{c}$.

NEGATIVE VECTORS

If two vectors $\mathbf{a}$ and $\mathbf{b}$, have the same magnitude but are in opposite directions we say that $\mathbf{b} = -\mathbf{a}$

i.e. $-\mathbf{a}$ is a vector of magnitude $\mathbf{a}$ and in the direction opposite to that of $\mathbf{a}$.

We say that $\mathbf{a}$ and $\mathbf{b}$ are equal and opposite.

Similarly it follows that

$$\overrightarrow{AB} = \overrightarrow{CD} = \overrightarrow{EF}$$

and $\overrightarrow{AB} = -\overrightarrow{HG}$

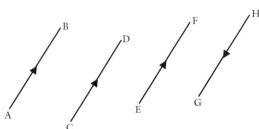

MULTIPLICATION OF A VECTOR BY A SCALAR   If k is a positive number then $k\mathbf{a}$ is a vector in the same direction as $\mathbf{a}$ and of magnitude $k|\mathbf{a}|$.

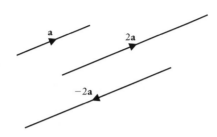

In the diagram, for example, the line representing $2\mathbf{a}$ is twice the length of the line representing $\mathbf{a}$ and parallel to it.

It follows that $-k\mathbf{a}$ is a vector in the opposite direction with magnitude $k|\mathbf{a}|$.

EXERCISE 6B   **1**

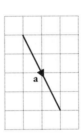

Copy the vector $\mathbf{a}$ onto squared paper, and on the same grid draw line segments to represent the vectors

a $2\mathbf{a}$ **c** $\frac{3}{2}\mathbf{a}$ **e** $-3\mathbf{a}$

b $\frac{1}{2}\mathbf{a}$ **d** $4\mathbf{a}$ **f** $-\mathbf{a}$

2

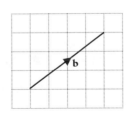

Copy the vector $\mathbf{b}$ onto squared paper, and on the same grid draw line segments to represent the vectors

a $-\mathbf{b}$ **c** $-\frac{1}{2}\mathbf{b}$ **e** $-2\mathbf{b}$

b $3\mathbf{b}$ **d** $\frac{5}{2}\mathbf{b}$ **f** $4\mathbf{b}$

3

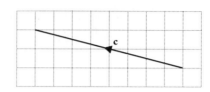

Copy the vector **c** onto squared paper, and on the same grid draw line segments to represent the vectors

a $3\mathbf{c}$ **b** $-2\mathbf{c}$ **c** $-\frac{3}{2}\mathbf{c}$ **d** $\frac{3}{4}\mathbf{c}$

4

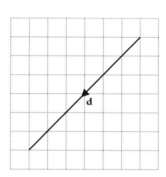

Copy the vector **d** onto squared paper, and on the same grid draw line segments to represent the vectors

a $-\frac{1}{2}\mathbf{d}$ **b** $-\frac{5}{2}\mathbf{d}$ **c** $-3\mathbf{d}$ **d** $\frac{2}{3}\mathbf{d}$

**EQUIVALENT
DISPLACEMENTS**

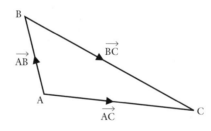

The displacement from A to B, followed by the displacement from B to C, is equivalent to the displacement from A to C.

This is written as the vector equation

$$\overrightarrow{AB} + \overrightarrow{BC} = \overrightarrow{AC}$$

Note that, in vector equations like the one above,

$+$ means 'together with' and $=$ means 'is equivalent to'

All vector quantities can be combined in this way.

**ADDITION OF
VECTORS**

We can see on page 88 that, in a triangle ABC, the displacement from A to B followed by the displacement from B to C, is equivalent to the displacement from A to C.

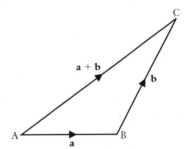

If the sides $\overrightarrow{AB}$ and $\overrightarrow{BC}$ of a triangle ABC represent the vectors **a** and **b** then the third side $\overrightarrow{AC}$ is equivalent to **a** followed by **b**. We say that $\overrightarrow{AC}$ represents the *vector sum* or *resultant*, of **a** and **b**.

We write $\overrightarrow{AC} = \mathbf{a} + \mathbf{b}$.

Note that **a** and **b** follow each other round this triangle (they go anticlockwise) whereas the resultant **a** + **b** goes the opposite way round (i.e. clockwise).

This is known as the *triangle law* for the addition of vectors and can be extended to the addition of more than two vectors.

In the diagram

$$\overrightarrow{AB} + \overrightarrow{BC} = \mathbf{a} + \mathbf{b} = \overrightarrow{AC}$$

and $\overrightarrow{AC} + \overrightarrow{CD} = (\mathbf{a} + \mathbf{b}) + \mathbf{c} = \overrightarrow{AD}$

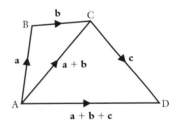

Note again that **a**, **b** and **c** go one way round the quadrilateral (clockwise) while the resultant, **a** + **b** + **c**, goes in the opposite sense.

In the parallelogram ABCD, if $\overrightarrow{AB}$ represents the vector **a** then so does $\overrightarrow{DC}$ (the opposite sides of a parallelogram are parallel and equal in length).

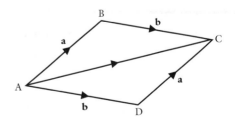

Likewise, if $\overrightarrow{BC}$ represents the vector **b**, so does $\overrightarrow{AD}$.

Using the triangle law for addition in triangle ABC

$$\mathbf{a} + \mathbf{b} = \overrightarrow{AC}$$

and using the law again in $\triangle ADC$ we have

$$\mathbf{b} + \mathbf{a} = \overrightarrow{AC}$$

Hence $\qquad\qquad\qquad\qquad \mathbf{a} + \mathbf{b} = \mathbf{b} + \mathbf{a}$

i.e. the order in which we add two vectors does not matter.

If the vector $\mathbf{c}$ is represented by the line $\overrightarrow{AC}$ then $\mathbf{c} = \mathbf{a} + \mathbf{b}$ and $\mathbf{c} = \mathbf{b} + \mathbf{a}$.

**SUBTRACTION
OF VECTORS**

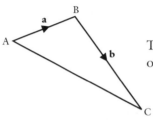

This diagram shows how to find the sum of two vectors $\mathbf{a}$ and $\mathbf{b}$.

The vector $-\mathbf{b}$ can be shown on a diagram by changing the direction of the arrow on BC, but we cannot use this diagram to find the sum of $\mathbf{a}$ and $-\mathbf{b}$ since the arrows on the lines representing the two vectors are not in the same sense.

We can overcome this problem by moving the line representing $-\mathbf{b}$ to a new position.

> Remember that a vector is not changed if the line representing it is moved to a different position, provided that its magnitude and direction are unchanged.

> The line representing $-\mathbf{b}$ has moved to a new position so that the end of $-\mathbf{b}$ originally marked C now coincides with the end of $\mathbf{a}$ marked B. In its new position the other end of $-\mathbf{b}$ is marked D.

This new position is shown opposite.

Then the resultant of $\mathbf{a}$ and $-\mathbf{b}$ is represented by the line $\overrightarrow{AD}$

so $\qquad \overrightarrow{AD} = \mathbf{a} + (-\mathbf{b})$

i.e. $\qquad \overrightarrow{AD} = \mathbf{a} - \mathbf{b}$

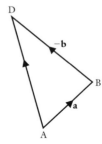

EXERCISE 6C

Find the single vector equivalent to $\overrightarrow{AC} + \overrightarrow{CD}$.

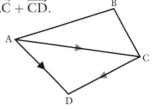

Marking $\overrightarrow{AC}$ and $\overrightarrow{CD}$ on the diagram we see that the displacement equivalent to $\overrightarrow{AC}$ followed by $\overrightarrow{CD}$, is $\overrightarrow{AD}$.

$\overrightarrow{AC} + \overrightarrow{CD} = \overrightarrow{AD}$

1 Find the single vector that is equivalent to

  a $\overrightarrow{PQ} + \overrightarrow{QR}$

  b $\overrightarrow{PR} + \overrightarrow{RQ}$

  c $\overrightarrow{RQ} + \overrightarrow{QP}$

  d $\overrightarrow{QP} + \overrightarrow{PR}$

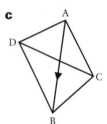

2 Give an alternative route for $\overrightarrow{AB}$.

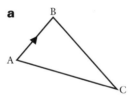 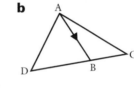 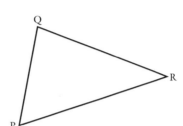

3 Find the single vector that is equivalent to

  a $\overrightarrow{AB} + \overrightarrow{BC}$

  b $\overrightarrow{BC} + \overrightarrow{CD}$

  c $\overrightarrow{AB} + \overrightarrow{BC} + \overrightarrow{CD}$

  d $\overrightarrow{AB} + \overrightarrow{BD}$

  e $\overrightarrow{DA} + \overrightarrow{AC}$

  f $\overrightarrow{AB} + \overrightarrow{BD} + \overrightarrow{DC}$

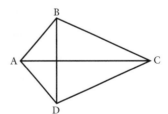

4 Find the single vector that is equivalent to

  a $\overrightarrow{AB} + \overrightarrow{BC} + \overrightarrow{CD}$

  b $\overrightarrow{BC} + \overrightarrow{CD} + \overrightarrow{DA}$

  c $\overrightarrow{AE} + \overrightarrow{EC} + \overrightarrow{CD}$

  d $\overrightarrow{DA} + \overrightarrow{AB} + \overrightarrow{BC}$

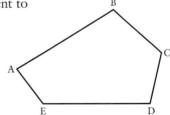

Which single vector is equivalent to

a $\overrightarrow{PS} - \overrightarrow{PQ}$

b $\overrightarrow{QR} - \overrightarrow{QX}$?

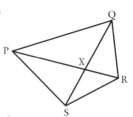

a Marking $\overrightarrow{PS}$ and $-\overrightarrow{PQ}$ on the diagram we see that, as the arrows follow round, $(-\overrightarrow{PQ}) + \overrightarrow{PS}$ is equivalent to the third side of $\triangle PQS$ in the direction Q to S.

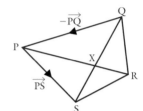

$$(-\overrightarrow{PQ}) + \overrightarrow{PS} = \overrightarrow{QS}$$

or $\quad \overrightarrow{PS} + (-\overrightarrow{PQ}) = \overrightarrow{QS}$

so $\quad \overrightarrow{PS} - \overrightarrow{PQ} = \overrightarrow{QS}$

b Similarly, marking $-\overrightarrow{QX}$ and $\overrightarrow{QR}$ on the diagram we see that

$$-\overrightarrow{QX} + \overrightarrow{QR} = \overrightarrow{XR}$$

i.e. $\quad \overrightarrow{QR} - \overrightarrow{QX} = \overrightarrow{XR}$

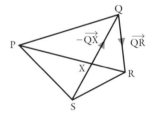

5 Which single vector is equivalent to

    a $\overrightarrow{AC} - \overrightarrow{BC}$

    b $\overrightarrow{BA} - \overrightarrow{CA}$

    c $\overrightarrow{BC} - \overrightarrow{BA}$?

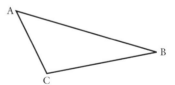

6 Which single vector is equivalent to

    a $\overrightarrow{AC} - \overrightarrow{AD}$

    b $\overrightarrow{DC} - \overrightarrow{BC}$

    c $\overrightarrow{AD} + \overrightarrow{DC} - \overrightarrow{BC}$

    d $\overrightarrow{AB} - \overrightarrow{DC} - \overrightarrow{CB}$?

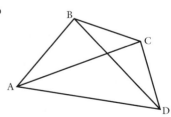

Copy the vectors **a** and **b** onto a grid.
On your grid draw line segments to
represent the vectors

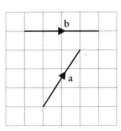

a $\mathbf{a} + \mathbf{b}$ **c** $2\mathbf{a} + \mathbf{b}$

b $\mathbf{a} - \mathbf{b}$ **d** $3\mathbf{b} - 2\mathbf{a}$

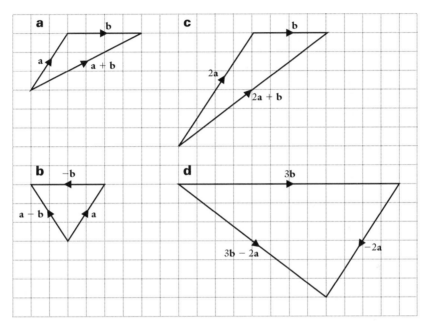

7 Copy the vectors **a** and **b** onto a grid.
On your grid draw line segments to represent
the vectors

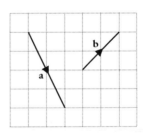

  a $\mathbf{a} + \mathbf{b}$ **c** $\mathbf{a} - \mathbf{b}$

  b $2\mathbf{a} + \mathbf{b}$ **d** $\mathbf{a} - \frac{1}{2}\mathbf{b}$

8 Copy the vectors **a** and **b** onto a grid.
On your grid draw line segments to represent
the vectors

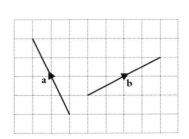

  a $\mathbf{a} + \mathbf{b}$ **d** $\mathbf{b} - 2\mathbf{a}$

  b $\mathbf{a} - 2\mathbf{b}$ **e** $\frac{5}{2}\mathbf{b} - \frac{3}{2}\mathbf{a}$

  c $2\mathbf{a} + 3\mathbf{b}$

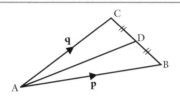

$\overrightarrow{AB} = \mathbf{p}$, $\overrightarrow{AC} = \mathbf{q}$ and D is the mid-point of BC.

Give $\overrightarrow{BC}$ and $\overrightarrow{AD}$ in terms of $\mathbf{p}$ and $\mathbf{q}$.

> Give an alternative route from B to C first.

$$\overrightarrow{BC} = \overrightarrow{BA} + \overrightarrow{AC}$$
$$= -\mathbf{p} + \mathbf{q}$$
$$= \mathbf{q} - \mathbf{p}$$
$$\overrightarrow{AD} = \overrightarrow{AB} + \overrightarrow{BD}$$
$$= \overrightarrow{AB} + \tfrac{1}{2}\overrightarrow{BC}$$
$$= \mathbf{p} + \tfrac{1}{2}(\mathbf{q} - \mathbf{p})$$
$$= \tfrac{1}{2}\mathbf{p} + \tfrac{1}{2}\mathbf{q}$$
$$= \tfrac{1}{2}(\mathbf{p} + \mathbf{q})$$

1

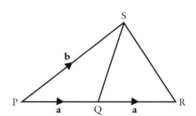

a Is PQR a straight line? Justify your answer.

b Give in terms of **a** and **b**

  i $\overrightarrow{QS}$ **ii** $\overrightarrow{SR}$ **iii** $\overrightarrow{RS}$

2

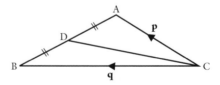

If D is the mid-point of AB, give in terms of **p** and **q**

  a $\overrightarrow{AB}$ **b** $\overrightarrow{AD}$ **c** $\overrightarrow{DB}$ **d** $\overrightarrow{CD}$

3

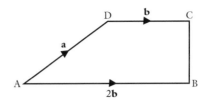

a What type of quadrilateral is ABCD ?

b Give in terms of **a** and **b**

  i $\overrightarrow{BC}$ ii $\overrightarrow{BD}$ iii $\overrightarrow{AC}$

4

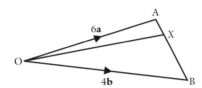

In the diagram X is the point on AB such that $AX = \frac{1}{3}XB$. Given that $\overrightarrow{OA} = 6\mathbf{a}$ and $\overrightarrow{OB} = 4\mathbf{b}$, express in terms of **a** and/or **b**

a $\overrightarrow{AB}$ b $\overrightarrow{AX}$ c $\overrightarrow{OX}$

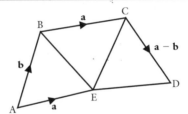

$\overrightarrow{AB} = \mathbf{b}$, $\overrightarrow{AE} = \mathbf{a}$, $\overrightarrow{BC} = \mathbf{a}$, and $\overrightarrow{CD} = \mathbf{a} - \mathbf{b}$.

a Find $\overrightarrow{BE}$ and $\overrightarrow{ED}$ in terms of **a** and **b**.

b What type of quadrilateral is BEDC ?

c Prove that A, E and D lie in a straight line.

a $\overrightarrow{BE} = \overrightarrow{BA} + \overrightarrow{AE}$ $\overrightarrow{ED} = \overrightarrow{EB} + \overrightarrow{BC} + \overrightarrow{CD}$
     $= -\mathbf{b} + \mathbf{a}$ $= -(\mathbf{a} - \mathbf{b}) + \mathbf{a} + (\mathbf{a} - \mathbf{b})$
     $= \mathbf{a} - \mathbf{b}$ $= \mathbf{a}$

b BEDC is a parallelogram because $\overrightarrow{BE} = \overrightarrow{CD}$
   i.e. CD is parallel to BE and CD = BE.

c Both $\overrightarrow{AE}$ and $\overrightarrow{ED}$ represent a.
   ∴ ED is parallel to AE.
   Also the point E is on both lines.
   ∴ A, E and D lie in a straight line.

5

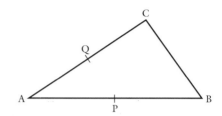

$\overrightarrow{AB} = \mathbf{b}$ and $\overrightarrow{AC} = \mathbf{c}$. P and Q are the mid-points of AB and AC respectively. Give in terms of $\mathbf{b}$ and $\mathbf{c}$

a $\overrightarrow{AP}$ **b** $\overrightarrow{AQ}$ **c** $\overrightarrow{BC}$ **d** $\overrightarrow{PQ}$

e Show that PQ is parallel to BC.

f What is the relationship between the lengths of PQ and BC? Justify your answer.

6

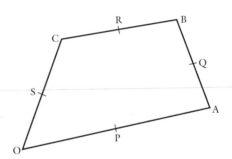

$\overrightarrow{OA} = \mathbf{a}$, $\overrightarrow{OB} = \mathbf{b}$ and $\overrightarrow{OC} = \mathbf{c}$. P, Q, R and S are the mid-points of OA, AB, BC and OC respectively. Give in terms of $\mathbf{a}$, $\mathbf{b}$ and $\mathbf{c}$

a $\overrightarrow{OP}$ **b** $\overrightarrow{AB}$ **c** $\overrightarrow{AQ}$ **d** $\overrightarrow{PQ}$ **e** $\overrightarrow{SR}$

f Show that PQ is parallel to SR.

g What type of quadrilateral is PQRS? Give reasons for your answer.

7

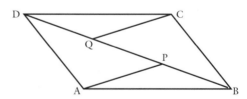

ABCD is a parallelogram. $\overrightarrow{AB} = \mathbf{a}$ and $\overrightarrow{AD} = \mathbf{b}$. P and Q are points on BD such that BP = PQ = QD. Give in terms of $\mathbf{a}$ and $\mathbf{b}$

a $\overrightarrow{BD}$ **b** $\overrightarrow{BP}$ **c** $\overrightarrow{BQ}$ **d** $\overrightarrow{AP}$ **e** $\overrightarrow{QC}$

f Show that APCQ is a parallelogram.

8

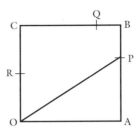

OABC is a square. $\overrightarrow{OA} = \mathbf{a}$ and $\overrightarrow{OC} = \mathbf{b}$.

P is the point on AB such that AP : PB = 2 : 1.

Q is the point on BC such that BQ : QC = 1 : 3.

R is the mid-point of OC.

Find in terms of **a** and **b**

a $\overrightarrow{AB}$ **b** $\overrightarrow{AP}$ **c** $\overrightarrow{OP}$ **d** $\overrightarrow{OR}$ **e** $\overrightarrow{CQ}$ **f** $\overrightarrow{RQ}$

g Show that RQ is parallel to OP.

h How do the lengths of RQ and OP compare?

9

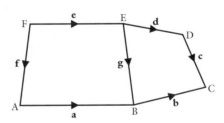

The diagram shows a rough sketch of two quadrilaterals.

a If $\mathbf{a} = 2\mathbf{b}$ what can you say about A, B and C?

b If $\mathbf{a} = \mathbf{b} = \mathbf{e} = \mathbf{d}$ what type of figure is ABCDEF?

c If $\mathbf{g} = 2\mathbf{c}$ what type of figure is BCDE?

d If $\mathbf{d} + \mathbf{c} = \mathbf{e} + \mathbf{g}$ name four points that are vertices of a parallelogram.

10

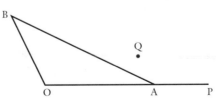

$\overrightarrow{OA} = 4\mathbf{a}$, $\overrightarrow{OB} = 2\mathbf{b}$, $\overrightarrow{AP} = \frac{1}{2}\overrightarrow{OA}$ and $\overrightarrow{OQ} = 3\mathbf{a} + \mathbf{b}$.

Give in terms of **a** and **b**

a $\overrightarrow{BP}$ **b** $\overrightarrow{BQ}$

c Show that B, Q and P lie in a straight line.

d Find BQ : BP.

11

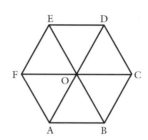

ABCDEF is a regular hexagon whose diagonals intersect at O.
$\overrightarrow{OA} = \mathbf{a}$ and $\overrightarrow{OB} = \mathbf{b}$.

a Find, in terms of **a** and **b**

    i $\overrightarrow{OC}$ **ii** $\overrightarrow{OD}$ **iii** $\overrightarrow{OE}$ **iv** $\overrightarrow{OF}$

b Give in terms of **a** and **b** the vectors

    i $\overrightarrow{AB}$ **ii** $\overrightarrow{BC}$ **iii** $\overrightarrow{CD}$ **iv** $\overrightarrow{DE}$ **v** $\overrightarrow{EF}$ **vi** $\overrightarrow{FA}$

12

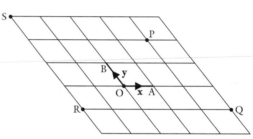

In the diagram $\overrightarrow{OA} = \mathbf{x}$ and $\overrightarrow{OB} = \mathbf{y}$. Express each of the
following vectors in the form $h\mathbf{x} + k\mathbf{y}$. For each vector give the
values of h and k.

    a $\overrightarrow{OP}$ **b** $\overrightarrow{OQ}$ **c** $\overrightarrow{OR}$ **d** $\overrightarrow{OS}$

13

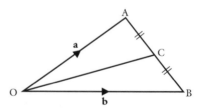

In the diagram $\overrightarrow{OA} = \mathbf{a}$, $\overrightarrow{OB} = \mathbf{b}$ and C is the mid-point of AB.
Give in terms of **a** and **b**

    a $\overrightarrow{BA}$ **b** $\overrightarrow{AB}$ **c** $\overrightarrow{BC}$ **d** $\overrightarrow{AC}$

e $\overrightarrow{OC}$ can be given as $\overrightarrow{OB} + \overrightarrow{BC}$ or $\overrightarrow{OA} + \overrightarrow{AC}$. Use each of
these two versions to find $\overrightarrow{OC}$ in terms of **a** and **b**. Are your two
answers the same?

14

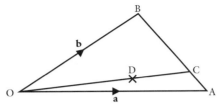

$\overrightarrow{OA} = \mathbf{a}$ and $\overrightarrow{OB} = \mathbf{b}$. C is on AB such that AC : CB = 1 : 3.
D is on OC such that OD : DC = 2 : 1.
Give the following vectors in terms of **a** and **b**.

a $\overrightarrow{AB}$ **b** $\overrightarrow{BA}$ **c** $\overrightarrow{AC}$ **d** $\overrightarrow{BC}$ **e** $\overrightarrow{OC}$ **f** $\overrightarrow{OD}$ **g** $\overrightarrow{DC}$

15 $\overrightarrow{OP} = \mathbf{p}$ and $\overrightarrow{OQ} = \mathbf{q}$.
R is the point on PQ
such that PR = kPQ.
Give in terms of **p** and **q**.

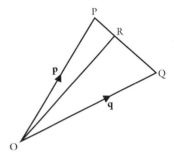

a $\overrightarrow{PQ}$ **c** $\overrightarrow{RQ}$

b $\overrightarrow{PR}$ **d** $\overrightarrow{OR}$

**USING VECTORS
TO MODEL
PRACTICAL
SITUATIONS**

So far we have visualised vectors as displacements but there are many
other quantities that have magnitude and direction, e.g. velocities,
accelerations, forces, magnetic fields. Problems involving vector
quantities can often be solved by representing the vector quantities by
lines.

If an athlete runs 100 m with a wind immediately behind him the time
will be faster than it would be without the wind. On the other hand a
wind blowing into his face will give a slower time than a wind blowing on
his back. Between these two extremes any wind blowing across the track
has an effect on the actual speed of the runner and hence on the time
taken. It is for this reason that, for a time to be acceptable as a record,
the wind speed must be very low. The effect of a wind on a moving
object can be found by using vectors.

For example, when an aeroplane is being steered due east but a
crosswind is blowing from the south-east, the actual flight path of the
aircraft will be somewhere between the two directions.

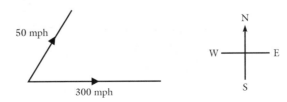

The actual flight path and speed are found using vector addition to combine the direction and speed of the wind with that of the aircraft due to its own power. The speed of the aircraft due to its own power is called 'its speed in still air'. We refer to the two velocities that combine to give the resultant velocity of the aircraft as *component velocities* or simply *components*.

If $\overrightarrow{AB}$ represents the velocity of the aircraft and $\overrightarrow{BC}$ the velocity of the wind then $\overrightarrow{AC}$ represents the resultant or actual velocity, that is it gives the actual speed and direction of the aircraft over the ground below.

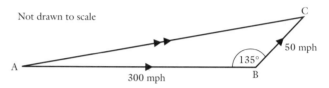

EXERCISE 6E

1   The motion of a body is made up of two velocities, **p** and **q**. Draw a diagram to show the resultant velocity, **p** + **q**.

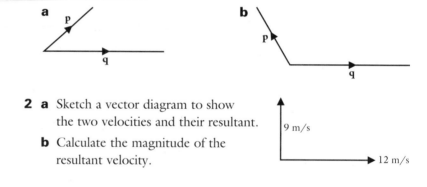

2 a Sketch a vector diagram to show the two velocities and their resultant.

   b Calculate the magnitude of the resultant velocity.

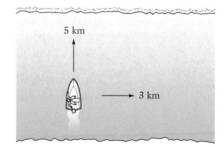

3   The diagram shows a boat crossing a river. Its motion is made up of two velocities. One is the current in the river. The other is the speed and direction set by the boatman.

   a Draw an accurate vector diagram to show the resultant velocity.

   b Find the angle between the resultant velocity and the direction of the current.

   c Check the accuracy of your drawing by calculation.

4 a Use the information given on page 100 to make an accurate drawing of ABC.

  b Hence find the direction in which the aircraft moves.
    Give your answer as a three-figure bearing.

  c Find the speed of the aircraft over the ground.

  d Check your answer to part **c** by calculation.

A plane needs to travel due north at 300 km/h. There is a cross wind of 90 km/h blowing from the west. Find the direction which the pilot should set.

We know the resultant velocity (300 km/h due north) of the plane and one of the two component velocities. First sketch the two known velocities separately.

Now we can draw a second combined diagram.

$\overrightarrow{CB}$ is the resultant,

$\overrightarrow{AB}$ is one component velocity

so $\overrightarrow{CA}$ is the second component velocity.

To find the direction of $\overrightarrow{CA}$ we need to calculate $\widehat{ACB}$.

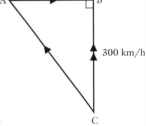

In $\triangle ABC$, $\tan \widehat{C} = \dfrac{90}{300} = 0.3$

$\therefore$ $\widehat{C} = 16.69\ldots°$

          $= 16.7°$ (correct to 1 decimal place)

The pilot has to set course on a bearing

$360° - 16.69\ldots° = 343.30\ldots°$

          $= 343.3°$ (correct to 1 decimal place)

Notice that, because of the wind, the speed of the plane due to its engine power has to exceed 300 km/h.

---

5 A ball rolls, at a speed of 4 m/s, across the floor of a carriage of a train which is travelling due south at 20 m/s. Find the speed and direction of the movement of the ball in relation to the ground.

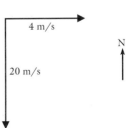

6 The resultant of two forces is 9 Newtons at an angle of 45° to the larger force. If the larger force is 8 Newtons find, by drawing an accurate diagram, the magnitude and direction of the smaller force.

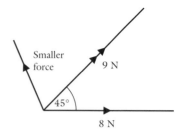

7 A pilot sets course due south at 120 km/h but, because of the wind, the plane actually flies at 130 km/h on a bearing of 150°. Find the speed and direction of the wind

a by drawing an accurate diagram   **b** by calculation.

8 A pallet with bricks loaded on it is pulled by two ropes, each inclined at 30° to the direction in which the pallet moves. The force in one rope is 10 Newtons. Find the force in the other rope.

9 A helicopter tries to fly at 60 km/h on a bearing of 085°, but is blown off course by a wind blowing at 30 km/h in the direction 175°.

a Make a scale drawing, and use it to estimate the magnitude and direction of the resultant velocity.

b Show by calculation that the magnitude of the resultant velocity is $30\sqrt{5}$ km/h.

c Calculate the bearing on which the helicopter actually flies, giving the answer correct to the nearest degree.

10 A boat, whose speed is set to 3 m/s (this is the speed it would have in still water) is driven across a river which is flowing at 1.5 m/s. Use scale drawings to answer the following questions.

a If the boat is pointed straight across the river, in what direction does it actually move?

b The boat needs to go across the river at right angles to the bank. In which direction should it be pointed?

c The speed of the current changes so that when the boat is pointed upstream at 70° to the bank, it actually moves downstream at 80° to the bank. What is the speed of the current?

d Check your answers to parts **a**, **b** and **c** by calculation.

11 Solve the problem stated at the beginning of the chapter.

FURTHER
PROBLEMS

If **a** and **b** are two non-parallel vectors, a vector **c** that is made up from **a** and **b** can be expressed in the form $h\mathbf{a} + k\mathbf{b}$, where h and k are numbers. If information is given that allows **c** to be found in another form, e.g. $3\mathbf{a} - \frac{1}{2}\mathbf{b}$, then we know that $h = 3$ and $k = -\frac{1}{2}$ because the coefficients of **a** and **b** must be the same in both forms.
In the following exercise you will see how this idea can be used in solving problems.

EXERCISE 6F

$\overrightarrow{OQ} = \mathbf{q}$, $\overrightarrow{OP} = \mathbf{p}$, $\overrightarrow{OR} = \frac{1}{3}\mathbf{p} + k\mathbf{q}$, $\overrightarrow{OS} = h\mathbf{p} + \frac{1}{2}\mathbf{q}$ and R is the midpoint of QS. Find h and k.

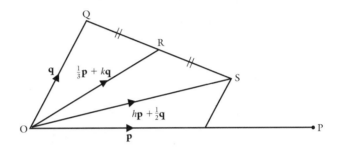

> We will find $\overrightarrow{QR}$ and $\overrightarrow{RS}$ in terms of **p**, **q**, h and k.

$$\overrightarrow{QR} = \overrightarrow{QO} + \overrightarrow{OR}$$
$$= -\mathbf{q} + \tfrac{1}{3}\mathbf{p} + k\mathbf{q}$$
$$= \tfrac{1}{3}\mathbf{p} + (k-1)\mathbf{q}$$

$$\overrightarrow{RS} = \overrightarrow{RO} + \overrightarrow{OS}$$
$$= -\left(\tfrac{1}{3}\mathbf{p} + k\mathbf{q}\right) + h\mathbf{p} + \tfrac{1}{2}\mathbf{q}$$
$$= \left(h - \tfrac{1}{3}\right)\mathbf{p} + \left(\tfrac{1}{2} - k\right)\mathbf{q}$$

But $\overrightarrow{QR} = \overrightarrow{RS}$ > R is the midpoint of QS.

$\therefore$ $\quad \tfrac{1}{3}\mathbf{p} + (k-1)\mathbf{q} = \left(h - \tfrac{1}{3}\right)\mathbf{p} + \left(\tfrac{1}{2} - k\right)\mathbf{q}$

Comparing coefficients of **p**, $\tfrac{1}{3} = h - \tfrac{1}{3}$ i.e. $h = \tfrac{2}{3}$

Comparing coefficients of **q**, $k - 1 = \tfrac{1}{2} - k$

$$2k = 1\tfrac{1}{2} \;\Rightarrow\; k = \tfrac{3}{4}$$

So $h = \tfrac{2}{3}$ and $k = \tfrac{3}{4}$

1

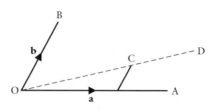

$\overrightarrow{OA} = \mathbf{a}$, $\overrightarrow{OB} = \mathbf{b}$, $\overrightarrow{OC} = \frac{2}{3}\mathbf{a} + \frac{1}{3}\mathbf{b}$.

D is the point such that $\overrightarrow{OD} = k\overrightarrow{OC}$.

a Find $\overrightarrow{OD}$ and $\overrightarrow{BD}$ in terms of **a** and **b**.

b If BD is parallel to OA, find the value of k.

c Find the ratio OC : CD.

2

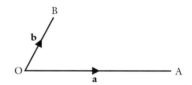

$\overrightarrow{OA} = \mathbf{a}$, $\overrightarrow{OB} = \mathbf{b}$,

C and D are the points such that $\overrightarrow{OC} = \mathbf{a} - \frac{1}{2}\mathbf{b}$ and $\overrightarrow{OD} = k\,\mathbf{a} + \frac{3}{4}\mathbf{b}$.

a Find $\overrightarrow{BD}$ in terms of **a** and **b**.

b If BD is parallel to OC find the value of k.

c Find $\dfrac{BD}{OC}$.

3

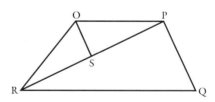

OPQR is a trapezium with OP parallel to RQ. $\overrightarrow{OP} = \mathbf{p}$, $\overrightarrow{OR} = \mathbf{r}$,

RQ = hOP and PS = kPR. Express in terms of **p** and **r**

a $\overrightarrow{RQ}$ **b** $\overrightarrow{PR}$ **c** $\overrightarrow{PQ}$

d $\overrightarrow{PS}$ **e** $\overrightarrow{OS}$

f If OS is parallel to PQ, find h in terms of k.

g If, in addition, $\dfrac{PS}{PR} = \dfrac{1}{2}$, find k and h.

4

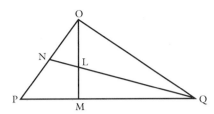

$\overrightarrow{OP} = \mathbf{p}$, $\overrightarrow{OQ} = \mathbf{q}$, M is a point on PQ such that
PM : MQ = 1 : 2. N is the mid-point of OP. LQ = hQN.

Give in terms of $\mathbf{p}$, $\mathbf{q}$ and h

a $\overrightarrow{PQ}$ **b** $\overrightarrow{PM}$ **c** $\overrightarrow{OM}$ **d** $\overrightarrow{ON}$ **e** $\overrightarrow{QN}$ **f** $\overrightarrow{QL}$ **g** $\overrightarrow{OL}$

h If OL = kOM, express $\overrightarrow{OL}$ in terms of $\mathbf{p}$, $\mathbf{q}$ and k.

i Using the two versions of $\overrightarrow{OL}$, find the values of h and k.

5 ABCDEF is a regular hexagon,
$\overrightarrow{AB} = \mathbf{a}$ and $\overrightarrow{AC} = \mathbf{b}$.

G is the point such that $\overrightarrow{CG} = \mathbf{b}$
and H is the point such that

$\overrightarrow{CH} = 2\mathbf{a} - \mathbf{b}$.

Find, in terms of $\mathbf{a}$ and $\mathbf{b}$.

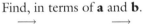

a $\overrightarrow{AD}$ **c** $\overrightarrow{EG}$

b $\overrightarrow{BE}$ **d** $\overrightarrow{HG}$

e Show that HG is parallel to EF.

f What type of quadrilateral is ADGH ?

6

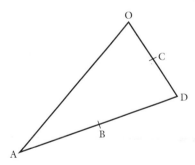

$\overrightarrow{OA} = \mathbf{a}$, $\overrightarrow{OB} = \mathbf{b}$ and $\overrightarrow{OC} = \mathbf{c}$.
B and C are the mid-points of AD and OD.

a Give $\overrightarrow{OD}$ and $\overrightarrow{AD}$ in terms of $\mathbf{a}$ and $\mathbf{c}$.

b Find $\mathbf{b}$ in terms of $\mathbf{a}$ and $\mathbf{c}$.

c E is a point on OA produced such that $\overrightarrow{OE} = 4\overrightarrow{AE}$.
   If $\overrightarrow{CB} = k\overrightarrow{AE}$ find the value of k.

7 O, A and B are the points (0, 0), (3, 4) and (4, −6) respectively. C is the point such that $\overrightarrow{OA} = \overrightarrow{OC} + \overrightarrow{OB}$. Find the coordinates of C.

MIXED EXERCISE

EXERCISE 6G

1

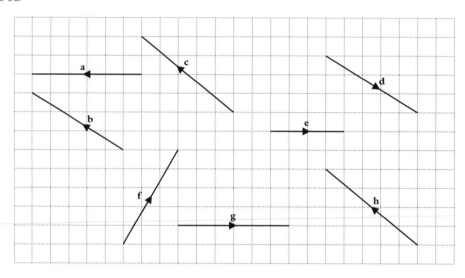

Look at the vectors in the diagram.

a Find the magnitude of the vector
  i a   **ii h**

b Which two vectors are equal?

c Which two vectors are equal in magnitude but opposite in direction?

d Which vector is 50% larger than **e**?

2 Copy the vectors **p** and **q** onto a grid.
On your grid draw line segments to represent the vectors

  a p + q

  b p − q

  c $3p - \frac{1}{2}q$

  d 2p + 3q

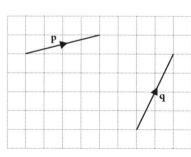

3 Find the single vector that is equivalent to

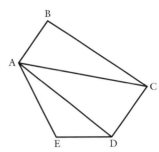

a $\overrightarrow{AE} + \overrightarrow{ED}$

b $\overrightarrow{AC} + \overrightarrow{CD} + \overrightarrow{DE}$

c $\overrightarrow{AC} - \overrightarrow{BC}$

d $\overrightarrow{AE} + \overrightarrow{ED} - \overrightarrow{CD}$

4

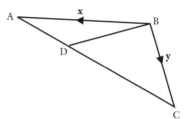

$\overrightarrow{BA} = \mathbf{x}$ and $\overrightarrow{BC} = \mathbf{y}$. D is the point on AC such that AD : DC = 1 : 2. Give in terms of $\mathbf{x}$ and $\mathbf{y}$

a $\overrightarrow{AC}$ **b** $\overrightarrow{AD}$ **c** $\overrightarrow{CA}$ **d** $\overrightarrow{CD}$

e $\overrightarrow{BD}$ can be given as $\overrightarrow{BA} + \overrightarrow{AD}$ or $\overrightarrow{BC} + \overrightarrow{CD}$.

Find $\overrightarrow{BD}$ in terms of $\mathbf{x}$ and $\mathbf{y}$ by the two different ways.

5 A ferry boat, which needs to sail due west, is affected by a current of 6 km/h flowing from a direction of 20° east of north. The speed of the ferry in still water is 24 km/h.

a On what course should the ferry be steered to travel in the desired direction?

b What is the actual speed of the ferry through the water?

6

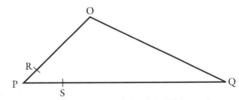

In the diagram $OR = \frac{4}{5}OP$, $\overrightarrow{OP} = \mathbf{p}$, $\overrightarrow{OQ} = \mathbf{q}$ and PS : SQ = 1 : 4.

a Express $\overrightarrow{OR}$, $\overrightarrow{RP}$ and $\overrightarrow{PQ}$ in terms of $\mathbf{p}$ and $\mathbf{q}$.

b Express $\overrightarrow{PS}$ and $\overrightarrow{RS}$ in terms of $\mathbf{p}$ and $\mathbf{q}$.

c What conclusion do you draw about RS and OQ?

d What type of quadrilateral is ORSQ?

e The area of $\triangle$PRS is 5 cm². What is the area of ORSQ?

Baldrick has enjoyed a relaxing day on the river and is rowing downstream back towards his starting point. As he passes under a bridge, unknown to him his hat falls into the river. Luckily his hat floats. He continues to row downstream and after 20 minutes realises what has happened. The river is flowing at 4 km/h and Baldrick can row in still water at 5 km/h. How long will it take him to row upstream to recover his hat and how far will he be downstream from the bridge at this time?

A river flows at v m/s and is 100 m wide. The greatest speed at which a boatman can row in still water is $2v$ m/s. The boatman wishes to cross the river, starting at A and landing at B, the point on the other bank immediately opposite A.

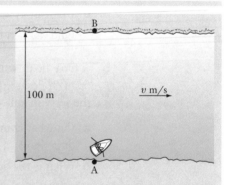

a Investigate whether or not he can travel directly from A to B. If he can, must he row at one particular speed or is there a range of speeds that he can choose from? Justify your answer.

b If the speed of the river increases to $1.5v$ m/s can he still travel directly across the river from A to B? What happens if the speed of the river increases to $2v$ m/s?

c Investigate the maximum speed of the river that allows him to travel directly from A to B.

d He points his boat towards B and rows at the same speed as the water is flowing in the river. How far downstream from B will he land?

e Will your answers to the previous parts change if the width of the river is 150 m? Justify your answers.

SAMPLES

The town planners proposed making the centre of a small town a traffic-free zone between the hours of 8 a.m. and 10 p.m.

For a statistics project, Raj decided to investigate attitudes to this plan. The question he posed was

> All traffic should be banned from the town centre from 8 a.m. to 10 p.m. Do you
>
> STRONGLY AGREE   AGREE   HAVE NO OPINION
>   DISAGREE   STRONGLY DISAGREE?

Raj was at school during the day so he put this question to people between 8 a.m. and 8.30 a.m. and between 5 p.m. and 5.30 p.m. The place he chose to find people was at the pedestrian exit of the main town centre car park.

His results showed that the overwhelming majority disagreed or strongly disagreed with the plan.

One problem associated with collecting information is how to make sure that the questions posed are easily understood, unambiguous, likely to be answered truthfully and are not phrased so that they lead to a 'correct' response.

This example illustrates some of the other problems associated with collecting useful data.

- For the results to be 100% reliable, Raj would have to ask everyone who uses the town centre. This is impossible for at least two reasons; not all the people who use the town centre can be identified and, even if they could, the time needed to question them all and analyse the results would take far more time than Raj has available.
- Members of interested groups, such as parents with young children, are not likely to be included among the people questioned because of the times when Raj gathered the evidence. The place he chose to find people means that the majority questioned are likely to be car drivers and others who have an interest in keeping car access.

SAMPLES

From the conclusion above we see that we may have to use only part of all the relevant information for a statistical investigation.

The selected information taken from the possible total is called a *sample*.

BIAS

The conclusion also shows that the way a sample is chosen can lead to conclusions that are probably not the same as those that would be reached if all the information could be analysed. This could be because the sample does not represent the make-up of the total pool of information from which it is drawn. Such a sample is called *unrepresentative*.

A sample can be unrepresentative by chance. For example, if a sample of four cards is selected at random from an ordinary pack of playing cards, the four cards could all be aces; these are not representative of the complete pack but this is because of chance.

The sample that Raj used is also unlikely to reflect the spread of views that all the people using the town centre are likely to hold. This is not because of chance but is a consequence of the way he chose the sample.

A questionnaire can also be worded in such a way as to produce unrepresentative replies. For example, Raj could have worded his question so that it was loaded towards agreement, e.g.

> Traffic pollution causes high levels of respiratory illness in this town, so the plan to make the centre traffic free is beneficial. Do you

> STRONGLY AGREE AGREE HAVE NO OPINION
>   DISAGREE STRONGLY DISAGREE

A sample that is unrepresentative for reasons other than chance, is called a *biased* sample.

EXERCISE 7A

1 The information for an opinion poll on voting intentions in a local election was collected between the hours of 9 a.m. and 5 p.m. by telephoning people chosen at random from the local telephone directory.

   a Do all people on the electoral register have a telephone?

   b Are all telephone numbers in the telephone directory?

   c Is everyone listed in a telephone directory eligible to vote?

   d Are all members of a household reached by telephone equally likely to be questioned?

   e Are all the members of the household equally likely to be at home between 9 a.m. and 5 p.m.?

   f Is this method of collecting information likely to give an unbiased sample?

2 A questionnaire was distributed by a bank to all its personal banking customers, by post, seeking their opinions on the satisfaction with services offered by the bank. The recipients were asked to complete the forms and return them.

   a Are all the questionnaires likely to be returned?

   b Are some customers more likely than others to return the forms? If so, what distinguishes these customers from others?

   c What advantages are there for this way of collecting information?

3 A group of students undertook a survey and each needed to obtain information from 10 men between the ages of 20 and 60 as part of the sample. The students all went to different locations to get their quota. Give brief reasons why each of the following locations might introduce bias.

   a The exit from the local supermarket between 11 a.m. and 12 a.m.

   b The bus stop between 8 a.m. and 9 a.m.

   c The local cinema in the half hour before the start of the evening programme.

4 Decide whether a sample *has* to be used to obtain information about each of the following cases. Give reasons for your decision.

   a The weights of grey squirrels.

   b The mass of nitrates per litre of water in a river.

   c The sizes of pebbles on a beach.

   d The life of light bulbs sold under a supermarket brand name.

   e The time spent on the mathematics coursework by all the candidates entering GCSE this year.

5 The population of a village is 4500. A group of activists in the village wanted to stop the local council removing the recreation ground in the centre of the village. They telephoned 50 people in the village one evening and asked 'Have you been to the recreation ground this week?' and 24 people said yes. They concluded that 48% of the people that lived in the village used the recreation ground. Give at least three reasons why this result is unreliable.

6 The group described in question **5** want to find out how many people living in the village use the recreation ground and how often. Write down two questions that could be asked with at least three alternative responses to choose from that will give the group reasonably reliable information.

CHOOSING A
SAMPLE

As we have seen, to avoid bias we need to choose a sample that is likely to reflect the information that we would get if we could use all the items.

We must first decide on the size of the sample; common sense says that it should be as large as possible. However the time and resources available, as Raj found, usually place a limit on the size of a sample.

The other consideration is how to choose the items that make up the sample. To make sure that bias does not affect the choice, we try to ensure that every item of information has an equal chance of selection.

RANDOM
SAMPLES

A sample that is chosen in such a way that every possible item that could be part of the sample has an equal chance of being selected, is called a *random sample*.

A *simple random sample* is chosen by first allocating a number to each item that could be chosen.

Suppose that we want to select a random sample of 5 people from the 13 members of a dining club. We start by allocating a number, from 1 to 13, to each member.

The people chosen for the sample can then be selected by any method that produces numbers at random.

One method is to use tickets (small pieces of paper) numbered 1 to 13. These can be folded, put in a box and thoroughly mixed; five tickets are then drawn out. The members whose numbers correspond to those on the tickets withdrawn form the sample.

The main disadvantage of the 'raffle' technique is that it is very time-consuming unless the total number of items involved is really small. Another problem is that we cannot be certain that the choice is really random; it may be influenced by subtle personal preferences such as the shape of a particular folded ticket.

USING RANDOM
NUMBERS

Using a mechanical process to choose the numbers avoids any chance of human preference introducing bias.

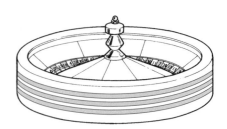

Most scientific calculators have a function that generates random numbers. On a Casio calculator, the key is marked Ran # and produces numbers between 0 and 1, each to 3 decimal places. Using this function five times gave the numbers 0.719, 0.664, 0.423, 0.618, 0.115.

As these are random numbers between 0 and 1, we can think of each of them as representing the item which is that fraction of the way through the list of 13 cards,

i.e. 0.719 represents the 0.719×13th member on the list, and so on. Now $0.719 \times 13 = 9.347$ and, as there is no such thing as a 9.347th card, we round up to the next integer, 10.

These calculations are always rounded up because this ensures that 1 is the lowest number that will occur and 13 is the highest.

Using the other numbers generated gives

$$0.664 \times 13 = 8.632 \Rightarrow 9$$
$$0.423 \times 13 = 5.499 \Rightarrow 6$$
$$0.618 \times 13 = 8.034 \Rightarrow 9 \quad \text{(this is a repeat, so we need another random number)}$$
$$0.115 \times 13 = 1.495 \Rightarrow 2$$
$$0.003 \times 13 = 0.039 \Rightarrow 1$$

We now have 5 different numbers; 10, 9, 6, 2, 1. Therefore the members allocated the numbers 1, 2, 6, 9 and 10 form the sample.

Note that using a 3 decimal place random number generator in this way will not produce a random sample if the selection is from a list of more than 1000 items, for example $0.001 \times 3000 = 3$, so item 2 out of a list of 3000 has no chance of selection. (The random number generator in a computer usually produces 7 decimal places so the problem will not arise unless the list is greater than 10 million.)

The advantage of using random number generators is that they are quick and easy to operate, particularly if a computer program is used.

There is one major disadvantage; the numbers produced are not really random numbers because the sequence repeats after a given number of steps. For this reason they are called pseudorandom numbers.

USE OF RANDOM NUMBER TABLES

If the use of a random number generator is not considered acceptable, or is unavailable, published tables of random numbers can be used. These tables consist of lists of the digits 0 to 9 in random order. The digits are arranged in rows and columns,

e.g.

5956	0827	8682	7958	0381	3185
2608	4397	4522	6363	2032	4740
0550	9879	3274	1666	7375	6793
6743	1502	6553	2651	2757	9540
7372	3469	4466	2211	0605	5948
0556	4336	2795	2035	6840	6863...

There is no significance in any particular division of the digits into rows and columns; it simply makes the tables easier to read.

By choosing a starting point at random, we can read either across the rows or down the columns to get a string of random digits.

To get our five random numbers from the list of 13, we need two-figure numbers between 01 and 13. Starting at the fourth figure in the second row, and reading across in groups of two, gives

84, 39, 74, 52, 26, 36, 32, 03, 24, 74, 00, 55, 09, 87, 93, ...

Discarding duplicates, and all numbers larger than 13, leaves 03, 09, ...

We have to continue to nearly the end of the second row in the lower block of numbers before finding five numbers between 1 and 13, so although this method is easy and direct, it can be lengthy (as this example shows); also it does not use all the numbers.

These problems can be overcome by using the first five numbers and treating them as two-figure decimals in the same way as the calculator numbers are used, i.e. $0.84 \times 13 = 10.92 \Rightarrow 11$, $0.39 \times 13 = 5.07 \Rightarrow 6$, and so on.

Note that if random number tables are used to select more than one sample, it is important that a *different* starting point is used each time.

EXERCISE 7B

Explain how you would select a random sample of four bottles of wine from a rack of 123 bottles using the following random numbers.

4 3 8 8 2 3 9 6 7 3 2 9 2 0 0

First number the bottles in the rack from 001 to 123.

> We cannot select four three-figure numbers in the range 001 to 123 from this set, so we will use the decimal number method. We can start anywhere in this list.

Use groups of three digits as the three decimal places of a number between 0 and 1.

Multiply each by 123 and then round up to the next integer. Starting at the beginning of the list, the sample is those bottles with the following numbers:

$$0.438 \times 123 = 53.8 \ldots \Rightarrow 54$$
$$0.823 \times 123 = 101.2 \ldots \Rightarrow 102$$
$$0.967 \times 123 = 118.9 \ldots \Rightarrow 119$$
$$0.329 \times 123 = 40.4 \ldots \Rightarrow 41$$

1 Use the list of random numbers on page 113 to select two numbers

   a between 1 and 50, starting with the second digit in the first row

   b between 1 and 670, starting with the fourth digit in the first row

   c between 1 and 2500, starting with the third digit in the first row.

2 There are 450 students enrolled at colleges of further education in Wessex for a one-year evening course in keyboard skills. Describe how you would select 10% of these students to make a random sample.

3 Use the random numbers given below to select a sample of three children from a class of 30 children and explain how you do it.

6 9 0 7 4 0 9 1 9 7 6

4 This diagram shows twenty rods of varying lengths.

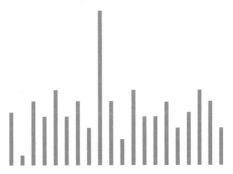

a Choose a sample of three rods which you think are representative of the twenty rods. Measure and write down their lengths. Find the mean length of the rods in your sample. (*Reminder* The mean length is the sum of the individual lengths divided by the number of rods.)

b Repeat part **a** 4 times. (It does not matter if a rod is included in more than one sample.)

c Number the rods 01 to 20 and then repeat parts **a** and **b** but this time use random numbers (from a table or calculator) to select the samples.

d Compare the sample means obtained from the random samples with those of the samples that you selected.

e Measure all the rods and find the mean length of the twenty rods. Which method of selecting samples do you consider to be better? Justify your answer.

5 The response to a competition resulted in 76 correct entries but there were only 10 prizes. The correct entries were numbered 01 to 76 and the 10 prize winning entries were selected by taking two-digit random numbers from a random number table and, if the number was greater than 76, subtracting 76 and using the resulting number. Explain why this method does not give each entry an equal chance of being selected.

One of the disadvantages of simple random sampling is that it may not be representative. Suppose that there are 200 girls and 50 boys who are members of a junior tennis club. A random sample of these members will not necessarily contain numbers of girls and boys in the same proportion as the membership. This does not matter if the gender of a member has no effect on the problem being investigated, eye colour, for example. If the problem has any relationship to gender, such as height, a sample that contains girls and boys in the same proportion as the members of the club is more likely to reflect that aspect of the members.

We now define exactly what is meant by a representative sample.

> A representative sample has numbers of each distinct group in the same proportion as their numbers in the complete set.

A stratified sample produces a representative sample by first dividing the set into groups. Then random numbers are used to select members from each group in numbers proportional to their occurrence in the whole set.

To produce a stratified sample of 25 from our 200 girls and 50 boys, we need first to find the number of girls and the number of boys required.

The sample size as a fraction of the total group is $\frac{25}{250}$, i.e. $\frac{1}{10}$

so we need $\frac{1}{10}$ of the 200 girls, i.e. 20 girls

and $\frac{1}{10}$ of the 50 boys, i.e. 5 boys

We then use random numbers to select this number of girls and boys from the respective lists.

There are many ways of dividing a set into groups and which particular groups are chosen will depend on the nature of the investigation. For the members of the junior tennis club, division into age groups might be appropriate for investigating attitudes towards pocket money, while division into groups according to height might be appropriate for an investigation into provision of club strip.

The advantage of stratified sampling is that it ensures proper representation of each group and still keeps an equal chance of selection for each member of the population.

The disadvantage is that division of a population into groups is not always easy or clearcut. For example, if the population is all people living in Greater London, the proportion of this population that is under the age of 20 can only be, at best, an informed guess. If the members of the tennis club are divided into good and average players, human judgment is needed to place an individual in one group.

Situations like these can introduce bias into the sample.

EXERCISE 7C

A consignment of 3560 bottles of water is delivered to a restaurant.

The number of bottles of each type of water is given in the table.

Category	Still	Carbonated	Lemon-flavoured	Lime-flavoured
Number	1280	1980	55	245

How many bottles of each type of water should be selected to give a representative sample that is 5% of the consignment?

For a 5% sample, we need 5% of the bottles of each type.

5% of 1280 = 64

5% of 1980 = 99

5% of 55 = 2.75 = 3 to the nearest whole number

5% of 245 = 12.25 = 12 to the nearest whole number

Rounding each calculation to the nearest whole number gives a total of 178 bottles for the sample. We need to check that this is 5% of the total number of bottles delivered.

The number delivered = 3560, and $\frac{178}{3560} = 0.05$
$= 5\%$ of the number delivered.

Sometimes the rounding process does not give the correct total. If this happens, and the total is too small, the largest decimal part that would normally be rounded down is rounded up instead, and so on until the correct total is reached. If the total is too large then the opposite procedure is adopted.

We need 64 bottles of still water,
99 bottles of carbonated water,
3 bottles of lemon-flavoured water
and 12 bottles of lime-flavoured water.

1 The table shows the number of students enrolled in each department of a college.

Department	Catering	Academic	Engineering	Sport	Technology
Number of students	590	280	770	60	520

How many students should be chosen from each department to give a 10% representative sample?

2 There are 180 pupils on the register at Lower Hill Primary School. The table shows the number of pupils in each year.

Year	1	2	3	4	5	6
Number	34	33	29	28	35	21

How many students need to be selected from each year to give

a a 10% representative sample

b a representative sample of 30 pupils?

3 A farmer grew the same variety of potatoes in four different fields. The table shows the yield, in tonnes, from each of these fields.

Field	A	B	C	D
Yield (tonnes)	250	220	140	90

a What weight should be selected from each field to give a representative sample of one tonne?

b The purpose of the sample is to try to quantify the number of damaged potatoes in the whole crop. What other method of sampling could be used? Give reasons for your choice.

4 For some coursework, Gary decides to investigate the pupils' attitudes to wearing school uniform in his school, which is coeducational. He decides to choose a sample. How should he choose his sample so that his results are reasonably reliable? Give as many reasons for your answer as possible.

5 A food producer has three machines that each fill 100 ml cartons with juice. Machine A fills 200 cartons an hour, Machine B fills 150 cartons an hour and Machine C fills 180 cartons an hour.

a Each hour a sample of 5% of all the cartons filled are examined to check the quantity of juice in each carton. How many cartons are examined?

b How many cartons from each machine should be examined to give a stratified sample?

c Why is a stratified sample necessary?

d When one of the samples was examined, 20 cartons were not correctly filled. Estimate how many of the cartons produced in the previous hour were incorrectly filled. Explain why your answer is an estimate.

6 At the beginning of this chapter we described a statistical investigation that Raj undertook. The way in which he chose his sample introduced bias.

a Ideally, when and where should Raj ask people for their views to avoid obvious bias?

b Given that Raj cannot go into the town centre during school hours, suggest some practical ideas that Raj could use to make his sample more representative.

7 A university education department wants some information about pupils in the first year of secondary schools. It is decided to gather the information from 10 per cent of the pupils. Here are some ways in which those pupils can be selected. There are no definitive answers to the questions that follow; they invite you to consider some of the reasons why the sample may not be representative.

A list of all secondary schools in the country can be obtained and the pupils in the schools selected using a simple random sample of 10% of the schools.

a Does this method of selecting a sample take account of independent schools?

b How do the differing arrangements for the age at which children change school affect this method of selection?

c Is this sample likely to be representative of the different types of school, i.e. single sex, selective, comprehensive and so on?

Instead of using a simple random sample, every tenth school in the list can be selected.

d Explain why this method is also unlikely to give a sample that is representative of the different types of school.

Alternatively the list of schools could be sorted into categories such as single-sex, church schools, etc. One tenth of the schools in each category could then be selected.

e This method of selection is likely to give a more representative sample than the first method, but what can be done about a school which is in more than one category?

f What other drawbacks can you think of?

g Another method of selection that overcomes the difficulty of categorising schools is to select one tenth of the pupils from every secondary school in the country. Can you suggest ways in which one tenth of the pupils in the first year of a school can be easily chosen so that any one pupil is as likely as any other to be selected?

This question investigates the reliability of sample size and is best done by working in a group of 3 or 4. You will need an ordinary pack of playing cards.

a If four cards are chosen at random, how many hearts should there be for these four cards to be representative of the proportion of hearts in the full pack?

b Shuffle a full pack of playing cards thoroughly and then deal out the top 4 cards. This is a sample of size 4. Write down the number of hearts.

c Replace the cards dealt for the first sample and then repeat part **b**.

d Repeat the process several more times.
(You will need at least 30 samples.)
Keep track of your results in a table like this one:

Sample	1	2	3	4	. . .
Number of hearts					

e Use your table to copy and complete this frequency table.

Number of hearts	0	1	2	3	4
Frequency					

f From your frequency table, find the relative frequency
(i.e. frequency divided by total number of samples) of samples that do represent the proportion of hearts in the whole pack.
Hence give an estimate of the probability that a sample of size 4 represents the proportion of hearts in the pack.

g Now repeat parts **b** to **e** using a sample of 20 cards.

h A sample of 20 cards should contain 5 hearts to represent exactly the proportion of hearts in the complete pack. If we allow 4, 5 or 6 hearts in a sample as 'good enough' to be representative, find the relative frequency of such samples. Hence estimate the probability that a sample of size 20 gives a representative proportion of hearts in the pack.

INVESTIGATION

This investigation will give you some insight into the problems involved in sampling. Your answers should be supported with reasons. Find some books on the subject to help research your ideas.

A company that distributes fresh fruit, imports 50 000 peaches. The company needs to have some idea of the quality of the peaches, particularly how many are likely to be damaged or bad. Suppose that they are prepared to accept a consignment if they think that not more than 5 per cent of the peaches are damaged. There are various ways in which they can come to a decision.

a They could make sure that every peach is inspected. The obvious advantage of doing this is that they would be certain about the proportion of bad peaches. What are the disadvantages? List as many as you can think of.

b They could inspect a sample of, say, 50 peaches. Suppose that this sample contained 2 damaged peaches, and on this evidence they decided to accept the whole consignment. Is it possible that their decision resulted in accepting a consignment in which considerably more than 5 per cent were damaged? On the other hand suppose that the sample had 20 damaged peaches; assuming that they rejected the consignment on this evidence, could they have rejected a 'good' consignment?

c Based on the evidence from a sample, is there likely to be any cost involved in either accepting a 'bad' consignment or rejecting a 'good' consignment?

d If consignments are delivered every day for four months, can you give an intuitive idea of the proportion of consignments that are incorrectly accepted or rejected on the evidence from a sample of about 50 peaches? (By intuitive, we mean an intelligent guess based on experience – it is not intended that you should try to understand the theory involved.)

BOX PLOTS AND HISTOGRAMS

As part of a survey, Rhona wrote to several companies asking for information about part-time earnings. One company, Company A, supplied this table.

Pay (£)	0–1999	2000–3999	4000–5999	6000–9999	10 000–19 999
Frequency	20	36	25	14	5

This table was provided by another company, Company B.

Pay (£)	0–1999	2000–3999	4000–5999	6000–7999	8000–9999	10 000–11 999
Frequency	72	85	81	93	74	62

Rhona wanted to analyse this information so that she could compare the two sets of figures.

She started by drawing cumulative frequency polygons and then used them to find the medians and the upper and lower quartiles.

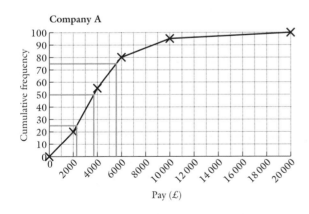

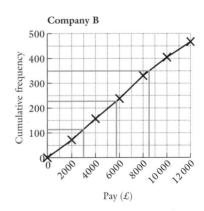

Reading from the diagrams we see that, while the range of pay for Company A is greater than the range for Company B, both the median and the middle 50% of the range (i.e. the interquartile range) is lower for Company A than for Company B.

The cumulative frequency polygons do not give a clear visual impression of these differences but there is a way of illustrating just the range and interquartile range that can be used to compare these aspects of the data. These are called box-and-whisker plots, or simply box plots.

BOX-AND-WHISKER PLOTS

A box-and-whisker plot is constructed by drawing a number line showing the values of the variable (£ in this case) then drawing a rectangle between the upper quartile and lower quartile (the box) to represent the middle 50% of the distribution. A line is drawn across the box at the median value. Then lines (the whiskers) are drawn from the edges of the box to show the bottom 25% and the top 25% of the range.

This diagram shows box-and-whisker plots for the distribution of the part-time earnings of employees in the two companies.

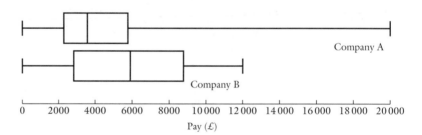

Notice that the box plot shows the range of each quarter of the earnings in each distribution but it tells us nothing about the number of earnings involved.

Now you can see clearly that the range of part-time earnings in Company A is nearly twice that in Company B but the bulk of earnings in Company A are lower than the bulk of earnings in Company B. This means that, with the exception of a small proportion, the part-time earnings in Company A are lower than those in Company B (the long right-hand whisker on the top box plot shows that there are a few employees with considerably larger part-time earnings than most). It is tempting to go on to say that Company A pays less per hour than Company B, however you *cannot* draw such conclusions because you know nothing about the nature of the work or the numbers of hours worked by each employee.

EXERCISE 8A

This cumulative frequency curve illustrates the distribution of the masses of 1000 apples.

Draw a box-and-whisker plot and explain why the interquartile range may be better than the range as an indication of the spread of the masses.

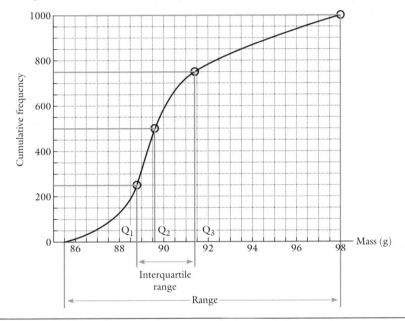

To draw a box-and-whisker plot we need the median and the quartiles. There are 1000 apples so the lower quartile is the mass of the 250th apple, the median is the mass of the 500th apple and the upper quartile is the mass of the 750th apple.

From the diagram, the lower quartile is 88.8 g, the median is 89.6 g and the upper quartile is 91.4 g.

Now we can draw the box-and-whisker plot.

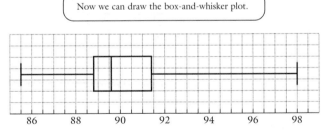

The heaviest quarter of the distribution has a range greater than the remaining three-quarters of the apples. This means that there are a few apples that are considerably heavier than the rest. Using the interquartile range as a measure of spread ignores these few heavier apples and is probably more representative of the spread of the masses of the majority of the apples.

1 Pupils in Year 9 were all given the same test.
The diagram shows box plots for the girls' results and the boys' results.

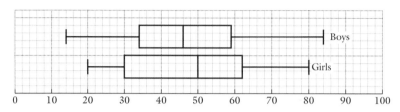

a What is the median mark for the boys?

b What is the highest mark achieved by a pupil in the test?

c Find the interquartile range for the girls' results.

d Describe two differences between the boys' marks and the girls' marks.

2 This box plot summarises the prices achieved at a car auction.

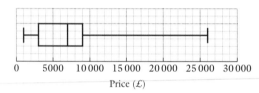

Price (£)

a What was the median price achieved?

b What was the highest price achieved?

c Is it true that 25% of the cars sold for £9000 or more? Justify your answer.

d 500 cars were sold at this auction.
How many cars were sold for the lower quartile price or less?

3 David weighed each of the potatoes in two 2.5 kg bags. One bag was labelled King Edwards. The other bag was labelled Reds. These two box plots summarise his results.

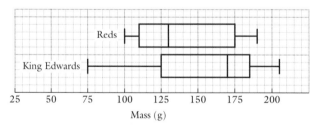

Mass (g)

a Which variety has the greatest range of masses?

b What is the interquartile range of the masses of the King Edward potatoes?

c What is the median mass of the Red potatoes?

d Write down two comparisons between the masses of potatoes in the two bags.

e Which of these two bags would you choose if you wanted potatoes that were of a fairly uniform size?

4 These figures summarise the lengths of some cucumbers.

Range: 22 cm to 36 cm
Interquartile range: 28 cm to 32 cm
Median: 30 cm

Draw a box plot to illustrate this information.
(Start your number line at 20 cm.)

5 All the pupils in Year 10 in a school were given the same literacy
test. A summary of the results is: lowest mark: 18
                                      highest mark: 84
                                      lower quartile: 44
                                      median: 59
                                      upper quartile: 65

a Draw a box plot to illustrate these results.
Use graph paper with 1 cm to represent 10 marks.

b Any pupil with a mark of less than 40 is offered special lessons.
Is it true that less than one quarter of the pupils are offered
special lessons ? Justify your answer.

6 Avril compared the number of words in each sentence in the leader
columns in two national newspapers. One paper was a tabloid and
the other was a broadsheet.
This is a summary of the results for the broadsheet:
range: from 5 to 25 words
interquartile range: from 12 to 20 words
median: 16 words.

This box plot summarises the results for the tabloid.

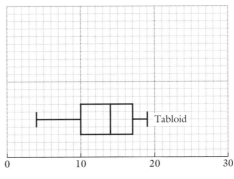

Number of words
in a sentence

a Copy the diagram and add a box plot to illustrate the results for
the broadsheet.

b Make two comparisons between the number of words in the
sentences in the tabloid and the number of words in the
broadsheet.

7 Ali measured the heights of two groups of tomato seedlings two weeks after they were sown. One group was from seeds sown outdoors in the ground. The other group was from seeds in pots in a greenhouse.

This is a summary of the results for the outdoor group.

Range: from 3 cm to 8.5 cm
Lower quartile: 5.4 cm
Upper quartile: 6.2 cm
Median: 5.9 cm

This box plot illustrates a summary of the greenhouse group.

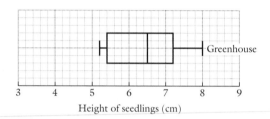

Height of seedlings (cm)

a Copy the diagram and add a box plot to illustrate the distribution of the heights of the outdoor group.

b Give two comparisons between the heights of the two groups of seedlings.

8 This cumulative frequency curve illustrates the distribution of the masses of the parcels delivered by a courier one morning.

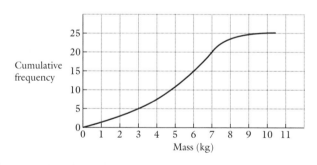

Mass (kg)

a Use the curve to estimate the median mass and the interquartile range of the masses.

b Draw a box plot to illustrate this distribution.

9 A group of people were asked to estimate the length of a brass tack. This cumulative frequency curve illustrates the distribution of those estimates.

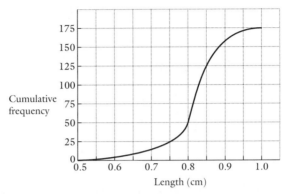

Cumulative frequency

Length (cm)

a Estimate the median length and the interquartile range.

b Illustrate this distribution with a box plot.

c The actual length of the tack is 0.85 cm.

Is it true that more people overestimated its length than underestimated its length?
Justify your answer.

10 Some screws were selected from a box of mixed screws and their lengths were measured. The results are recorded in the table.

Length, l cm	Frequency	Cumulative frequency
$0 \leqslant l < 0.5$	8	
$0.5 \leqslant l < 1.5$	20	
$1.5 \leqslant l < 2.5$	15	
$2.5 \leqslant l < 3.5$	9	
$3.5 \leqslant l < 4.5$	4	
$4.5 \leqslant l < 5.5$	2	
$5.5 \leqslant l < 6.5$	1	
$6.5 \leqslant l < 7.5$	1	

a Copy and complete the table.

b Construct a cumulative frequency curve.

c Construct a box plot to illustrate this information.

d Would you use the range or the interquartile range to describe the spread of lengths?
Justify your answer.

Box plots give a simple way of showing how values are spread through the range and they are particularly useful for comparing the spread of two or more sets of information but they give no indication about the number of values involved. However, bar charts do include information about numbers of values as well as giving an impression of the spread of those values.

Rhona decided she needed to use bar charts as well as box plots.

This is the information that Company A supplied.

Pay (£)	1–1999	2000–3999	4000–5999	6000–9999	10 000–19 999
Frequency	20	36	25	14	5

To illustrate the information, Rhona drew this bar chart.

Figure I

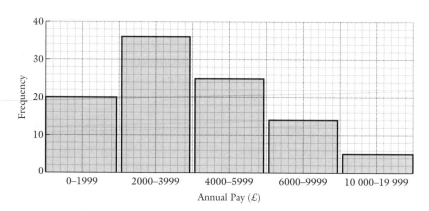

She realised that it was unsatisfactory because the width of the bars gives the impression that the spread of incomes in the last group is the same as the spread in the first group; it is in fact five times larger.

She then drew this chart but was not happy with the impression it gave.

Figure II

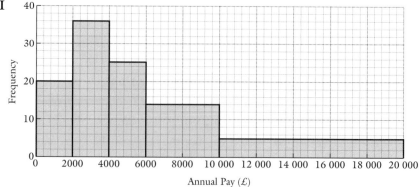

EXERCISE 8B

1 What causes problems about the way in which the information is given to Rhona?

2 Look at the bar chart in Figure I. How is the horizontal axis scaled? What is the disadvantage of doing it this way?

3 Look at the first bar and the last bar in Figure II. What impression do they give about the number of incomes in each of these groups?

4 From Figure II, how many people appear to have a part-time income of between £10 000 and £12 000 a year? Is this justified?

5 What aspect of the bars give the visual impression of the numbers represented by those bars?

6 This is the table that was provided by company B.

Pay (£)	0–1999	2000–3999	4000–5999	6000–7999	8000–9999	10 000–11 999
Frequency	72	85	81	93	74	62

Gavin illustrated the information with this chart.

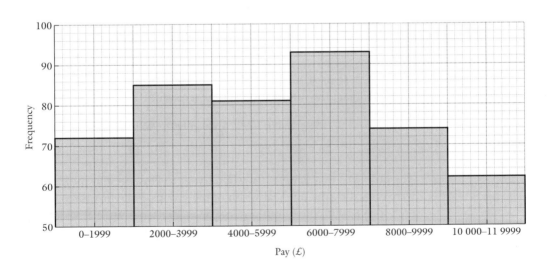

What impression does this chart give of the number of people earning below £2000 a year compared with the number of people earning £10 000 a year or more? What aspect of the bars leads to this impression? Do the facts support this impression? Justify your answer.

HISTOGRAMS

Discussion from the last exercise indicates that, when we look at a bar chart, it is the *area* of the bar that gives a visual impression of the number of items represented by that bar.

Therefore, to give a visual impression that reflects the true nature of the data, we need to construct a bar chart so that the area of each bar is proportional to the frequency of items in the group represented by that bar.

A bar chart drawn this way is called a *histogram*.

A standard bar chart is a histogram when the groups are all the same width, the frequency scale starts from zero, and the scale on both axes is uniform. For example, when the data given in question **6, Exercise 8B**, is represented by this bar chart then the area of each bar is proportional to the height of the bar (which is the frequency) because the width of all the bars is the same, so this bar chart is a histogram.

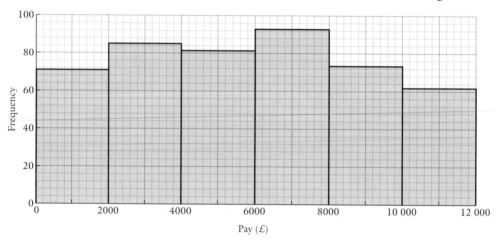

UNEQUAL WIDTH GROUPS

Consider again the information Rhona received.

Pay (£)	0–1999	2000–3999	4000–5999	6000–9999	10 000–19 999
Frequency	20	36	25	14	5

The problems that Rhona had when trying to illustrate this information were caused by the unequal width of the groups. The first three groups each span £2000, the fourth group spans £4000 so is twice as wide as the first three and the last group spans £10 000 so it is five times as wide as each of the first three groups.

To construct a histogram we must make the *area* of each bar represent the frequency of items in the group. We do this by making

- the width of a bar the same as the width of the group it represents

- the height of a bar equal to $\dfrac{\text{frequency}}{\text{width of group}}$;

this fraction is called the *frequency density*.

We can choose the unit that denotes the width of the bars. We might, for example, use £1 as the unit of width, but this would give very small values for the frequency densities. As the width of the narrowest group is £2000, it is better to choose this as our unit of width.

Adding two more rows to the table helps organise the calculations to find the frequency density for each group.

Pay (£)	0–1999	2000–3999	4000–5999	6000–9999	10 000–19 999
Frequency	20	36	25	14	5
Group width (in units of £2000)	1	1	1	2	5
Frequency density, i.e. frequency per £2000 of income	20	36	25	7	1

Now we can draw the histogram.

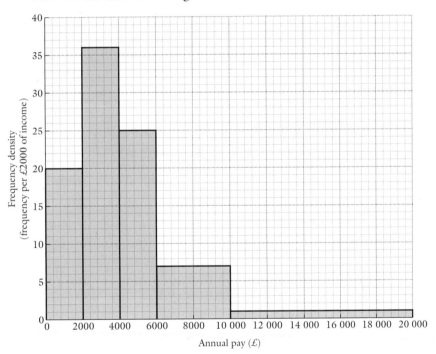

Notice that labelling the vertical axis 'frequency density' is not specific enough; for someone else to be able to obtain information from the histogram we must also give the unit in which the group is measured. Alternatively we can give the area that represents a given number of items; in this case we can add this key:

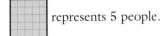 represents 5 people.

1 The frequency table shows the distribution of the ages of 200 people attending a film show.

Age, n years	$0 \leqslant n < 20$	$20 \leqslant n < 30$	$30 \leqslant n < 40$	$40 \leqslant n < 50$	$50 \leqslant n < 60$	$60 \leqslant n < 100$
Frequency	20	40	50	60	10	20

A histogram is to be constructed to illustrate the data.

a Using 1 year as the unit of group width, what is the frequency density of

   i the first group   **ii** the sixth group   **iii** the fifth group?

b Repeat part **a** using 10 years as the unit of group width.

For each question from **2** to **6**, copy and complete the table and draw a histogram to illustrate the data. Add rows for the width and frequency density for questions **5** and **6**.

2 This table shows the distribution of the number of words per sentence on one page of a particular book.

Number of words	1–3	4–6	7–10	11–14	15–20	21–30
Frequency	2	7	11	16	18	13
Width of group	3	3	4			
Frequency density						

Draw the bars so that the first starts at 0.5 and ends at 3.5, the second starts at 3.5 and ends at 6.5, and so on.

3 This table gives the distribution of the marks of 100 pupils in an English examination.

Mark	0–29	30–39	40–49	50–59	60–99
Frequency	9	15	24	36	16
Width of group	30	10	10	10	40
Frequency density	0.3				

Draw the bars so that the first starts at −0.5 and ends at 29.5, the second starts at 29.5 and ends at 39.5, and so on.

4 The table shows the distribution of the heights of 65 fourteen-year-old boys.

Height, h cm	$100 \leqslant h < 130$	$130 \leqslant h < 140$	$140 \leqslant h < 145$	$145 \leqslant h < 155$	$155 \leqslant h < 165$	$165 \leqslant h < 185$
Frequency	6	11	8	14	14	12
Width of group	30					
Frequency density						

5   The table shows the distribution of the times taken by 78 pupils to get to school.

Time, t min	$0 \leqslant t < 5$	$5 \leqslant t < 10$	$10 \leqslant t < 15$	$15 \leqslant t < 18$	$18 \leqslant t < 21$	$21 \leqslant t < 26$	$26 \leqslant t < 30$	$30 \leqslant t < 40$
Frequency	6	1	18	9	12	19	9	4

6   The table gives the results of a survey, which was conducted in a region of France, to find the areas of 185 farms.

Area, A ha	$0 \leqslant A < 4$	$4 \leqslant A < 8$	$8 \leqslant A < 12$	$12 \leqslant A < 17$	$17 \leqslant A < 24$	$24 \leqslant A < 28$	$28 \leqslant A < 32$	$32 \leqslant A < 42$
Frequency	34	22	38	30	28	13	10	10

INFORMATION
FROM
HISTOGRAMS

When we are given information in a histogram then, provided that it is properly labelled, we can construct a frequency table. That table can then be used to find other information about the distribution such as the mean.

EXERCISE 8D

Make a frequency table from the histogram below which shows the distribution of the ages of people boarding buses at the bus station between 0830 and 0900 one morning.

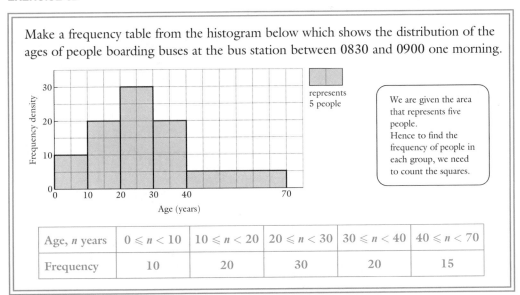

represents 5 people

We are given the area that represents five people.
Hence to find the frequency of people in each group, we need to count the squares.

Age, n years	$0 \leqslant n < 10$	$10 \leqslant n < 20$	$20 \leqslant n < 30$	$30 \leqslant n < 40$	$40 \leqslant n < 70$
Frequency	10	20	30	20	15

In each question from **1** to **3**, make a frequency table from the histogram.

1

Age of houses on estate

2

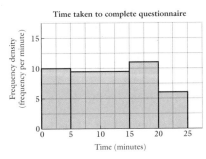

Time taken to complete questionnaire

3

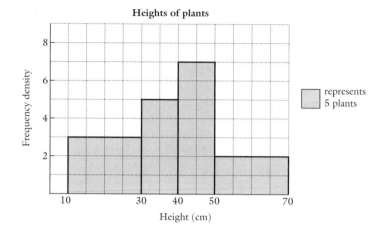

4 The histogram below shows the distribution of the heights of plants of the same variety grown by a horticulturalist.

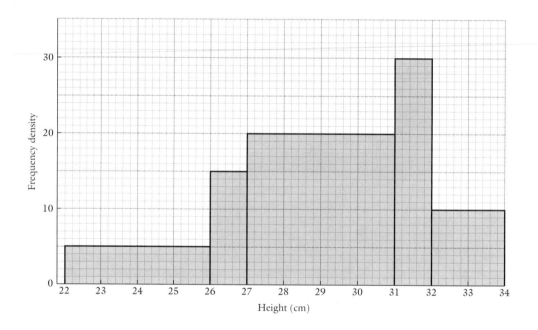

There are 20 plants in this sample whose heights are at least 32 cm but less than 34 cm.

a What key could have been given instead of this information?

How many plants in the sample are there

b with heights at least 31 cm but less than 32 cm?

c with heights at least 27 cm but less than 31 cm?

The polygon illustrates the distribution of the area covered by moss per square metre in a meadow. Find the mean area of moss per square metre.

A frequency density polygon is constructed in the same way as a frequency polygon: the midpoints of the bars representing the groups are joined by straight lines. The groups are therefore 0–2 cm², 2–4 cm², 4–6 cm², 6–10 cm². This information can be used to construct a frequency table, from which we can calculate an estimate of the mean.

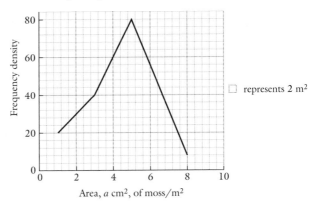

□ represents 2 m²

Area, a cm², of moss/m²	$0 \leqslant a < 2$	$2 \leqslant a < 4$	$4 \leqslant n < 6$	$6 \leqslant n < 10$
Frequency, f	50	100	200	40
Mid-class value, x	1	3	5	8

$$\text{Mean area of moss/m}^2 = \frac{\sum fx}{\sum f} = \frac{50 + 300 + 1000 + 320}{50 + 100 + 200 + 40} \text{ cm}^2$$

$$= 4.28 \text{ cm}^2 \text{ (correct to 3 s.f.)}$$

5 The polygon opposite shows the distribution of the ages of passengers alighting from buses at Camberley Crescent during the course of one week.

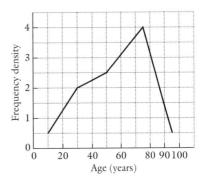

Copy this frequency polygon on squared paper and use it to complete the following table.

Age, n years	$0 \leqslant n < 20$	$20 \leqslant n < 40$	$40 \leqslant n < 60$	$60 \leqslant n < 90$	$90 \leqslant n < 100$
Frequency density	0.5				
Width of bar	20				
Frequency	10				

How many passengers were there altogether?

6 Similar information about the ages of passengers collected in the same week, at Bramberdown bus stop, gave the following table.

Age, n years	$0 \leqslant n < 20$	$20 \leqslant n < 40$	$40 \leqslant n < 60$	$60 \leqslant n < 90$	$90 \leqslant n < 100$
Frequency	60	85	40	30	0

a Superimpose the polygon for this data on a copy of the polygon for question **5**.

b Compare the two polygons.

c Give a possible explanation for their differences.

7

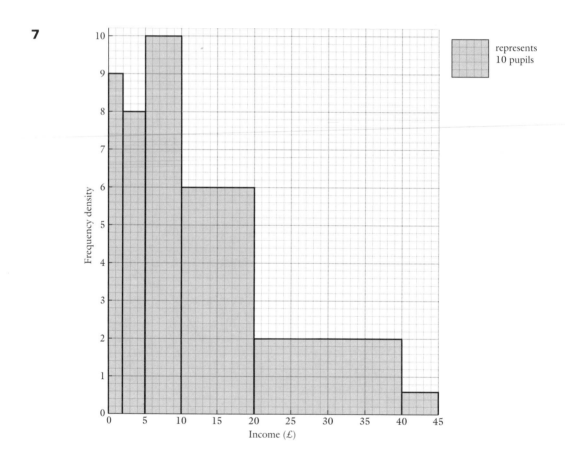

The histogram illustrates the results of a survey in which a number of pupils were asked how much money they received each week.

a Construct a frequency table.

b Estimate the mean.

8 This cumulative frequency polygon illustrates the distribution of the heights of 90 five-year-old children.

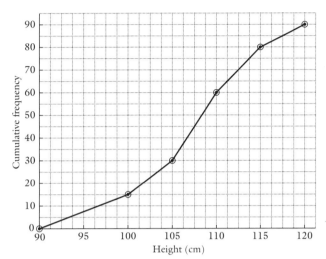

a Use the graph to estimate the median and the interquartile range.

b Construct a frequency table from the graph.

c Illustrate the information with a histogram.

d On your histogram, draw a vertical line through the median height. Compare the areas of the parts of the histogram on each side of the line.
Explain why your answer will be the same for any histogram.

e Deduce the values of the upper and lower quartiles from the histogram. Check these values from the cumulative frequency curve.

9 Information about the quantity of alcohol in the blood stream of a group of 18-year-old men is given by the graph.

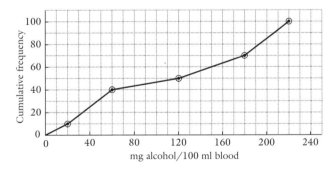

a The legal limit for driving is 80 mg/100 ml of blood. What percentage of this group are over the limit?

b Illustrate this information with a histogram.

c Estimate the median from   **i** the cumulative frequency polygon   **ii** the histogram.
Compare your answers.

USING THE AREA UNDER A HISTOGRAM

As the area of a bar in a histogram represents the number of items in the group,

> the total area of the histogram represents the total number of items in the distribution.

This histogram illustrates the distribution of the number of hours worked in March by the part-time members of staff in one company.

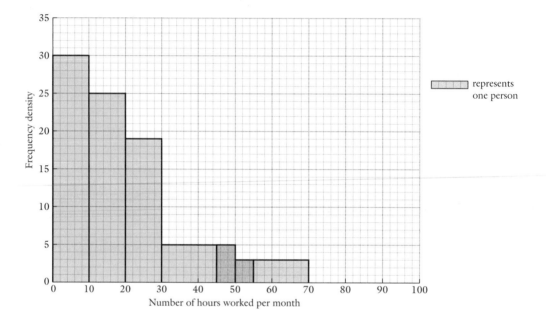

represents one person

The area of the histogram is 450 small squares.
5 small squares represents 1 person.
Therefore the area represents 90 people.

We can also use the given unit of area to estimate, for example, the number of people who worked between 45 and 55 hours in March. To do this we need to find the area shaded grey.

The required area $= (2.5 \times 5) + (2.5 \times 3)$ small squares.

$$= 20 \text{ small squares}$$

Therefore 4 people worked between 45 and 55 hours in March.

As the median is the value of the middle ranked item, it follows that

> the vertical line through the median divides the area in half.

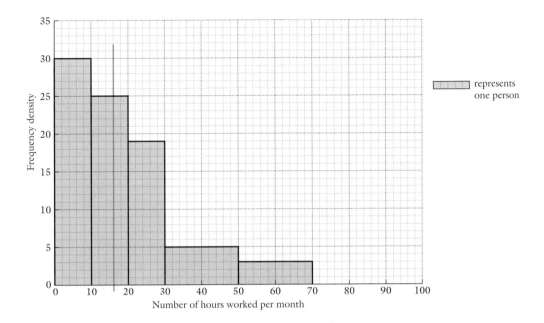

The vertical line that divides this area into two equal parts, each representing 45 people, gives an estimate of the median number of hours worked.
Counting units of area, this line goes through 16 on the horizontal axis, i.e. the median is 16 hours.

EXERCISE 8E

1 Explain why the line that divides the area covered by a histogram into two equal parts gives an *estimate* of the median.

2 The histogram shows the distribution of midday temperatures in degrees Celsius in Northampton for the first 120 days of a year (the months January, February, March and April).

    a Why does this chart not need a vertical scale?

    b On how many days was the temperature less than $0\,°C$?

    c Estimate the number of days on which the temperature was greater than $12\,°C$.

    d Estimate the median temperature for this period.

    e Make a frequency table and use it to find the mean daily temperature.

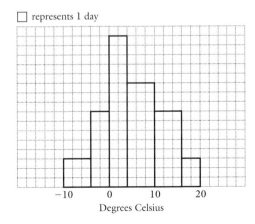

3 The histogram illustrates the distribution of goals scored in one match by a hockey team over the course of one season.

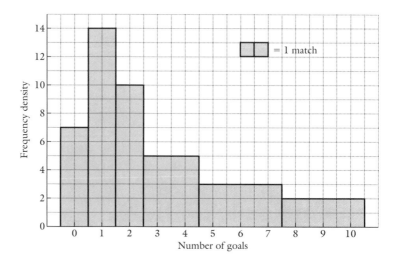

a How many matches were played by the team?

b In how many matches did the team score 3 or more goals?

c Estimate the median number of goals scored.

4 The histogram shows the distribution of scores from the schools in one area in the Key Stage 2 tests.

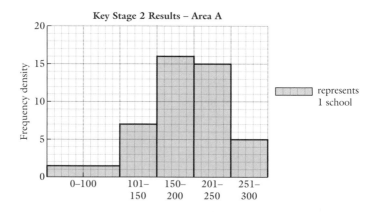

a How many schools have their results included in this histogram?

b Estimate the median score.

c Estimate the number of schools that have a score of
   i more than 270   **ii** less than 120.

5 The distribution of the Key Stage 2 results of schools in another area are shown in this histogram.

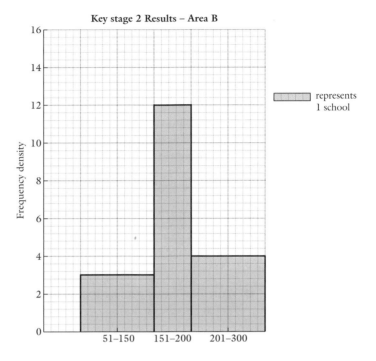

Key stage 2 Results – Area B

a How many schools have their results included in this histogram?

b Estimate the median score.

c Compare the results from area A in question **4** with those from area B using only the information given and the median scores.

d What other information can be found that would help with the comparison and in what way would it help?

6 The distribution of the masses of individual new potatoes dug from Mr Shah's allotment are summarised in the frequency table.

Mass (grams, correct to the nearest gram)	0–20	21–30	31–40	41–60	more than 60
Frequency	10	40	45	20	5

a Explain why it would be difficult to draw a histogram or a cumulative frequency curve to illustrate this distribution.

b Explain why estimates for the median mass and the interquartile range can be found and find them.

c Say, with reasons, whether it is possible to estimate the mean mass.

7 The distribution of the ages of the 103 staff in a large comprehensive school are given in the table.

Age (years)	22–24	25–29	30–39	40–49	50–59	60–69
Frequency	9	10	42	32	7	3
Frequency density	3			3.2		

a i Copy and complete the table of frequency densities.
ii On graph paper, draw a histogram to illustrate the age distribution of the comprehensive school staff.

In a nearby private school the staff has the following age distribution.

Age (years)	25–29	30–39	40–49	50–69
Percentage of staff	5	26	44	25

b Write down two comparisons about the age distributions of the staff at the two schools.

8

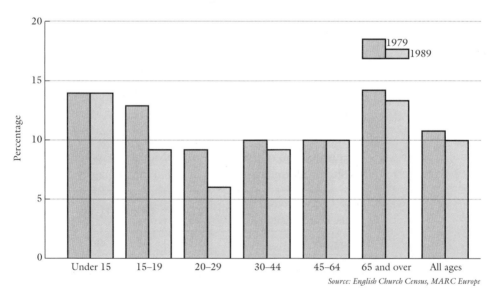

Source: English Church Census, MARC Europe

The chart, from *Social Trends 22*, shows the percentage of the population, by age, attending church services in England in 1979 and 1989.

Explain why it is not possible to work out the number of people under 15 attending church in each of these two years.

Is it reasonable to say that the number of people aged 45 to 64 attending church in 1989 had not fallen when compared with 1979?

Explain your answer.

Write down two comparisons that you can make about church attendances in the two years.

This chart gives the structure of the population, by age, in 1979 and 1989.

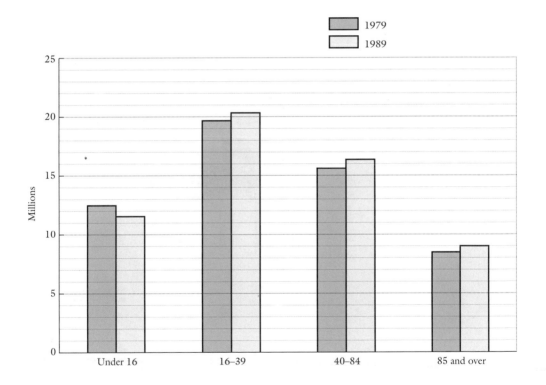

Use this information together with the chart opposite to construct two histograms showing an estimate for the age distribution of numbers of people attending church in 1979 and 1989.

Use the histograms to make two comparisons of the distributions that you could not make using just the first chart.

A chart-drawing package is included with most spreadsheet programs. Such a package can produce a variety of graphs from data entered in the spreadsheet.

This information, shows the number of office heating systems installed in each quarter of one year in the four regions in which a company operates.

	1st quarter	2nd quarter	3rd quarter	4th quarter
North	65	45	34	71
Southeast	40	30	36	58
Midlands	16	35	28	14
West	9	22	29	24

These charts were produced using the data in a spreadsheet.

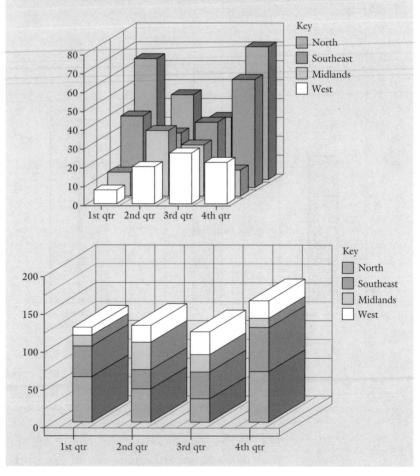

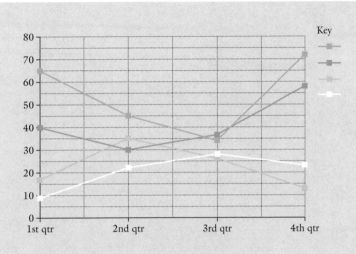

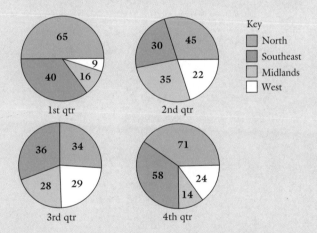

Discuss the advantages and disadvantages of each of these charts. Use this data, or some of your own, to experiment with a chart drawing package.

Revision for Higher Tier

1 Number

WHOLE NUMBERS

An **integer** is a positive or a negative whole number.

A **factor** of a number will divide into that number exactly,

e.g.   3 is a factor of 6 because 3 divides into 6.

The **highest common factor** of two or more whole numbers is the largest number that divides into each of them exactly.

A **multiple** of a number has that number as a factor,

e.g.   12 is a multiple of 3.

The **lowest common multiple** of two or more given numbers is the smallest number that has each of the given numbers as a factor.

A **prime number** has only two distinct factors, 1 and itself; for example, 7. Hence 1 is not a prime number.

A number can be expressed as a **product of prime factors** by starting with the lowest prime number that will divide into it and repeating this process until no more divisions are possible. For example, to express 180 as a product of its prime factors, the working can be set out as follows.

$$
\begin{array}{c|c}
2 & 180 \\ \hline
2 & 90 \\ \hline
3 & 45 \\ \hline
3 & 15 \\ \hline
5 & 5 \\ \hline
& 1
\end{array}
$$

This shows that $180 = 2 \times 2 \times 3 \times 3 \times 5$ which, using indices, can be expressed as $2^2 \times 3^2 \times 5$.

Expressing numbers as products of prime factors can help to find their HCF or LCM when these are not obvious.

For example, $180 = 2^2 \times 3^2 \times 5$ and $588 = 2^2 \times 3 \times 7^2$

Choosing the factors that are common to both, i.e. $2^2 \times 3$, gives the HCF.

Choosing the factors that include all those of both numbers, i.e. $2^2 \times 3^2 \times 5 \times 7^2$, gives the LCM.

Positive and negative numbers are collectively known as **directed numbers**; they are not necessarily whole numbers.

The **rules for multiplying and dividing directed numbers** are

> when the signs are the same, the result is positive
>
> when the signs are different, the result is negative.

The same rules apply when adding and subtracting directed numbers, since for example, $3 - (-2) = 3 - 1(-2)$

$$= 3 + (-1)(-2) = 3 + 2 = 5$$

The sign in front of a number refers to that number only.

When a calculation with any type of number involves a **mixture of operations**, start by calculating anything inside brackets, then follow the rule 'do the multiplication and division first'.

EXERCISE 1.1

Do not use a calculator for this exercise.

1 a Express 504 as a product of its prime factors in index form.

  b Find the lowest common multiple of 2, 9 and 12.

  c Find the highest common factor of 156 and 252.

  d Find the lowest common multiple of 18 and 21.

2 The frequency table gives the number of syllables in the words from a passage in a book.

Number of syllables	1	2	3	4	5
Frequency	402	355	159	47	9

How many words are there in the passage?

3 Which of the numbers

$$1, \quad 2\tfrac{1}{2}, \quad 3, \quad 6, \quad 7, \quad 9, \quad 13, \quad 15$$

are **a** not integers

  b prime

  c not integer multiples of 3

  d integer factors of 30?

4 Find the value of

  a $240 + 20 \times 56$ **d** -25×-30 **g** $8 - (-10)$

  b $64 - (20 - 14)$ **e** $390 \div 130$ **h** $\dfrac{-3 \times 4}{-6}$

  c $14 - 6 \div (-2)$ **f** $484 \div 11 \times 2$ **i** $\dfrac{1 - 2(-4)}{-3}$

5 Calculate the number of seconds in one week.

6 Write down the integers which satisfy the inequality illustrated on this number line.

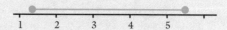

7 A car hire firm charges £8 a day plus 15p a mile for a Ford Focus. Zaid hires a Ford Focus for 2 days and drives 380 miles. What is the total hire charge?

8 Find the highest common factor of 112 and 120.
Several square mats are to be used to cover a rectangular floor measuring 11.2 m by 12 m without having to cut any mats. Find the size of the largest square mat that can be used.

9 A newsagent employs 14 people to deliver daily newspapers. They are all paid the same amount. The total weekly payroll for these 14 people is £238.

a How much is each person paid?

b Each delivery person takes papers to 34 households each day. Each household is charged 75 p a week for having their newspapers delivered; this is to cover wages for the delivery person and the newsagent's overheads. How much are the newsagent's weekly overheads?

10 a Calculate
$$(-3) \times (-3) \times (-3) - 2 \times (-3) \times (-3) - 4 \times (-3) + 3$$

b The table shows some values of y for values of x between -3 and 2 where
$$y = x^3 - 2x^2 - 4x + 3$$
Copy and complete the table.

x	-3	-2	-1	0	1	2
y		-5		3		

11 a Find the lowest common multiple of 12, 15 and 20.

b There are three buoys each with a flashing light at the entrance to a harbour. The light on one buoy flashes every 12 seconds, the light on another buoy flashes every 15 seconds and the light on the third buoy flashes every 20 seconds. They flash together at 12 noon. When do they next flash together?

FRACTIONS

The word **fraction** means part of a whole. When we refer to a fraction in mathematics we mean part of a whole expressed as a number in the form $\frac{a}{b}$ where a and b are integers; a is called the **numerator** and b is called the **denominator**.

Equivalent fractions are the same part of a whole;
e.g. $\frac{2}{4}$ and $\frac{3}{6}$ are equivalent fractions (they are both half of a quantity).
Equivalent fractions are formed by multiplying (or dividing) both the top and bottom of a fraction by the same number.

Fractions are added or subtracted by first changing them into equivalent fractions with the same denominators, then adding or subtracting the numerators. For example, $\frac{1}{2} + \frac{2}{3} = \frac{3}{6} + \frac{4}{6} = \frac{7}{6}$.

An **improper fraction** has a numerator larger than its denominator, i.e. $\frac{7}{6}$ is an improper fraction. An improper fraction can be changed into a **mixed number**, which is the sum of a whole number and a fraction.

For example, $\frac{7}{6} = \frac{6}{6} + \frac{1}{6} = 1 + \frac{1}{6}$, which is written as $1\frac{1}{6}$.

When one number is divided by another, we can write the division as a fraction,

i.e. $1 \div 4$ can be written as $\frac{1}{4}$.

In general, $\frac{a}{b}$ means $a \div b$ and conversely $a \div b$ means $\frac{a}{b}$.

The **reciprocal** of a number is 1 divided by that number,

for example, the reciprocal of 4 is $\frac{1}{4}(= 0.25)$

the reciprocal of $\frac{2}{3}$ is $\frac{3}{2}$

and   the reciprocal of 0.8 is $1 \div 0.8(= 1.25)$

To **multiply one fraction by another fraction**, we multiply their numerators and multiply their denominators,

e.g. $\frac{1}{2} \times \frac{5}{3} = \frac{1 \times 5}{2 \times 3} = \frac{5}{6}$

To **divide by a fraction**, we multiply by the reciprocal of the fraction,

e.g. $\frac{1}{2} \div \frac{5}{3} = \frac{1}{2} \times \frac{3}{5} = \frac{3}{10}$

To multiply or divide with **mixed numbers**, for example $1\frac{3}{4}$, first change the mixed numbers to improper fractions,

e.g. $2\frac{2}{3} \div 1\frac{3}{4} = \frac{8}{3} \div \frac{7}{4} = \frac{8}{3} \times \frac{4}{7} = \frac{32}{21} = 1\frac{11}{21}$

To multiply or divide a fraction by a whole number, first express the whole number as a fraction whose denominator is 1,

e.g. $\frac{2}{3} \times 6 = \frac{2}{3} \times \frac{6}{1}$

To find **a fraction of a quantity**, we multiply that fraction by the quantity,

e.g. $\frac{1}{2}$ of $\frac{3}{4}$ lb is $\frac{1}{2} \times \frac{3}{4}$ lb, and $\frac{3}{8}$ of £24 $= £\left(\frac{3}{8} \times 24\right)$

(Remember that 'of' means '×'.)

To express **one quantity as a fraction of another**, first express both quantities in the same unit, then place the first quantity over the second,

e.g. 24 p as a fraction of £2 is $\frac{24}{200}\left(= \frac{3}{25}\right)$,

i.e. $24\,\text{p} = \frac{3}{25}$ of £2.

When working with fractions, look for opportunities for reducing the size of the numbers involved by cancelling before multiplying,

e.g. $\frac{\overset{1}{\cancel{25}}}{\underset{9}{\cancel{36}}} \times \frac{\overset{4}{\cancel{16}}}{\underset{3}{\cancel{78}}}\left(= \frac{1}{9} \times \frac{4}{3}\right) = \frac{4}{27}$

EXERCISE 1.2

Do not use a calculator for this exercise.

1 Find the value of

  a $\frac{2}{3} \times 12$ **c** $1\frac{2}{3} + \frac{3}{4}$ **e** $1\frac{2}{3} \div \frac{3}{4}$ **g** $\dfrac{1\frac{1}{2}}{\frac{3}{4}}$

  b $\frac{2}{3} \div 8$ **d** $1\frac{2}{3} - \frac{3}{4}$ **f** $1\frac{2}{3} + \frac{3}{4} \times \frac{1}{6}$ **h** $1\frac{3}{5} - \frac{3}{8} \div 15$

2 A metal pin is $3\frac{1}{8}$ inches long. A piece of length $1\frac{2}{3}$ inches is cut off one end of this pin. What length is left?

3 Find

  a $\frac{3}{4}$ of £26   **b** $\frac{2}{3}$ of 12.6 metres   **c** $\frac{5}{8}$ of $2\frac{1}{2}$ hours

4 Find

  a 25 p as a fraction of £2

  b 20 seconds as a fraction of 1 minute

  c $1\frac{1}{2}$ feet as a fraction of 2 yards.

5 A book costs £5.40 from the publisher. A bookseller wants to make a gain of $\frac{1}{3}$ of this cost. How much should the book be sold for?

6 The mass of a silver bead is 32 grams and the mass of a gold bead is 56 grams. What fraction of the mass of the gold bead is the mass of the silver bead?

7 Given that $y = \dfrac{2x - 3}{4x}$, find y when $x = 1\frac{2}{3}$.

8 The frequency table gives the number of syllables in the words from a passage in a book.

Number of syllables	1	2	3	4	5
Frequency	402	355	159	47	9

What fraction of the words in this passage have one syllable?

DECIMALS

A decimal fraction (usually called simply a **decimal**) expresses part of a quantity by using figures to the right of the units (separated by a point), where the first figure represents tenths, the second figure represents hundredths, and so on,

e.g. 2.56 means 2 and $\frac{5}{10}$ plus $\frac{6}{100}$.

Decimals are added and subtracted in the same way as whole numbers; when written in columns it is important that the decimal points are in line.

To **multiply decimals** without using a calculator, first ignore the decimal point and multiply the numbers. Then add the number of decimal places in each of the decimals being multiplied together; this gives the number of decimal places in the answer,

e.g. $7.5 \times 0.5 = 3.75$ $(75 \times 5 = 375)$

$[(1) + (1) = (2)]$

To **multiply a decimal by 10, 100, ...**, move the decimal point 1, 2, ... places to the right,

e.g. $1.562 \times 100 = 156.2$

To **divide by a decimal**, move the point in *both* numbers to the right until the number we are dividing by is a whole number,

e.g. $2.56 \div 0.4 = 25.6 \div 4 = 6.4$

To **divide a decimal by 10, 100, ...**, move the decimal point 1, 2, ... places to the left, and fill empty spaces with zeros,

e.g. $1.562 \div 100 = 0.01562$

The first **significant figure** in a decimal is the first non-zero figure (reading from the left), the second and subsequent significant figures are the next and subsequent figures regardless of whether or not they are zero, for example in the number 0.0205, the first significant figure is 2, the second significant figure is 0.

To **correct a number** to a specified place value or number of significant figures, look at the figure in the next place: if it is 5 or more, add 1 to the previous figure, otherwise leave the previous figure as it is.

When a number has been corrected, its true value lies within a range that can be shown on a number line. For example, if a nail is 23.5 mm long correct to 1 decimal place, then the length is from 23.45 mm up to, but not including, 23.55 mm as shown on this number line.

23.45 mm is the **lower bound** and 24.55 mm is the **upper bound** of this length.

Estimating the result of a calculation

Whether working with or without a calculator, an estimate should be made of any calculation so that you are immediately aware of any mistake in your working. There is no hard and fast rule for how to estimate; it involves simplifying the numbers to some that you can work with easily in your head. Correcting each number to one significant figure will give reasonable estimates, but use common sense to decide how far to take the simplification. If the number is close to the borderline between being rounded up or down, it may be better to correct to two significant figures depending on context.

For example, $\dfrac{12.6 - 9.554}{0.357} \approx \dfrac{13 - 10}{0.4} = 7.5$

Using a calculator

Most calculations can be keyed directly into a calculator using brackets where necessary.

For example, to find $\dfrac{12.6 - 9.554}{0.357}$, key in

$$(\; 1 \; 2 \; \cdot \; 6 \; - \; 9 \; \cdot \; 5 \; 5 \; 4 \;) \; \div \; 0 \; \cdot \; 3 \; 5 \; 7$$

The display shows 8.532 212 885. There is no need to write all these figures down; to give an answer correct to 3 significant figures it is sufficient to write down the first four significant figures, i.e. 8.532 . . .

Then $\dfrac{12.6 - 9.554}{0.357} = 8.53$ correct to 3 significant figures.

EXERCISE 1.3

SECTION A Do not use a calculator for this section.

1 Find   **a** $2.15 + 0.034$ **b** $0.035 - 0.0035$ **c** $6 - 0.27$

2 Find   **a** 2.1×0.3 **c** 4×0.015 **e** 0.06^2
          b 2.5×1.6 **d** 2.36×20 **f** 300×1.4

3 Find   **a** $2.1 \div 0.3$ **c** $0.035 \div 0.01$ **e** $12.5 \div 50$
          b $40 \div 2.5$ **d** $0.3 \div 1.2$ **f** $12.5 \div 4000$

4 Find, correct to 3 significant figures,

   a 3.6×2.7 **b** $3.3 \div 0.7$ **c** $0.02 \div 0.3$

5 a The height of a tree is $15\,\text{m}$ correct to the nearest metre.
   Write down the range in which the height of the tree lies.

   b The weight of a pill is 0.140 grams correct to 3 significant figures.
   Write down the range in which the weight of the pill lies.

6 A screw weighs 6 grams correct to the nearest gram. Find

   a the lowest possible mass of 500 of these screws

   b the largest possible mass of 1000 of these screws.

7 A plank is $240\,\text{cm}$ long. A length of $45\,\text{cm}$ is cut off the plank.
   Both measurements are correct to the nearest centimetre. Find the
   greatest possible length of the remaining plank.

8 Estimate the value of

   a 12.07×15.9 **d** $\sqrt{83.7}$ **g** $\dfrac{\sqrt{200}}{(2.3)^2}$

   b $\dfrac{1.267 + 0.944}{0.513}$ **e** $\dfrac{4.86}{(9.87)^2}$ **h** $\sqrt{\dfrac{1.87 - 0.03}{(0.45)^2}}$

   c $\dfrac{356}{297 - 159}$ **f** $\sqrt{127.3 - 6.4}$

   In each case, state whether your estimate is greater or less than the
   actual value.

SECTION B You may use a calculator for this section.

9 Find, correct to three significant figures, the area of a rectangle
   measuring $126\,\text{cm}$ by $97\,\text{cm}$.

10 Find, correct to three significant figures, the greatest possible area of
   a rectangle measuring $126\,\text{cm}$ by $97\,\text{cm}$, where each measurement is
   correct to the nearest centimetre.

11 Evaluate, correct to 3 significant figures, each part of question **8**.

12 Kate put £20 worth of petrol priced at 64.5 p per litre into her car. How many litres did she buy?

13 Calculate, giving your answer correct to 3 significant figures,

$$\textbf{a} \quad \frac{59.7 - 4.63}{9.25^2 + 1.04} \qquad\qquad \textbf{c} \quad \sqrt{\frac{48.5}{5 - 2.6 \tan 51°}}$$

$$\textbf{b} \quad \frac{2.56^2 - 1.38 \sin 31°}{4.37^2} \qquad\qquad \textbf{d} \quad \frac{4.8 \tan 36°}{\sqrt{1 - \tan 36°}}$$

PERCENTAGES, DECIMALS AND FRACTIONS

A **percentage** gives a fraction of a quantity as a number of hundredths,

e.g. 20% means 20 out of 100, i.e. $\frac{20}{100}$.

A **percentage can be expressed as a fraction** by placing the percentage over 100,

e.g. $33\% = \frac{33}{100}$,

and a **percentage can be expressed as a decimal** by dividing the percentage by 100, i.e. by moving the decimal point two places to the left,

e.g. $33\% = 0.33$

A **fraction can be expressed as a percentage** by multiplying the fraction by 100,

e.g. $\frac{2}{5} = \frac{2}{5} \times 100\% = \frac{2}{5} \times \frac{100}{1}\% = 40\%$

and a **decimal can be expressed as a percentage** by multiplying the decimal by 100,

e.g. $0.325 = 32.5\%$

A **fraction can be changed to a decimal** by dividing the denominator into the numerator,

e.g. $\frac{3}{8} = 3 \div 8 = 0.375$

A **decimal can be changed to a fraction** by placing the figures after the decimal point over 10, 100, ... when there are 1, 2, ... decimal places,

e.g. $0.025 = \frac{25}{1000} \left(= \frac{1}{40} \right)$

Recurring decimals arise when some fractions are changed to decimals and the result is a recurring pattern of numbers that repeats indefinitely

For example, $\frac{1}{11} = 1 \div 11 = 0.090\,909 \ldots = 0.0\dot{9}$

A recurring decimal can be expressed as a fraction by multiplying the decimal by 10, 100, ... when there are 1, 2, ... figures in the repeating pattern,

e.g. $0.\dot{1}0\dot{5} = 0.105\,105\,105\ldots$ becomes

then subtracting the
original decimal

$$1000 \times 0.\dot{1}0\dot{5} = 105.105\,105\,105\ldots$$
$$-\underline{0.\dot{1}0\dot{5} = 0.105\,105\,105\ldots}$$

to give

$$999 \times 0.\dot{1}0\dot{5} = 105$$

so

$$0.\dot{1}0\dot{5} = \frac{105}{999} = \frac{35}{333}$$

To find **one quantity as a percentage of another quantity**, we place the first quantity over the second quantity and multiply this fraction by 100,

e.g. 24 p as a percentage of £2 is $\frac{24}{200} \times \frac{100}{1} = 12\%$,

i.e. 24 p = 12% of £2.

To find a **percentage of a quantity**, change the percentage to a decimal and multiply it by the quantity,

e.g. 32% of £18 = $0.32 \times £18\,(\,= £5.76\,)$

Percentage change

A change is expressed in percentage terms as a percentage of the original quantity, that is, before any changes are made.

To **increase** a quantity by 15%,

we find *the increase* by finding 15% of the quantity,

we find *the new quantity* directly by finding $100\% + 15\%$, i.e. 115%, of the original quantity so we multiply by 1.15.

To **decrease** a quantity by 15%,

we find *the decrease* by finding 15% of the quantity,

we find *the new quantity* directly by finding $100\% - 15\%$, i.e. 85%, of the original quantity so we multiply by 0.85.

To **find the original quantity** when the changed quantity is known, we divide the changed quantity by the factor that changed it.
For example, if a shirt is sold for £25 after allowing for a discount of 15%, the discount is 15% of the price before it is reduced,

i.e. £25 = 85% of the original price = $0.85 \times$ original price.

Therefore the original price = £25 ÷ 0.85.

Compound percentage change is an accumulating change. If, for example, the value of a house is initially £P and then increases by 5% p.a. of its value at the start of each year;

its value after one year is 105% of its initial value,
i.e. £$P \times 1.05$,
after another year its value is 105% of its value at the start of that year,
i.e. £$P \times 1.05 \times 1.05 = £P \times (1.05)^2$, and so on,
so its value after n years is £$P \times (1.05)^n$.

Interest

When a sum of money is borrowed or lent, interest is usually charged on a yearly basis and is given as a percentage of the sum borrowed, for example 3% p.a.

When the interest is calculated on the initial amount only it is called **simple interest**. When the interest is added to the amount each year it is called **compound interest**.

Percentage error is found by expressing the error as a percentage of the correct value. For example, if a garden is 21.5 m long and we use 20 m as the length, we have introduced an error of 1.5 metres.

The percentage error is $\dfrac{1.5}{21.5} \times 100 = 7.0\%$ to 2 significant figures.

EXERCISE 1.4

SECTION A   Do not use a calculator for this section.

1 Copy and complete the following table.

Fraction	$\frac{2}{5}$			$1\frac{3}{4}$		$\frac{3}{8}$		
Decimal		0.15		0.03			0.045	
Percentage			36%		210%	95%		312%

2 Express each fraction as a recurring decimal.

 a $\frac{2}{3}$ **b** $\frac{1}{9}$ **c** $\frac{3}{11}$ **d** $\frac{5}{6}$

3 Express as a fraction in its simplest form

 a $0.\dot{7}$ **b** $0.0\dot{7}$ **c** $0.0\dot{1}4\dot{2}$

4 Find   **a** 12 p as a percentage of £3   **b** 3 ml as a percentage of 5 cl.

5 Find   **a** 30% of £12.50   **b** 32% of 500 kg.

6 A bottle of perfume is sold for £24 which is a gain of 20% of the wholesale price. Find the wholesale price.

7 Find the value of $\dfrac{x-y}{x}$ when $x = \frac{2}{3}$ and $y = 0.8$.

SECTION B   You may use a calculator for this section.

8 Sally earns £205 per week now. At the end of the month she will receive a rise of 3.5%. How much will her new weekly pay be?

9 At the end of last year, Tim's pay increased from £14 500 p.a. to £15 200 p.a. What percentage rise did he receive?

10 After a 5% salary increase, Amin's salary was £25 540. What was his salary before the increase?

11 A computer costs £957 including VAT at 17.5%. What is the price of the computer excluding VAT?

12 Candidates in a general election lose their deposits if they get less than 5% of the total votes cast in their constituency. In the last election, the votes cast in the constituency of Upper Haven were as follows.

Peter Smith	12 437
Joanne Brown	9 533
Eric White	10 204
Ann Jones	524
Rudy Williams	835

a Which candidates lost their deposits?

b What was the minimum number of votes necessary for a candidate to avoid losing the deposit?

13 Jim invests £500 in a fixed interest rate bond paying compound interest at 4.5% p.a. How much is the bond worth after 3 years?

14 The value of an industrial sewing machine depreciates each year by 25% of its value at the start of the year. It is now worth £110. What was its value 2 years ago?

15 Anthea Jones uses bottles of mineral water in her café. She pays 36 p per bottle when it is delivered. If she collects the bottles from a cash-and-carry, it costs 27 p per bottle.
What is the percentage saving if she collects the bottles?

16 A bottle holds 25.4 cl of water. Jim uses 25 cl as an estimate for the capacity of the bottle. What percentage error is this?

17 Mary estimates the area of this rectangular rug by correcting each measurement to 1 significant figure.

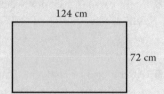

124 cm

72 cm

a Find the difference between Mary's estimate for the area of the rug and the area calculated from the measurements given.

b What is the percentage error in Mary's estimate?

18 Cleo invested £5000 in a savings bond that guaranteed an annual rate of growth of 6.5%.

a What was Cleo's investment worth after 15 years?

b What annual rate of growth is needed for the bond to double in value after 15 years?

19 The population of Archester is now 156 000 to the nearest thousand.
The population is declining at the rate of 2.5% a year.

a If this rate continues, what will the population be in 10 years' time?

b What annual percentage rate of decline would result in the population halving in 10 years?

RATIONAL AND IRRATIONAL NUMBERS

We often need to generalise arithmetic results, that is, to give a result in a form that represents all numbers. We do this by using letters to represent numbers. These are some of the conventions used with letters.

Terms such as $5n$ mean $5 \times n$, i.e. $n + n + n + n + n$.

Similarly ab means $a \times b$.

Like terms such as $2x + 5x$ can be simplified to $7x$.

A **square root** of a number is another number which, when multiplied by itself, gives the first number.
A number has two square roots, one positive and one negative,

e.g. if $x^2 = 4$, then $x = \pm\sqrt{4} = \pm 2$.

Note that $\sqrt{4}$ means the positive square root of 4.

In general, $\sqrt{x}$ means the positive square root of x.

A **rational number** can be written in the form $\frac{a}{b}$ where a and b are integers.

An **irrational number** cannot be expressed in the form $\frac{a}{b}$ where a and b are integers, nor can it be expressed exactly as a decimal; it has an infinite number of decimal places with no pattern. Examples of irrational numbers are $\sqrt{2}$ and π.

The square root of an integer that is not a perfect square is an irrational number and when it is written in square root form, it is called a **surd**; $\sqrt{2}$ and $\sqrt{3}$ are examples of surds.

When a fraction contains a surd in its denominator, it can be expressed as an equivalent fraction with a denominator which is a whole number. We do this by multiplying top and bottom by that surd,

e.g. $\qquad \dfrac{2}{\sqrt{3}} = \dfrac{2 \times \sqrt{3}}{\sqrt{3} \times \sqrt{3}} = \dfrac{2\sqrt{3}}{3}$.

This process is called **rationalising the denominator**.

The **nth root of a number** is the factor such that n of these factors multiplied together give the number. The nth root of x is written as $\sqrt[n]{x}$.

For example, $\sqrt[4]{16}$ means the 4th root of 16,
and, as $16 = 2 \times 2 \times 2 \times 2$, $\sqrt[4]{16} = 2$.

INDICES

When a number is written in the form 3^4, 3 is called the base and the raised 4 is called the **index** or **power** and 3^4 means $3 \times 3 \times 3 \times 3$. A **negative sign in an index** means 'the reciprocal of',

so $\qquad 3^{-4}$ means $\dfrac{1}{3^4}$,

Zero index means that the expression is identical to 1,

i.e. $3^0 = 1$, in fact $a^0 = 1$ whatever number a represents, except zero.

Fractional indices

$x^{\frac{1}{n}}$ means the nth root of x, i.e. $x^{\frac{1}{n}} = \sqrt[n]{x}$.

For example, $8^{\frac{1}{3}} = \sqrt[3]{8} = 2$

Laws of indices

We can multiply different powers of the same base by adding the indices,

e.g. $3^4 \times 3^2 = 3^{4+2} = 3^6$

We can divide different powers of the same base by subtracting the indices,

e.g. $3^4 \div 3^2 = 3^{4-2} = 3^2$

A power of a number written in index form can be simplified by multiplying the powers, e.g. $(3^3)^4 = 3^{12}$.

In general terms, these laws are

1 $a^x \times a^y = a^{x+y}$, e.g. $2^3 \times 2^4 = 2^7$

2 $a^x \div a^y = a^{x-y}$, e.g. $2^3 \div 2^4 = 2^{-1}$

3 $(a^x)^y = a^{xy} = (a^y)^x$, e.g. $(2^2)^3 = 2^6 = (2^3)^2$

A number written in **standard form** is a number between 1 and 10 multiplied by a power of ten,

e.g. 1.2×10^5, 2.4×10^{-8}

EXERCISE 1.5

SECTION A   Do not use a calculator for this section.

1 Which of these numbers are rational?

   a $(1 + \sqrt{2})^2$ **b** $(1 + \sqrt{2})(1 - \sqrt{2})$ **c** π^2

2 Find the value of

   a $4^{\frac{1}{2}}$ **d** $2^0 + 2^{-1}$ **g** $\dfrac{x^5 \times y^2}{x^3 y}$ **j** $(16)^{-\frac{3}{4}}$

   b 6^{-1} **e** $\left(\dfrac{2}{3}\right)^{-2} \times \left(\dfrac{1}{4}\right)^{\frac{1}{2}}$ **h** $p^3 q^{-1} \div \left(\dfrac{p}{q}\right)^2$ **k** $\dfrac{1}{8^{-\frac{1}{3}}}$

   c $8^{\frac{2}{3}}$ **f** $\left(\dfrac{9}{16}\right)^{-\frac{1}{2}} \times \left(\dfrac{1}{3}\right)^{-1}$ **i** $\dfrac{1}{3^{-1}}$ **l** $\left(\dfrac{1}{125}\right)^{-\frac{2}{3}}$

3 Write the numbers in standard form   **a** 105 000 000   **b** 0.000 026

4 Find the value of x if

   a $x^{\frac{1}{2}} = 6$ **b** $x^{-2} = 4$ **c** $x^{\frac{1}{3}} = \frac{1}{2}$ **d** $81^{\frac{1}{4}} = 3^x$ **e** $64^x = 2$

5 Simplify

   a $\sqrt{10} \times \sqrt{5}$ **b** $\sqrt{18} + \dfrac{1}{\sqrt{2}}$ **c** $(1 - \sqrt{2})^2$

   d $\sqrt{2}(1 - \sqrt{6})$ **e** $(2 - \sqrt{3})(1 + \sqrt{2})$ **f** $(4 - \sqrt{3})(4 + \sqrt{3})$

6 Write those of the following numbers that are rational in the form $\dfrac{p}{q}$ where p and q are integers.

$$(1 - \sqrt{3})(1 + \sqrt{3}), \quad \sqrt{4\tfrac{1}{9}}, \quad \sqrt{1\tfrac{7}{9}}, \quad \sqrt{6} \times \sqrt{24}, \quad (\sqrt{3})^3, \quad 0.\dot{5}$$

7 Find, in standard form, the value of

a $(2 \times 10^{-4})^2$

c $(16 \times 10^4)^{\frac{1}{2}}$

b $\dfrac{8.1 \times 10^{-5}}{8.1 \times 10^{-5} + 2.43 \times 10^{-4}}$

d $\dfrac{4.9 \times 10^{-2} + 2.1 \times 10^{-4}}{7 \times 10^{-5}}$

8 The mass of one grain of sand is $2.5 \times 10^{-3}\,\text{g}$
Find the mass of 5 million grains of sand.

9 The island of Torrina has an area of 1.05×10^6 hectares of which 4.9×10^5 hectares is desert.
What fraction of the land is desert?

SECTION B   You may use a calculator for this section.

10 Find the value of x when

a $x^{20} = 10\,000$

b $x^{-15} = 0.0038$

11 1 nanometre is 10^{-9} metres.
Find, in metres, the value of 42×157 nanometres, giving your answer in standard form.

12 The mass of the Earth is about $5.98 \times 10^{24}\,\text{kg}$.
The mass of the Sun is about $1.9 \times 10^{30}\,\text{kg}$.
Giving your answer in standard form to an appropriate degree of accuracy, find

a the mass of the Earth as a fraction of the mass of the Sun

b the difference between the masses of the Sun and the Earth.

13 Water running down a rock face eats into the vertical surface at the rate of $1.4 \times 10^{-3}\,\text{m}$ a year. How far will the water have eaten into the rock face after two millennia?

RATIO

Ratios are used to give the relative sizes of quantities.
For example, if a car is 400 cm long and a model of it is 20 cm long, we say that their lengths are in the ratio $400 : 20$.
Ratios can be simplified by dividing the parts of the ratio by the same number,

e.g. $400 : 20 = 20 : 1$ (dividing 400 and 20 by 20).

A **map ratio** is the ratio of a length on the map to the length it represents on the ground. When expressed as a fraction (or sometimes as a ratio), it is called the **representative fraction**.

Division in a given ratio

To divide £200 into three amounts of money in the ratio 2 : 5 : 3, means that £200 has to be divided into $2 + 5 + 3$, i.e. 10, equal parts; so the first amount is 2 of these parts, i.e. $\frac{2}{10}$ of £200,

the second amount is $\frac{5}{10}$ of £200 and the third amount is $\frac{3}{10}$ of £200.

DIRECT AND INVERSE PROPORTION AND VARIATION

When two quantities are related so that when one of them trebles, say, the other also trebles, the quantities are **directly proportional** (that is, they are always in the same ratio). If x and y represent the numerical values of the quantities, then x and y are related by $y = kx$, where k is a constant.

Another way to describe this relationship is to say that y varies with x,

i.e.   'y varies as x' means that 'x and y are directly proportional'

and $y = kx$ where k is a constant.

It follows that if, for example, y varies as x^2, then y and x^2 are related by the equation $y = kx^2$.

When two quantities are related so that when one of them trebles, say, the other becomes a third of its original size, the quantities are **inversely proportional** and their product is constant.

If x and y represent the numerical values of the quantities, then x and y are related by $y = \dfrac{k}{x}$ ($\Rightarrow$ $xy = k$)

This relationship can also be described by saying that y **varies inversely** as x and $y = \dfrac{k}{x}$ where k is a constant.

It follows that if we are told that y varies inversely as, say, the square root of x, then $y = \dfrac{k}{\sqrt{x}}$.

Once we know how x and y are related we can find the constant k when we are given a pair of corresponding values of x and y.
For example, if y varies as x^3 and $y = 6$ when $x = 2$,

then we know that $\hspace{4cm} y = kx^3$

and, as $y = 6$ when $x = 2$, $\hspace{1.5cm} 6 = k \times 8$, i.e. $k = \frac{3}{4}$,

so $\hspace{6cm} y = \frac{3}{4}x^3$

EXERCISE 1.6

SECTION A Do not use a calculator for this section.

1 Simplify the ratios.

   a 1 minute : 20 seconds   **c** 200 cm : 200 mm

   b 40 ml : 1 litre   **d** 2 kg : 500 g : 1.5 kg

2 An orange drink is made by mixing concentrated squash with water in the ratio 2 : 5 by volume.

   a How much water is needed to mix with 25 ml of squash?

   b How much squash, to the nearest millilitre, is there in 500 ml of the drink?

3 A special purpose compost is made by mixing sand, peat and fertiliser in the ratio 3 : 4 : 1 by weight. What quantities of peat and fertiliser are needed to mix with 12 kg of sand?

4 The lengths of the corresponding sides of the two triangles are in the same ratio. Find the length of AB.

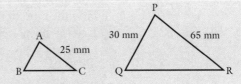

5 On a map with a map ratio of 1 : 10 000, a road is 12 cm long. How long is the road on the ground?

6 A diagram of a garden plan is drawn to a scale of 1 : 400. The end fence of the garden is 30 m long. How long should the line be drawn on the diagram to represent this fence?

7 A ream (500 sheets) of A4 file paper is 54 mm thick. How many sheets would you expect there to be in a pile of similar file paper that is 30 mm thick. Comment on the accuracy of your answer.

8 It takes 5 hours to fill a swimming pool when 3 hoses are turned on. Assuming that all the hoses deliver water at the same rate, how long would it take to fill the pool if 8 hoses were used?

9 Given that y varies as the square of x, and that $y = 2$ when $x = 3$, find y when $x = 5$.

10 Given that y is inversely proportional to the square root of x and that $y = 5$ when $x = 9$, find

   a the relationship between x and y

   b the value of x when $y = 12$.

SECTION B You may use a calculator for this section.

11 A bonus of £5000 is divided between Ann Jones, Colin Smith and Glyn Brown in the ratio of their salaries. Ann earns £12 000 a year, Colin earns £14 400 a year and Glyn earns £16 000 a year. What bonus does each of them receive?

12 When a stone is dropped from a height of h metres, it hits the ground with speed v m/s. Given that v varies as the square root of h, and that a stone dropped from a height of 20 m hits the ground at 20 m/s, find

a the relationship between v and h

b the speed at which a stone dropped from a height of 100 m hits the ground.

13 The mass of one atom of compound A is 2.5×10^{-24} kg. The mass of one atom of compound B is 7.6×10^{-22} kg. Find the ratio of the mass of compound A to the mass of compound B in the form $1{:}n$, giving the value of n correct to two significant figures.

14 David's salary is £36 500. Abdul's salary is £44 500. A bonus of £100 000 is shared between David and Abdul in the ratio of their salaries.

a What percentage of the bonus does David receive?

b What sum of money does Abdul receive?

MIXED EXERCISE 1.7

1 a i Write down a value of x for which $\sqrt{2} \times \sqrt{x}$ is an irrational number.

ii Write down a value of y other than $y = 3$, for which $\sqrt{3} \times \sqrt{y}$ is a rational number.

b i Write down and add together two irrational numbers which will give an answer that is also an irrational number.

ii Write down and add together two irrational numbers which will give an answer that is a rational number.　　　　　　　　　　　　　　　　　(OCR)

2 A full jar of coffee weighs 750 g. The empty jar weighs 545 g. Both weights are accurate to the nearest 5 g. Calculate the maximum and minimum possible values of the weight of coffee in the jar.　　　　　　　　　　　　　　　　　(OCR)

3 The formula $f = \dfrac{uv}{u+v}$ is used in the study of light.

   a Calculate f when $u = 14.9$ and $v = -10.2$.
     Give your answer correct to 3 significant figures.

   b By rounding the values of u and v in part **a** to 2 significant figures, check whether your
     answer to part **a** is reasonable. Show your working.   (OCR)

4

Four Star Leaded Petrol	Unleaded Petrol
62 p per litre	54 p per litre

Susan calls at a petrol station and puts £23.87 worth of four-star leaded petrol into her car.

   a How many litres of four-star petrol did she buy?

   b How much would the same quantity of petrol have cost her if she had put the
     unleaded petrol in her car?

   c Susan is thinking of converting her car to be able to use unleaded petrol. What would
     be the percentage drop in the cost of her petrol bills, if prices remained as they are
     now?   (WJEC)

5 **a** A particular CD costs a shopkeeper £9.20 from the wholesaler. The shopkeeper wants
     to make a profit of 35% on his costs. What price will the CD be in the shop?

   b Another CD is offered in a sale at a discount of 15%. The sale price is £13.26.
     What was the price of the CD before the sale discount?   (WJEC)

6 Given that y is inversely proportional to the square of x, and that y is 12 when $x = 3$

   a find an expression for y in terms of x

   b calculate   **i** y when $x = 6$ **ii** x when $y = 27$.   (WJEC)

7 Sarah uses the formula $t = \dfrac{2s}{u+v}$

She has to calculate the value of t when $s = 623.25$, $u = 11.37$ and $v = 87.22$. Sarah
estimates the value of t **without using her calculator**.

   a **i** Write down the numbers Sarah could use in the formula to estimate the value of t.
     ii Work out the estimate for the value of t that these numbers would give.

   b Use your calculator to work out the actual value of t. Give your answer to an
     appropriate degree of accuracy.   (Edexcel)

8 Nesta invests £508 in a bank account paying compound interest at a rate of 8.5% per
annum.
Calculate the total amount in Nesta's bank account after 2 years.   (Edexcel)

9 When a stone is thrown upwards with an initial speed s metres per second, it reaches a maximum height, h metres.

Given that h varies directly as the square of s and that $h = 5$ when $s = 10$,

a work out a formula connecting h and s.

b calculate the value of s when $h = 20$.

Two stones are thrown up. The ratio of their initial speeds is $3 : 1$.

c Work out the ratio of the maximum heights achieved. (Edexcel)

10 In Cooper's Store BRITE colour television sets are priced at £390 each in 1996. Lyn pays cash and is given a discount of 12%.

a Calculate the amount that Lyn pays.

The price of a BRITE colour television set has risen by 4% since 1995 to £390.

b Calculate the price of the television in 1995. (Edexcel)

11 The dimensions of the rectangle are given correct to one decimal place.

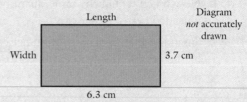

Length

Diagram *not* accurately drawn

Width

3.7 cm

6.3 cm

a Calculate the lower bound of the area of the rectangle.

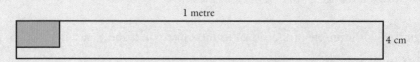

Diagram *not* accurately drawn

1 metre

4 cm

Rectangles with the same dimensions as the one in part **a** are to be cut from a large piece of card of length 1 metre, correct to the nearest centimetre and width exactly 4 centimetres.

b Calculate the maximum number of rectangles that could be cut from the large piece of card. (Edexcel)

12 a Simplify the expression $\sqrt{12} \times \sqrt{6}$ leaving your answer in surd form.

b **i** Find, as a single integer, the highest common factor of 216 and 168.

ii A rectangular field measures 21.6 m by 16.8 m. Fencing posts are placed along its sides at equal distances apart so that fence panels of equal size can be fastened between them. The posts are as far apart as possible. What is the distance between them?

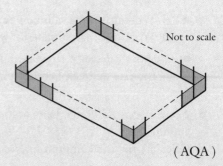

Not to scale

(AQA)

13 At the end of 1993 there were 5000 members of a certain rare breed of animal remaining in the world.
It is predicted that their number will decrease by 12% each year.

   a How many will be left at the end of 1996?

   b By the end of which year will the number first be less than 2500?   (OCR)

14 A number x is such that $x^{\frac{1}{2}} = \frac{1}{5}$

   a Find the value of $x^{\frac{3}{2}}$

   b Find the value of x^{-1}.   (OCR)

15 a Use your calculator to work out the value of $\dfrac{6.08 \times (9.72)^2}{581 + 237}$.

     Write down the full calculator display.

   b **i** Write down a calculation that could be done mentally to check the answer to part **a** using numbers rounded to one significant figure.

     ii Write down the answer to your calculation in part **b i**.   (OCR)

16 On a motorway there are three lanes, an inside lane, a middle lane and an outside lane.
One day, at midday, the speed of the traffic on these three lanes was in the ratio 3 : 4 : 5.
The speed on the outside lane was 70 miles per hour.
Calculate the speed on the inside lane.   (AQA)

17 a Light takes about 12 minutes and 40 seconds to reach the planet Mars from the Sun.
Light travels at approximately 299 800 kilometres per second.
Calculate the approximate distance of the Sun from Mars.
Give your answer in standard form correct to 2 significant figures.

   b The distance from the Earth to the Moon is approximately 384 400 km. The distance from the Earth to the Sun is approximately 1.496×10^8 km.
Use these approximations to express the ratio

        distance of Earth to Moon : distance of Earth to Sun

     in the form $1 : n$ where n is a whole number.

   c Light travels at the rate of approximately 186 000 miles per second. Light takes 12 years to reach Earth from a particular star.
Find the approximate distance, in miles, of this star from the Earth. Give your answer in standard form correct to 3 significant figures.   (AQA)

18 a What is a rational number?   **b** Convert $0.1\dot{5}$ to a fraction.   (AQA)

19 Solve the following equations

   a $8^{\frac{1}{x}} = 2$ **b** $16^{\frac{1}{4}} = 2^x$ **c** $32^y = 2$ **d** $125^x = \frac{1}{5}$ (AQA)

20 A transport plane can carry a maximum of 23 000 kg of cargo.
The owners of the plane wish to transport 60 crates of food to a famine stricken area.
The crates are labelled 'weight 380 kg to the nearest 10 kg'.
Could 60 crates exceed the maximum cargo load? Show the calculation on which you
base your decision. (OCR)

21 This year there were 4.27×10^6 listeners to Radio 5 which was an increase of 80 000
on the total for last year.
Find the number of listeners to Radio 5 last year. Give your answer in standard form.
(OCR)

22 a State whether the following numbers are rational or irrational.
    i 5.252 525   **ii** 5.2̇5̇

b State, with reasons, whether the following lengths are rational or irrational.

    i The circumference of a circle, radius 2 cm.

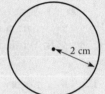

    ii The hypotenuse length of a right-angled triangle
       whose other two sides are $\sqrt{5}$ cm and 2 cm.

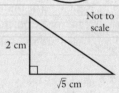

    iii The perpendicular height of an equilateral
        triangle of side 2 cm.

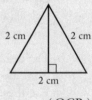

(OCR)

23 The number 10^{100} is called a googol.

   a Write the number 50 googols in standard index form.

A nanometre is 10^{-9} metres.

   b Write 50 nanometres in metres. Give your answer in standard index form.
   c How many nanometres are there in 10 metres? (Edexcel)

24 n is a positive integer such that $\sqrt{n} = 15.4$ correct to 1 decimal place.

   a **i** Find a value of n.   **ii** Explain why $\sqrt{n}$ is irrational.
   b Write down a number between 10 and 11 that has a rational square root. (Edexcel)

25 a Express $81^{\frac{1}{2}}$ as a fraction in the form $\dfrac{a}{b}$ where a and b are integers.

b Simplify $a^{\frac{1}{6}} \div a^{-2}$.

c Find the value of y for which $2 \times 4^y = 64$ (Edexcel)

26 The surface area of the Earth is approximately 1.971×10^8 square miles.
The surface area of the Earth covered by water is approximately 1.395×10^8 square miles.

a Calculate the surface area of the Earth not covered by water.
Give your answer in standard form.

b What percentage of the Earth's surface is not covered by water?   (AQA)

27 The length of each side of a square, correct to 2 significant figures, is 3.7 cm.

a Write down the least possible length of each side.

b Calculate the greatest and least possible perimeters of this square.

c **i** When calculating the perimeter of the square how many significant figures is it appropriate to give in the answer?
  ii Explain your answer.

d If this question had referred to a regular octagon, instead of a square, would your answer to part **c i** have been the same? Explain your answer.   (Edexcel)

28 a Write down a rational number between

  i 3 and 4   **ii** $\frac{3}{4}$ and 1   **iii** $\sqrt{3}$ and $\sqrt{5}$

b Write down an irrational number between

  i 3 and 4   **ii** $\frac{3}{4}$ and 1   **iii** $\sqrt{3}$ and $\sqrt{5}$ (OCR)

29 In each of the following cases simplify the expression, writing your answer in the form x^k.

a $\dfrac{x^2}{x^5}$ **b** $\sqrt{x^3}$ **c** $\sqrt{\dfrac{x^2}{x^6}}$ (OCR)

30 Change $0.4\dot{5}$ into a fraction in its lowest terms.   (Edexcel)

31 Square pegs are to be put into round holes.

Each peg is made 42 millimetre square,
measured to the nearest millimetre.
Each hole is drilled as a circle of radius
30.1 mm measured to three significant figures.
Will every peg fit into every hole?
Show all your working.

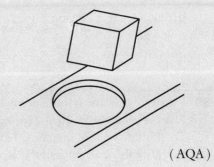

                         (AQA)

32 Phil has 80 birds; some are blue, the rest are yellow.

Phil sells 30% of his birds.
The new ratio of blue birds to yellow birds is 4:3.

How many blue birds has he got left? (AQA)

33 What is the area of a square of side $(1 + \sqrt{2})$ units?
Give your answer in the form $a + b\sqrt{2}$. (AQA)

34 a i Express 72 and 96 as products of their prime factors.
ii Use your answer to **i** to work out the Highest Common Factor of 72 and 96.
b Change the decimal $0.4\dot{5}$ into a fraction in its lowest terms. (Edexcel)

35 Calculate

$$\frac{3.81 \times 2.96^2 - 4.61\cos 20°}{\sqrt{7.83 + 0.593}}$$

(Edexcel)

36 The Andromeda Galaxy is 21 900 000 000 000 000 000 km from the Earth.

a Write 21 900 000 000 000 000 000 in standard form.

Light travels 9.46×10^{12} km in one year.

b Calculate the number of years that light takes to travel from the Andromeda Galaxy to Earth.
Give your answer in standard form correct to 2 significant figures. (Edexcel)

37 The population of the world is increasing at an **annual** rate of 1.6%.
In 1990, it was 5250 million.
Calculate an estimate for the population of the world in 2010. (Edexcel)

38 a Express the following in the form $p\sqrt{q}$ where p and q are integers and q is as small as possible.
i $\sqrt{75}$
ii $\sqrt{8} + \sqrt{18}$
b Simplify the following. Give your answer in the form $a + b\sqrt{2}$ where a and b are integers.

$$(1 + \sqrt{2})(3 - \sqrt{2})$$

(OCR)

39 I wish to paint the outside of my house.
A tin of paint covers $25\,\text{m}^2$, correct to the nearest $5\,\text{m}^2$.
The outside walls of my house have an area of $310\,\text{m}^2$, correct to the nearest $10\,\text{m}^2$.
Calculate the maximum number of tins of paint I may have to buy. (OCR)

40 Mrs Blake put £3000 in a building society account that offered 6% interest per year. Interest was added to the account at the end of each year.

   a How much did she have in her account 3 years later, after the final interest had been added?

   b What annual rate of interest would be required for a sum of money to double in ten years? Give your answer as a percentage to 1 decimal place. Show your calculations.

        (OCR)

41 **a** Calculate $81^{-\frac{1}{2}}$, giving your answer as a fraction.

   b Calculate $\dfrac{2^3}{2^{-2}}$ leaving your answer in the form 2^p.

   c Write $\dfrac{\sqrt{3}}{\sqrt{2}}$ in the form $\dfrac{\sqrt{a}}{b}$ where a and b are whole numbers.   (AQA)

42 Cashchem sells sun oil in special offer bottles which contain an extra 25% free. In a sale, Cashchem discounts all stock by 30%. What percentage is the reduction of cost of 100 ml of sun oil when bought in the special offer bottles in the sale?   (AQA)

43 The sets $\{7, 8, 9\}$ and $\{110, 111, 112\}$ are examples of sets of three consecutive numbers. Explore the validity of the following statements.

   i 'In any set of three consecutive numbers one of them is divisible by 3.'

  ii 'The product of any set of three consecutive numbers is divisible by 6.'

 iii 'The product of any set of three consecutive numbers is divisible by 4.'   (WJEC)

44 Show that the following statements are false.

   a If a is greater than b, then a^2 is always greater than b^2.

   b If $a > b$ then, for all values of c, $ac > bc$.

   c For all values of x, $x^2 > x$.   (OCR)

45 The sets $\{11, 13, 17\}$ and $\{23, 29, 31\}$ are examples of sets of three consecutive prime numbers.

For **each** of the following statements, investigate its validity giving an explanation, an example and a conclusion.

   a 'The sum of the numbers in any set of three consecutive prime numbers is always odd'.

   b 'In any set of three consecutive prime numbers the difference between the largest and the smallest is less than 10.'   (WJEC)

2 Algebra

The **coefficient** of a letter means the number (including the sign) it is multiplied by. In the expression $3x - 4y$, for example, the coefficient of x is 3 and the coefficient of y is -4.

The same rules apply to fractions with letter terms as to fractions with numbers only,

i.e. **fractions can be simplified** by cancelling common factors of the numerator and denominator. You may need to factorise some expressions before you can spot the common factors.

e.g.
$$\frac{x^2}{2xy - x^2} = \frac{x^2}{x(2y - x)} = \frac{x}{(2y - x)}$$

to **add or subtract fractions**, first change them into equivalent fractions with a common denominator, then add or subtract the numerators,

e.g.
$$\frac{3}{2x} + \frac{5}{4y} = \frac{6y}{4xy} - \frac{5x}{4xy} = \frac{6y - 5x}{4xy}$$

to **multiply fractions**, multiply the numerators together and multiply the denominators together,

e.g.
$$\frac{2x}{3y} \times \frac{5y^2}{7x} = \frac{2 \times 5y}{3 \times 7} = \frac{10y}{21}$$

and to **divide by a fraction**, turn it upside down and multiply,

e.g.
$$\frac{x}{x-2} \div \frac{2x^2}{3x-1} = \frac{x}{x-2} \times \frac{3x-1}{2x^2} = \frac{3x-1}{2x(x-2)}$$

Expansion of brackets

$x(2x - 3)$ means $x \times 2x + (x) \times (-3)$.

Therefore $x(2x - 3) = 2x^2 - 3x$

$(a + b)(c + d)$ means $a \times (c + d) + b \times (c + d) = ac + ad + bc + bd$,

i.e. each term in the second bracket is multiplied by each term in the first bracket. The order in which the terms are multiplied does not matter, but it is sensible to stick to the same order each time, e.g.

$$(2x - 3)(4x + 5) = (2x)(4x) + (2x)(5) + (-3)(4x) + (-3)(5)$$
$$= 8x^2 + 10x - 12x - 15$$
$$= 8x^2 - 2x - 15$$

In particular, $(x+a)^2 = (x+a)(x+a) = x^2 + 2ax + a^2$
$$(x-a)^2 = (x-a)(x-a) = x^2 - 2ax + a^2$$
$$(x+a)(x-a) = x^2 - a^2$$

The expression $x^2 - a^2$ is called the **difference of two squares**.

Factorising

Factorising is the reverse of the process of expanding (multiplying out) an algebraic expression.

A *common factor* of two or more terms can be seen by inspection and 'taken outside a bracket',

e.g. $2ab + 4bc = 2b(a + 2c)$

(This can be checked by expanding the result and remember to look at the expressions inside the brackets to ensure that you have not missed any factors.)

To factorise an expression such as $2x^2 + x - 10$, look for two brackets whose product is equal to the original expression.

Start by looking at the pattern of the signs;

when both are +, the sign in both brackets is +;

when the number term is + and the x term is −, the sign in both brackets is −;

when the number term is −, the sign in one bracket is + and the sign in the other is −.

In the case of $2x^2 + x - 10$, first note that, as the number term is negative, the signs in the brackets are different,

then the $2x^2$ term comes from $2x$ multiplied by x, so we start by writing $2x^2 + x - 10 = (2x \quad)(x \quad)$.

We then look for two numbers with a product of 10 and a difference of 1 between twice one of them and the other;

this gives $2x^2 + x - 10 = (2x + 5)(x - 2)$

We then check that the brackets are correct by expanding them.

Not all quadratic expressions factorise.

Completing the square

Expressions such as $x^2 + ax$ where a is a number can be made into a perfect square by adding a number; this number is found by halving the coefficient of x and then squaring it.

For example, $x^2 + 8x$ can be altered to $x^2 + 8x + \left(\frac{8}{2}\right)^2$ which is equal to $(x + 4)^2$

and $x^2 - 3x$ can be altered to $x^2 - 3x + \left(-\frac{3}{2}\right)^2$ which is equal to $\left(x - \frac{3}{2}\right)^2$.

Any quadratic expression can be written in the form $(x-a)^2 + b$.

For example, $x^2 + 8x - 3 = [x^2 + 8x + \left(\frac{8}{2}\right)^2] - \left(\frac{8}{2}\right)^2 - 3$

(adding $\left(\frac{8}{2}\right)^2$ to the first two terms to make them a perfect square and then subtracting it to keep the equality unchanged)

i.e. $\qquad x^2 + 8x - 3 = (x+4)^2 - 19$

Expressions, identities and equations

An **expression** is a collection of one or more algebraic terms, for example, $2x$, $x^2 - 3x$, $\sqrt{b^2 - 4ac}$ are expressions.

An **identity** is the equality between two forms of the same expression and it is true for all possible values of the unknown. for example, $2x = x + x$, $x^2 - 8x + 10 = (x-4)^2 - 6$ are identities.

This means that, if you know that $2x = x + ax$ is an identity, then a must be 1 since there must be the same number of xs on each side.

You can use this fact in less obvious situations, for example, given that $x^2 - 10x + a = (x-b)^2 + 8$ is an identity, the values of a and b can be found as follows.
Expanding the RHS gives $x^2 - 10x + a = x^2 - 2bx + b^2 + 8$
Then, as there must be the same number of xs on both sides, it follows that

$$-10 = -2b \quad \Rightarrow \quad b = 5$$

and the constant must be the same on each side so

$$a = b^2 + 8 \quad \Rightarrow \quad a = 25 + 8 = 33$$

An **equation** is an equality between two expressions which is true for some, but not all, values of the unknowns,
for example, $2x = 4$ is true only when $x = 2$
and $\qquad\qquad x + y = 1$ is true when $x = 2$ and $y = -1$
$\qquad\qquad\qquad\qquad$ (and for several other pairs of values of x and y)
$\qquad\qquad\qquad\qquad$ but it is not true when $x = 1$ and $y = 1$.

EXERCISE 2.1

Do not use a calculator for this exercise.

1 Simplify

a $\dfrac{x}{2} + \dfrac{2x}{5}$

b $\dfrac{x+1}{2} + \dfrac{2x}{5}$

c $\dfrac{x+1}{4} + \dfrac{2x-3}{6}$

d $\dfrac{2}{x} + \dfrac{3}{2x}$

e $\dfrac{2}{3x} + \dfrac{4}{y}$

f $\dfrac{1}{x} + \dfrac{2}{x-1}$

g $\dfrac{a}{2} - \dfrac{b}{3}$

h $\dfrac{2}{3p} - \dfrac{5}{q}$

i $\dfrac{1}{x+1} + \dfrac{2}{x-1}$

j $\dfrac{3}{x} - \dfrac{4}{2x-3}$

k $\dfrac{7}{x-1} + \dfrac{2x}{x^2+4}$

l $\dfrac{5x}{x^2+1} - \dfrac{4}{x}$

2 Simplify

a $\dfrac{x}{2} \times \dfrac{2x}{3}$ **e** $\dfrac{s}{2} \times \dfrac{3}{s^2}$ **i** $\dfrac{5a^4 b}{15ab^2}$

b $\dfrac{x^2}{3} \times \dfrac{6}{5x}$ **f** $\dfrac{x}{6} \div 2x^2$ **j** $\dfrac{3\pi r^2 \times h}{2\pi r}$

c $\dfrac{x}{2} \times 12$ **g** $\dfrac{x+2}{3} \times \dfrac{6}{2x-5}$ **k** $\dfrac{4xy \times 3y^3}{6x^2 y}$

d $\dfrac{x}{2} \div x^2$ **h** $\dfrac{a+2}{c^2} \div \dfrac{2a-3}{c}$ **l** $\dfrac{24a^5 b^3}{6a^2 \times 3ab}$

3 Multiply out and simplify

a $(x-4)(2x+1)$ **f** $(2s-t)(3s+2t)$

b $4y(7-3y)$ **g** $4a(a-b) - 6a^2(1-b)$

c $3x(x-3) - 2(4-x)$ **h** $5p^2 q(p^2 - q^2)$

d $(2a-7)(3-5a)$ **i** $2\pi r^2(r^2 - 3r)$

e $2pq(3p-2q)$ **j** $3mn(m^2 - 4n^2)$

4 Factorise

a $x^2 - 4$ **f** $6x^2 + 11x - 10$ **k** $2A^3 - 4A$

b $x^2 - 6x + 9$ **g** $4t^2 - 25s^2$ **l** $10x^2 - 19x + 6$

c $4st - 6t^2$ **h** $12 - 5x - 2x^2$ **m** $2p^2 q - 3pq^2$

d $x^2 + 6x + 8$ **i** $3x^2 - 12x + 9$ **n** $6x^3 y^2 - 3xy^3$

e $4x^2 + 5x + 1$ **j** $6y^2 + 7y - 3$ **p** $15\pi a^2 b - 10\pi ab^2$

5 Write down the number that needs to be added to make each of the following expressions a perfect square.

a $x^2 + 4x$ **c** $x^2 - 9x$

b $x^2 - 10x$ **d** $x^2 + 5x$

6 Write each expression in the form $(x+a)^2 + b$.

a $x^2 + 6x + 4$ **c** $x^2 + 12x - 3$ **e** $x^2 + 3x + 9$

b $x^2 - 8x + 1$ **d** $x^2 - 6x - 4$ **f** $x^2 - 5x + 1$

7 Find the values of p and q if, for all values of x,

$$x^2 - 14x + 2 = (x+p)^2 + q$$

8 Find the values of p and q if, for all values of x,

$$x^2 - px + 12 = (x-6)(x-q)$$

9 Find the values of a and b if, for all values of x,

$$x^2 + ax + 9 = (x+b)^2$$

10 Simplify

a $\dfrac{2x^2}{x^2 - 3x}$

c $\dfrac{a^2 + 2ab + b^2}{ab + b^2}$

b $\dfrac{x^2 - 9}{x^2 - 6x + 9}$

d $\dfrac{x^2 - 3x + 2}{2x^2 - 5x + 3}$

SOLVING EQUATIONS

An equation is a relationship between an unknown number, represented by a letter, and other numbers,

e.g. $2x - 3 = 5$

Solving the equation means finding the unknown number.
Provided that we do the same thing to both sides of an equation, we keep the equality; this can be used to solve the equation;
i.e. to rearrange the equation so that it is in the form $x = $ a number.

When an equation contains brackets, first multiply out the brackets.
When an equation contains fractions, multiply each term in the equation by the lowest number that each denominator divides into exactly (i.e. the lowest common multiple). This will eliminate all fractions from the equation.

Simultaneous equations

A pair of simultaneous equations in two unknowns can be solved algebraically by eliminating one of the letters.
It may be necessary to multiply one or both equations by numbers to make the coefficients of one of the letters the same in both equations.

For example, to solve $2y - x = 7$ [1]

and $3y + 4x = 5$ [2]

$[1] \times 4$ gives $8y - 4x = 28$ [3]

then $[2] + [3]$ eliminates x to give $11y = 33$

so $y = 3$
and, from [1], $x = -1$

It may also be necessary to rearrange one or both equations so that the letter terms are in the same position,

e.g. if we are given $y = 3x - 2$ [1]

and $2x - y = 4$ [2]

we can subtract $3x$ from both sides of equation [1] to give

$$-3x + y = -2 \quad [3]$$

The letters in equation [3] are now in the same order as those in equation [2].

Quadratic equations

To solve a quadratic equation, first arrange the equation in the form

$$ax^2 + bx + c = 0,$$

then, if the left-hand side factorises, use the fact that one or both of the factors must be zero,

e.g. $x^2 - 3x + 2 = 0$ becomes $(x-2)(x-1) = 0$

then either $x - 2 = 0$ in which case $x = 2$

or $x - 1 = 0$ in which case $x = 1$

If the left-hand side does not factorise, then use
either the method of **completing the square,**
e.g. $x^2 + 6x + 3 = 0$ can be solved by first expressing it
as $x^2 + 6x = -3$ then completing the square on the LHS, (remember
to add the same number to the RHS)
i.e. $x^2 + 6x + 9 = -3 + 9$ $\Rightarrow$ $(x+3)^2 = 6$
$\Rightarrow$ $x + 3 = \pm\sqrt{6}$ $\Rightarrow$ $x = -3 \pm \sqrt{6}$

Use this method when you need to give answers in an exact form
involving surds.

or use **the formula** $\qquad x = \dfrac{-b \pm \sqrt{b^2 - 4ac}}{2a}$

Simultaneous equations where one is quadratic and the other is linear

To solve a pair of equations such as $x^2 + y^2 = 5$ and $2x + y = 4$, use
the linear equation to express one letter in terms of the other,
($y = 4 - 2x$ is sensible here) then substitute for that letter in the
quadratic equation.
In this case we get $x^2 + (4 - 2x)^2 = 5$.
This can then be simplified to a quadratic equation in one unknown that
can be solved.

Polynomial equations in x contain terms involving powers of x,

e.g. $3x^3 - 3x + 1 = 0$ and $2x^2 = 5$ are polynomial equations.

Equations containing x^3 or higher powers of x can be solved by **trial
and improvement**, i.e. by trying possible values for x until we find a
value that 'fits' the equation.

Inequalities

An inequality remains true when the same number is added to, or subtracted from, both sides,

e.g.   if $x > 5$ then $x + 2 > 5 + 2$ and $x - 2 > 5 - 2$

An inequality also remains true when both sides are multiplied, or divided, by the same *positive* number,

e.g.   if $x > 5$ then $2x > 10$ and $\dfrac{x}{2} > \dfrac{5}{2}$

However, multiplication or division by a negative number must be avoided because this destroys the inequality.

Solving an inequality means finding the values of x for which the inequality is true.

e.g.   if $2x - 3 < 5$ then $2x < 8$ $\Rightarrow$ $x < 4$

Quadratic inequalities can be solved algebraically by first arranging the inequality in the form $ax^2 + bx + c < (\text{or} >)0$

then finding the values of x for which $ax^2 + bx + c = 0$.

If these values are p and q, a table can be constructed,

i.e.

value of x	$< p$	$p < x < q$	$> q$
value of $ax^2 + bx + c$			

from which it can be determined whether $ax^2 + bx + c$ is greater than or less than zero for each range of values of x.

Quadratic inequalities can also be solved graphically; this is covered on page 201.

EXERCISE 2.2

SECTION A   Do not use a calculator for this section.

1 Solve the equations

 a $2x - 7 = 3$

 b $2x - 7 = 3x - 10$

 c $3x - 4 = 5 - 6x$

 d $3(4 - 5x) = 1 + x$

 e $\frac{3}{4}x = 14 - x$

 f $\dfrac{x}{3} = \dfrac{1}{5} + x$

2 Solve the equations

 a $\dfrac{x - 1}{2} = \dfrac{x}{3}$

 b $\dfrac{2}{x - 1} = \dfrac{3}{x}$

 c $\dfrac{2}{x - 1} = \dfrac{5}{x + 1}$

 d $\dfrac{8}{x + 1} = \dfrac{8}{x - 1} + 25$

 e $\dfrac{5}{x + 1} = \dfrac{x}{4}$

 f $\dfrac{2x}{3} = \dfrac{1}{x - 1}$

3 Solve the simultaneous equations

a $x - y = 5$
$2x + y = 7$

c $3s + t = 7$
$2s - 3t = 1$

e $5x - 2y = 4$
$3x + 5y = 21$

b $2x + y = 6$
$2x - 3y = 4$

d $p = 3 - 2q$
$2p + 4 = q$

f $7x + 4y = -1$
$2x + 3y = 4$

4 A hotel has 54 bedrooms, some of which are single rooms and some are double rooms. If there are s single rooms and d double rooms

a write down an equation relating s and d.

The hotel can accommodate a total of 80 guests in these rooms.

b Write down another equation relating s and d.

c Solve your equations to find the number of single rooms.

5 **a** Find the range of values of n for which $3 - 2n < 4$.

b Give the smallest integer value of n for which $3 - 2n < 4$.

6 Find the integral values of n for which

a $2 < 3n - 4 < 11$ **b** $n < 2n - 1 < 7$

7 Solve the equations

a $x^2 - 6x + 5 = 0$

e $x^2 - 5x = 0$

b $2x^2 + 7x - 4 = 0$

f $x^3 = 4x - 3x^2$

c $2x = x^2 - 3$

g $\dfrac{15}{x-1} + \dfrac{15}{x+1} = 8$

d $\dfrac{2}{x-1} = \dfrac{3}{x} + 2$

h $\dfrac{2}{x+4} - \dfrac{1}{x+3} = 0$

8 The hypotenuse of a right-angled triangle is x cm long. The lengths of the other two sides are $(x - 1)$ cm and $(x - 2)$ cm.

a Use the information given to form an equation.

b Solve the equation and hence give the length of the longest side of the triangle.

9 Solve the following pairs of simultaneous equations.

a $x^2 + y^2 = 10$
$y - 2x = 1$

c $x^2 + y^2 = 13$
$2x - y = 1$

e $x + xy = 4$
$4x - 3y = 1$

b $x^2 + y^2 = 5$
$y - x = 3$

d $x^2 + xy = 10$
$3x - y = 3$

f $2x^2 - xy = 14$
$2x + y = 1$

10 Find the ranges of values of x for which $x^2 - x - 6 < 0$.

SECTION B   You may use a calculator for this section.

11 Solve the equations, giving answers correct to 3 significant figures

   a $x^2 - 3x + 1 = 0$ **c** $2 = x^2 - 2x$

   b $3x^2 - 7x - 5 = 0$ **d** $\dfrac{1}{x} = x + 1$

12 Find, correct to 1 decimal place, the value of t for which

$$2.3t - 3.7 = 2.2$$

13 Solve the equations $x^2 + y^2 = 7$ and $x + y = 3$ simultaneously, giving your answers correct to 2 decimal places.

14 Solve the equations $x^2 + xy = 3$ and $2x - y = 1$ simultaneously giving answers correct to 2 decimal places.

FORMULAS AND SEQUENCES

A **formula** is a general rule for finding one quantity in terms of other quantities,

e.g.   the formula for finding the area, A cm², of a rectangle measuring l cm by b cm, is given by $A = l \times b$

A is called the **subject of the formula**.

When the formula is rearranged to give $l = \dfrac{A}{b}$, l is the subject.

The process of rearranging $A = l \times b$ to $l = \dfrac{A}{b}$ is called **changing the subject of the formula**. It is achieved by thinking of $A = l \times b$ as an equation which has to be 'solved' to find l.

When the letter to be made the subject of a formula is squared, 'solve' the formula for the square and then remember that, if $x^2 = a$, $x = \pm\sqrt{a}$.

For example, to make a the subject of $b = a^2 + c^2$

            first 'solve' for a^2, giving $a^2 = b - c^2$,

then take the square root of both sides, i.e. $a = \pm\sqrt{b - c^2}$

When a formula contains square roots, first isolate the square root and then square both sides.

For example, to make m the subject of $h = 3 - \sqrt{m + n}$,

first 'solve' for $\sqrt{m + n}$ giving $\sqrt{m + n} = 3 - h$

then square both sides, i.e. $m + n = (3 - h)^2$, so $m = (3 - h)^2 - n$.

A **sequence** is an ordered set of terms, that is, there is a first term, a second term, and so on.

The **nth term of a sequence** is denoted by u_n and is sometimes expressed in terms of n, the position number of the term. Any term of the sequence can then be found by giving n a numerical value. For example, when the nth term is given by $u_n = 3n - 2$, the 10th term is given by substituting 10 for n,

i.e.   the 10th term is $3(10) - 2 = 28$

A **difference table** may help you to find the nth term in terms of n; a difference table is constructed by listing the given terms in the first row, then in the second row, listing $(u_2 - u_1)$, $(u_3 - u_2)$, $(u_4 - u_3)$, and so on. Subsequent rows are formed in the same way using the terms in the row above.

If the terms in the second row are constant (that is all equal to a) then the nth term of the sequence is a linear function of n,

i.e. $u_n = an + b$ where a and b are numbers.

If the terms in the second row are not constant but the terms in the third row are (all equal to $2a$) then the nth term of the sequence is a quadratic function of n, i.e. $u_n = an^2 + bn + c$ where a, b and c are numbers.

EXERCISE 2.3   **Do not use a calculator for this exercise.**

1 Given that $a = b(3 - c)$ find

    a a when $b = 3$ and $c = 2$ **c** a when $b = \frac{1}{2}$ and $c = -1.6$

    b a when $b = 4$ and $c = -3$ **d** b when $a = \frac{3}{4}$ and $c = -0.8$

2 If $R = \sqrt{d^2 - 2ef}$, find

    a R when $d = 3$, $e = 2$ and $f = -3$

    b R when $d = -4$, $e = -2$ and $f = -1$

3 Make the letter in brackets the subject of the formula.

    a $y = mx + c$ (c) **e** $s = \frac{1}{2}(a - b)^2$ (a)

    b $y = mx + c$ (x) **f** $T = \pi\sqrt{\dfrac{l}{g}}$ (l)

    c $A = \dfrac{PRT}{100}$ (R) **g** $\dfrac{1}{f} = \dfrac{1}{u} + \dfrac{1}{v}$ (u)

    d $C = \pi r^2$ (r) **h** $l = \sqrt{a^2 + b^2 + c^2}$ (b)

4 A square of side a units has a perimeter of P units and an area of A square units.

   a Find a formula for P in terms of a.

   b Find a formula for A in terms of a.

   c Hence find a formula for A in terms of P.

5 The formula $v = u + gt$ is used in science.

   a Make t the subject of this formula.

   The formula $s = ut + \frac{1}{2}gt^2$ is also used in science.

   b Find a relationship between s, u, v and g.

6 Write down the first five terms of the sequence whose nth term, u_n, is given by

   a $u_n = n(2 + 4n)$ **b** $u_n = n(n+1)(n+2)$

7 Find an expression for the nth term of each sequence.

   a $4, 7, 10, 13, 16, \ldots$ **c** $1, 3, 6, 10, 15, \ldots$

   b $-1, 2, 7, 14, 23, \ldots$ **d** $10, 8, 6, 4, 2, \ldots$

8 Find a formula for the nth term of the sequence

   a $7, 11, 15, 19, \ldots$ **b** $0, 7, 26, 63, 124, \ldots$

9 Use a difference table to find the next two terms in the sequence $5, 9, 16, 26, 39, \ldots$

10 Find a formula for the nth term of the sequence $-1, 0, 3, 8, 15, \ldots$ in terms of n. Use your formula to find the 12th term.

11 Bina is training for a marathon. She starts by running 500 metres on the first day. She then runs a further 500 m on each successive day.

   a Write down the distance she runs on each of the first 5 days.

   b Find, in terms of n, how far she runs on the nth day.

   c How many days does it take until she first runs 42 km?

12 The nth term of a sequence is 3^n.

   a Write down the first four terms of this sequence.

   b Show, using algebra, that the product of any two terms of this sequence is also a term of the sequence.

STRAIGHT LINE GRAPHS

A **straight line** graph gives the values of one quantity in terms of the other – it has meaning along the whole of its length.

This **conversion graph** is an example of a straight line graph; it gives the value in dollars of an amount in pounds sterling when the exchange rate is £1 ≡ $1.80. The graph can be used to convert between pounds sterling and US dollars, e.g. £75 ≡ $135

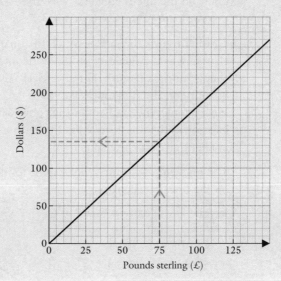

COORDINATES AND LINES

The **distance between two points** can be found using Pythagoras' Theorem with the help of a sketch. For example, the distance between A and B is given by

$$AB^2 = 7^2 + 6^2 = 85$$

so $AB = \sqrt{85}$

The x-coordinate of the **midpoint** of a line segment is half the sum of the x-coordinates of its end points. The y-coordinate is found in the same way.

For example, the coordinates of the midpoint of AB

are $\left(\dfrac{3 + (-4)}{2}, \dfrac{4 + (-2)}{2} \right) = (-\tfrac{1}{2}, 1)$

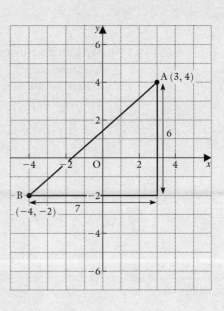

The **equation of a line gives the** y-coordinate of a point in terms of its x-coordinate. The relationship between the coordinates is true only for points on the line. In this case the equation of the line is $y = 2x - 3$; that is the y-coordinate of any point on the line is equal to twice the x-coordinate of the point minus 3.

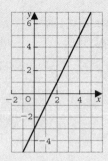

Gradient gives the rate at which the quantity on the vertical axis is changing as the quantity on the horizontal axis increases, that is the change in the quantity on the vertical axis per unit increase of the quantity on the horizontal axis.

For example, the gradient of the conversion graph on page 187 gives the increase in the dollar price for £1 increase in the sterling price, that is the gradient gives the exchange rate.

The gradient of a straight line can be found from any two points P and Q on the line, by calculating

$$\frac{\text{increase in } y \text{ in moving from P to Q}}{\text{increase in } x \text{ in moving from P to Q}} = \frac{y\text{-coordinate of Q} - y\text{-coordinate of P}}{x\text{-coordinate of Q} - x\text{-coordinate of P}}$$

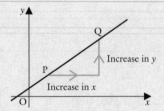

When the gradient is positive, the line slopes uphill when moving from left to right.
When the gradient is negative, the line slopes downhill when moving from left to right.

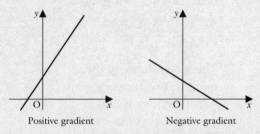

Positive gradient   Negative gradient

If **two lines are perpendicular**, the product of their gradients is -1, i.e. if a line has a gradient m, the **gradient of any perpendicular line** is $-\dfrac{1}{m}$.

For example, if a line has a gradient of $\frac{2}{3}$ then any line perpendicular to it has gradient $-\frac{3}{2}$.

The **equation of a straight line** is of the
form $y = mx + c$ where m is the gradient
of the line and c is the y-intercept,

e.g. the line whose equation is $y = \frac{1}{2}x - 5$

has gradient $\frac{1}{2}$ and its y-intercept is -5.

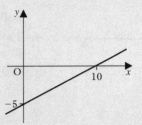

An equation of the form $y = c$
gives a line parallel to the x-axis.

An equation of the form $x = b$
gives a line parallel to the y-axis.

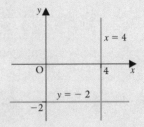

This straight line graph shows how the cost of a bag of potatoes is related
to the weight of potatoes in the bag.
For a given number of pounds we can find the cost and, conversely, we
can find how many pounds of potatoes we can buy for a given amount of
money. The gradient of the line gives the price of potatoes per pound
and, since the cost of 0 lb is 0 p, the line passes through the origin. The
price of the potatoes is directly proportional to the weight of the
potatoes.

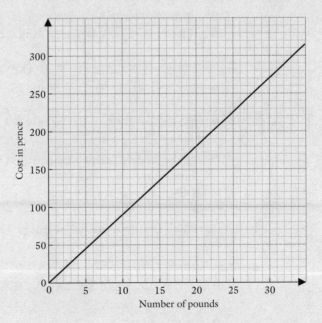

Any two quantities that can be represented by a straight line through the
origin are **directly proportional**.

Solving simultaneous linear equations using graphs

Two simultaneous equations can be solved graphically by drawing two straight lines and finding the coordinates of their point of intersection, that is the values of x and y that fit both equations.

To solve $\quad 3x - y = 4 \qquad$ [1]
$\qquad\qquad 2x + y = 5 \qquad$ [2]

we plot the lines represented by these equations.

It may help if each equation is first re-arranged in the form $y = \dots$

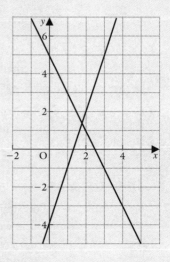

i.e. $\qquad y = 3x - 4 \qquad$ [3]
$\qquad\qquad y = -2x + 5 \qquad$ [4]

As accurately as we can read from the graph, the point of intersection is (1.8, 1.4).
The solution is therefore
$x = 1.8$ and $y = 1.4$.

---

1 For the line between the given points, find
  i the gradient
  ii the length in surd form
  iii the coordinates of the midpoint
  a $(1, 5)$ and $(3, 1)$ **b** $(3, 1)$ and $(2, 8)$ **c** $(-1, 6)$ and $(4, -2)$

2 Find the gradient of any line that is perpendicular to the line between
  a $(2, 6)$ and $(4, 2)$ **b** $(1, -3)$ and $(5, 7)$ **c** $(-1, -4)$ and $(3, -3)$

3 a Find the equation of the straight line that passes through the points A(2, 4) and B(8, 7).

  b For the straight line whose equation you found in part **a**, find
    i the value of y when $x = 6$
    ii the value of x when $y = 5.5$
    iii the coordinates of the mid-point of AB
    iv the length of the line segment AB.

4 On squared paper sketch the line whose equation is
  a $x = -3$ **b** $y = 2x + 5$ **c** $y = 6$ **d** $y = 3 - 2x$
  Label each line clearly.

5 Find the equation of the straight line that has a y-intercept of 3 and an x-intercept of 2. What is the gradient of this line?

6 Find the gradient and y-intercept of the straight line with equation

a $3y = x - 9$ **c** $\dfrac{x}{2} - \dfrac{y}{4} = 1$

b $3x + 4y = 12$ **d** $x + 3 = 4y$

7 Find the equation of the straight line that is

a parallel to the line $y = 4x - 3$ and passes through the point $(2, 9)$

b parallel to the line $3x + 4y = 1$ and cuts the x-axis where $x = -2$

c perpendicular to the line $y = 2x - 1$ and passes through the origin

d perpendicular to the line $y + x = 3$ and cuts the y-axis where $y = -1$.

8 The cost, $£C$, of repairing a washing machine on-site is made up of a fixed call-out charge plus an amount which depends on the time, t minutes, taken to complete the repair. This information is shown on the graph.

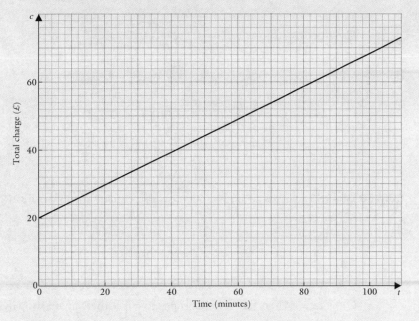

Find

a the fixed charge

b the total cost of a repair that takes 50 minutes

c the length of time that a repair takes if the total cost of the repair is £56

d the gradient of the line and what it means.

9 Pete was asked to draw the graph of $y = 3 - 2x$.
He plotted the six points shown in the diagram

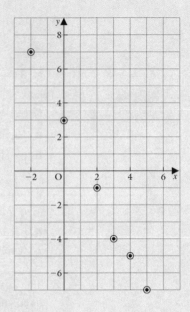

 a How can you tell from the shape of the graph
 that one of the points is in the wrong place?

 b Draw the graph of $y = 3 - 2x$ on your own
 set of axes using 1 cm as 1 unit.

 c By drawing another straight line on the
 same set of axes solve the simultaneous
 equations $y = 3 - 2x$ and $y = \frac{1}{2}x + 1$.

10 AC is perpendicular to AB. Find the equation of the line through A and C.

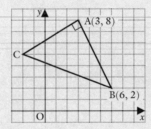

CURVED GRAPHS A **parabola** is a curve whose equation is in the form $y = ax^2 + bx + c$.

The shape of the curve looks like this:

When the x^2 term is positive it is the way up shown above.

When the x^2 term is negative the curve is upside down.

For example
the graph of
$y = x^2 - 2x$ is

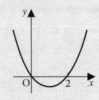

whereas
the graph of
$y = 2x - x^2$ is

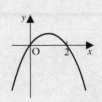

When the equation of a curve contains x^3 (and maybe terms involving x^2, x and a number), the curve is called a **cubic curve**.

The equations $y = x^3$, $y = 4x^3 - 2x + 6$ and $y = 5x^2 - x^3$ all give cubic curves.

A cubic looks like this ⟋ or ⩗ when the x^3 term is positive

and ⟍ or ⩗ when the x^3 term is negative.

The equation $y = \dfrac{a}{x}$ where a is a number, is called a **reciprocal equation**.

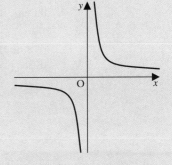

An equation of the form $y = \dfrac{a}{x}$,

where a is a constant (that is, a number), gives a two-part curve called a **hyperbola**.

Any two quantities, x and y, that are **inversely proportional**, are related by the equation $y = \dfrac{k}{x}$ and the graph representing them is a hyperbola.

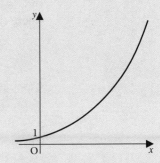

An equation of the form $y = a^x$ where a is a number greater than zero, is called an **exponential growth curve**.
This curve cuts the y-axis where $y = 1$.

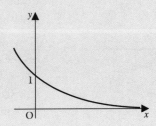

An equation of the form $y = a^{-x}$ where a is a number greater than zero, is called an **exponential decay curve**.
This curve cuts the y-axis where $y = 1$.

An equation of the form $x^2 + y^2 = a^2$ is the **equation of a circle** whose centre is the origin and whose radius is a.

If you plot this equation you need the same scale on both axes otherwise it will not look like a circle.

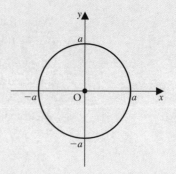

Solving equations using intersecting graphs

The equation $x^2 - 4x + 2 = 0$ can be solved by rearranging the equation as $x^2 = 4x - 2$ and drawing the graphs of $y = x^2$ and $y = 4x - 2$. The points where these graphs cross give the solutions to the given equation.

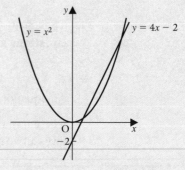

Similarly the equation $x^3 - 3x - 7 = 0$ can be solved by drawing the graphs of $y = x^3$ and $y = 3x + 7$ and finding where they cross.

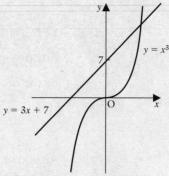

Sometimes a little more rearrangement is needed when using a given graph to solve an equation. For example, to solve the equation $x^2 - 3x - 9 = 0$ using the graph of $y = x^2 - x - 6$ rewrite the equation as $x^2 - x - 6 = 2x + 3$, i.e. so that the left-hand side is equal to the right-hand side of the equation of the given graph.

Now draw the graph of $y = 2x + 3$ on the same axes as the given graph. The values of x at the points of intersection of the two graphs give the solutions to the given equation.

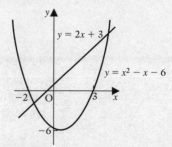

The coordinates of the points where two graphs intersect give the solutions to the simultaneous equations for the two graphs.

For example, the coordinates of the points A and B give the solution of the simultaneous equations

$$x^2 + y^2 = 9 \quad \text{and} \quad y = 2x - 1$$

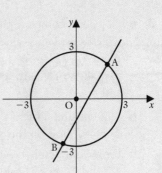

GRADIENT OF A CURVE AT A POINT

The **gradient of a curve** at a point is defined as the gradient of the tangent to the curve at that point, e.g. the gradient of the curve at P is the gradient of the tangent APB. Using a transparent ruler helps when drawing a tangent to a curve at a point. As a rough guide the tangent should be approximately at the same 'angle' to the curve on each side of the point on contact.

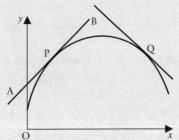

This gradient gives the rate at which the variable on the vertical axis is changing as values of the variable on the horizontal axis increase at the point, that is the increase in y for a unit increase in x. In the diagram the rate of change at P is positive so y is increasing in value but the rate of change at Q is negative, so y is decreasing in value.

EXERCISE 2.5

1 Use squared paper and the same scale on both axes to draw a *sketch* of the graph of $y = x^2$. On the same axes sketch the graph of $y = x + 1$.

a Use your sketch to estimate the solution of the equation $x^2 = x + 1$.

b Calculate the solution of the equation $x^2 = x + 1$, giving your answers to 2 decimal places.

2 a Draw the graph of $y = (x - 1)(2x + 3)$ for values of x between -2 and 2 using scales of 2 cm for one unit on both axes.

b Complete the table for values of y when $y = \dfrac{1}{x^2}$.

x	-2	-1.5	-1	-0.5	0.5	1	1.5
y		0.44					

Draw the graph of $y = \dfrac{1}{x^2}$ on the same set of axes used for part **a**.

c Write down the values of x at the points of intersection of the two graphs and the equation for which these values are solutions.

3 a Copy and complete the table which shows the value of
$y = x^3 - 2x^2 - 5x + 5$
for values of x between -3 and 4.

x	-3	-2	-1	0	1	2	3	4
y	-25	-1			-1		-1	

b Plot the graph of $y = x^3 - 2x^2 - 5x + 5$ for values of x from -3 to 4 using $2\,\text{cm} \equiv 1$ unit on the x-axis and $2\,\text{cm} \equiv 5$ units on the y-axis.

c From your graph, estimate the maximum and minimum values of y within the range -2 and 2, and the values of x for which they occur.

d By drawing a suitable straight line on the same axes, solve the equation
$x^3 - 2x^2 - 7x + 9 = 0$.

4 The graph shows the values of
$y = x^3 - 2x^2$ for values of x
from -1.5 to 3.

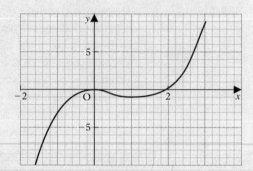

a Use the graph to find the values of x which satisfy the equation
$x^3 - 2x^2 + 2 = 0$.

b The equation $x^3 - 2x^2 = k$ is satisfied by one value only between -2 and 3.

Make a sketch copy of the graph and show how you would use your graph to find possible values for k.

5 a Copy and complete the table which gives values of $x^3 - 3x - 3$ for values of x from -3 to 3 at unit intervals.

x			-3	-2	-1	0	1	2	3
$x^3 - 3x - 3$			-21				-5		15

Plot these points on a grid using $2\,\text{cm} \equiv 1$ unit on the x-axis and $2\,\text{cm} \equiv 5$ units on the y-axis. Draw a smooth curve to pass through these points.

b By drawing a suitable straight line on the grid solve the equation
i $x^3 - 3x - 3 = -4$　　**ii** $x^3 - 4x - 5 = 0$

6 a Draw a set of axes for values of x and y from -8 to 8 using a scale of 1 cm for one unit on both axes. Draw the graph of $x^2 + y^2 = 36$.

b On the same axes draw the graph of $x + y = 7$.

c Hence estimate the solutions of the simultaneous equations $x^2 + y^2 = 36$ and $x + y = 7$.

7 a Sketch the graphs $x^2 + y^2 = 1$ and $x + y = 7$ on the same set of axes.

b What can you deduce from your sketch about the solutions of the simultaneous equations $x^2 + y^2 = 1$ and $x + y = 7$?

8 A set of five circular coins is made from the same alloy. Values of
the radius, r cm, and the mass, m g, are given in the table.

r	0.5	0.8	1	1.5	2
m	1.5	3.84	6	13.5	24

a Which of these graphs represents this information?

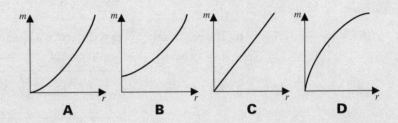

A   B   C   D

b Which of these equations describes the information given in the
table?

$$m = kr \qquad m = k\sqrt{r} \qquad m = a^r \qquad m = kr^2$$

c Would it appear from the data that all the coins are the same
thickness? Justify your answer.

d Find, in g/cm^3, the density of the alloy from which the coins are
made if each coin is 2 mm thick.

9 A car bought for £10 000 depreciates by 20% each year.
The value £V, of the car, after n years is given by the formula

$$V = 10\,000 \times 0.8^n.$$

a Copy and complete the table to show the value of the car over
the first 8 years of its life. Give the amounts correct to the nearest
£10.

n	0	1	2	3	4	5	6	7	8
V(£000s)	10	8				3.28			1.68

b Using a scale of 2 cm ≡ 1 year and 2 cm ≡ £1000, draw a
graph of V against n.

c Use your graph to find

i the value of the car after 6 months

ii how long it takes before the value of the car has halved.

10 The number of bacteria in a culture doubles every 2 hours.

a Copy and complete the table to show the number of bacteria at different times over a period of 12 hours.

Time, t hours	0	2	4	6	8	10	12
Number of bacteria, n	1	2	4				

b Plot the points on a graph, using a scale of $1\,\text{cm} \equiv 1\,\text{hour}$ and $1\,\text{cm} \equiv 5$ bacteria, and draw a smooth curve to pass through them.

c The relationship between t and n is given by the formula $n = 2^{\frac{1}{2}t}$. Copy and complete the following table.

t	3	7	11
n			

Verify that these points lie on your curve.

d Use your graph to find the value of n after
i 5 hours   **ii** 8 hours 24 minutes.

e From your graph find how long it takes before the number of bacteria is   **i** 10   **ii** 40.

11 The temperature, $T\,°C$, of some water t minutes after it was placed over a source of heat is given in the table.

Time, t minutes	0	1	2	4	5	7	9	11	13
Temperature ($°C$)	18	23	30	46	56	78	92	100	100

a On graph paper, using scales of 1 cm to 1 minute and 1 cm to $10\,°C$, draw a graph of temperature against time.

b Concurrently, in a similar experiment alcohol was heated instead of water. The following values were obtained.

Time, t minutes	0	1	2	4	5	7	9	11	13
Temperature ($°C$)	18	32	48	72	76	78	78	78	78

Plot a graph to represent this information on the same graph paper as you used for part **a**.

c Use your graph to find the rate of increase of the temperature after 3 minutes for
i the water   **ii** the alcohol.

d During the first 6 minutes of the experiments, is the temperature of the alcohol always increasing faster than the temperature of the water? Justify your answer.

12 Given below are five sketch graphs and five equations. Pair each equation with its graph.

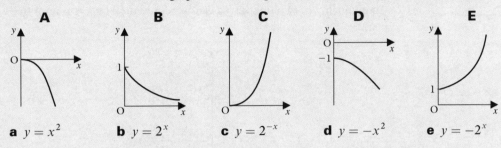

  A   **B**   **C**   **D**   **E**

a $y = x^2$ **b** $y = 2^x$ **c** $y = 2^{-x}$ **d** $y = -x^2$ **e** $y = -2^x$

13 It is known that N and t are related by an equation of the form $N = ka^t$ where k and a are constants. In an experiment the following values of N and t were obtained.

t	0	0.6	1.1	1.8	2.3	2.9	3.4
N	20	25.6	31.2	41.4	50.8	64.8	79.4

Plot these values on a graph and use your graph to write down the relationship between k and a when $t = 0$ and when $t = 1$. Solve these two simultaneous equations to find the values of k and a giving each value to 2 significant figures.

14 A curve passes through the points given in the table.

x	0	2	4	6	8	10	12	14	15
y	0	3	5.5	7.4	8.8	9.5	9.5	8.9	8.2

 a Using a scale of 1 cm for 1 unit on each axis plot these points on a graph and draw a smooth curve to pass through them.

 b Estimate the gradient of the tangent to the curve at the point where $x = 5$.

15 The table shows the number of houses in Ashton at 5-year intervals from 1930 to 1985.

Date, D	1930	1935	1940	1945	1950	1955	1960	1965	1970	1975	1980	1985
Number of houses, N	320	420	640	640	640	800	1060	1260	1380	1440	1480	1500

Using a scale of 2 cm ≡ 5 years and 1 cm ≡ 100 houses, draw the graph illustrating this information.
Find the gradient of the tangent to the curve when
 a $D = 1945$ **b** $D = 1965$
Interpret each result.

INEQUALITIES

Straight line inequalities

An inequality in two unknowns can be represented by a region of the x-y plane.

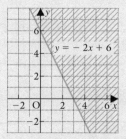

For example, the *unshaded area* in the diagram above represents the inequality $y \leqslant -2x + 6$. The boundary line, $y = -2x + 6$, is included in the region: this is indicated in the diagram by a solid line.

To find the inequalities that define an unshaded region, use a point as a test.

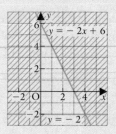

Obviously two inequalities that must be satisfied in this diagram are $x \geqslant 0$ and $y \geqslant -2$. The third boundary line is $y = -2x + 6$.
Test the point $(1, 1)$.
When $x = 1$, $-2x + 6 = 4$.
Since $y = 1$ and $1 < 4$,
the points in the region are where $y \leqslant -2x + 6$.

In this diagram, the unshaded region contains the points whose coordinates satisfy the three inequalities $y \leqslant -2x + 6$, $y \geqslant 1$ and $x > -1$.
The broken line shows that points on the line $x = -1$ are not included in the region.

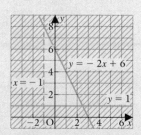

Quadratic inequalities

To find the range of values of x that satisfy the inequalities
$x^2 - 11x + 24 \leqslant 0$ and $x \geqslant 5$, first sketch the graph of
$y = x^2 - 11x + 24$, i.e. $y = (x - 3)(x - 8)$, which is shown below.

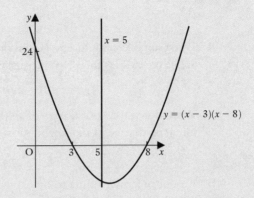

We see that $x^2 - 11x + 24 \leqslant 0$, that is y is negative, for $3 \leqslant x \leqslant 8$.
If we add the condition $x \geqslant 5$ we see that, to satisfy both the given
inequalities, $5 \leqslant x \leqslant 8$.

EXERCISE 2.6

1 Write down the three inequalities that define the triangular region
ABC.

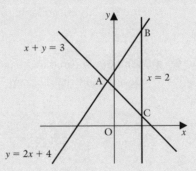

2 a A line has a gradient of 4 and passes through the point $(1, -2)$.
Find the equation of the line.

b Using a scale of 1 cm to one unit on each axis, draw the line for
values of x from -2 to 2.

c Find the gradient of the line $2y - x = 1$.

d Using the same axes as for part **b** draw the graph representing
$2y - x = 1$.
Leave unshaded the region that satisfies all the inequalities

$$2y - x \leqslant 1, \quad x > 0.5, \quad y > 4x - 6.$$

3 a Draw the graph of $y = x^2 - 4x$ for $-1 \leqslant x \leqslant 5$. Choose your own scales for the x- and y-axes.

b By drawing another straight line on the same diagram, solve the inequality $x^2 - 4x \geqslant 1$.

4 Draw suitable straight line graphs on the same diagram and shade the area that satisfies the three inequalities

$$2y - x \leqslant 1, \quad 2y + x \geqslant 4, \quad x < 4.$$

Write down the coordinates of all the points whose coordinates are integers, that lie in the region satisfied by all three inequalities.

5 A poultry farmer wants to buy at least 50 goslings and at least 100 ducklings.
He can afford to spend £450.
A gosling costs £1.50 and a duckling costs £1.

a If the farmer buys x goslings and y ducklings, express this information using inequalities.
Illustrate the inequalities on a graph and leave unshaded the area for which all the inequalities are satisfied.
Take $1\,\mathrm{cm} \equiv 25$ birds on each axis.

b Find the largest possible total number of birds the farmer can buy.

6 a Write down three inequalities needed to define the unshaded region, A, in the diagram.

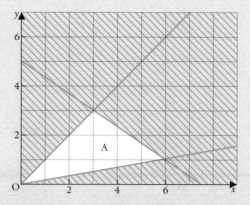

b The region A is reflected in the y-axis to give the region B. Write down the three inequalities that define the region B.

DISTANCE-TIME AND VELOCITY-TIME GRAPHS   The relationship between distance, speed and time when the speed is constant can be remembered from this triangle.

When the speed varies, average speed $= \dfrac{\text{total distance}}{\text{total time}}$.

A **distance-time graph** plots the distance from a starting point against the time taken to get there. It can be made up of straight lines and/or curves.

A distance-time graph is called a **travel graph**. The gradient of a sloping line gives the speed of the object. When the gradient is positive (such as from A to B) the object is moving away from the starting point and when the gradient is negative (from C to D) the object is moving towards the starting point. If a line is parallel to the horizontal axis (such as BC) the object is at a constant distance from the starting point, that is it is at rest.

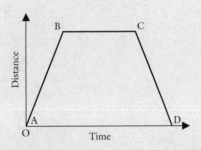

The **speed** of an object is the rate at which it is moving.
Velocity is the name given to the quantity that includes the speed and the direction of motion.

For a **curved distance-time graph** the gradient of a chord gives the average velocity over the time interval spanned by the chord, so gradient of PQ gives the average velocity from $t = 2$ to $t = 5$

which is $\dfrac{30}{3}$ m/s $= 10$ m/s.

The gradient of a tangent to the curve gives the velocity at that instant, e.g. when $t = 4$ the velocity is 8 m/s.

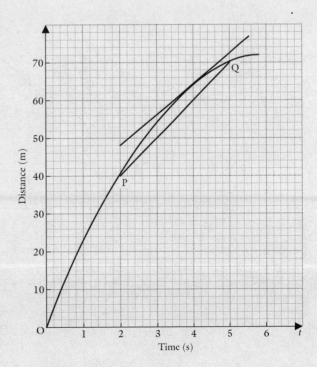

A **velocity-time graph** plots the velocity of an object against time, and can be made up of straight lines and/or curves.

This velocity-time graph is made up of straight lines.

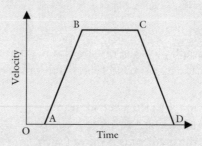

The gradient of a sloping line gives the **acceleration** of an object. When the gradient is positive (such as from A to B) the object is accelerating and when the gradient is negative (from C to D) the object is decelerating. If a line is parallel to the horizontal axis (such as BC) the object is travelling at a constant velocity.

For a curved velocity-time graph the gradient of the tangent at a point represents the acceleration at that instant.

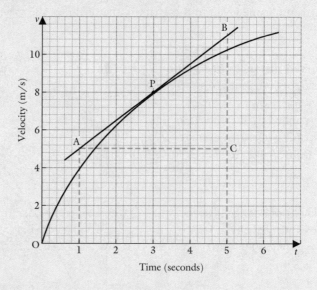

For this velocity–time graph,

the acceleration when $t = 3$ is given by $\dfrac{BC}{AC}$

i.e. acceleration when $t = 3$ is $\dfrac{6\,\text{m/s}}{4\,\text{s}} = 1.5\,\text{m/s}^2$

EXERCISE 2.7   **1**

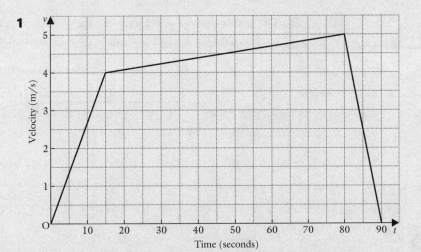

The graph shows the velocity of a cyclist during a short journey.

a How long did the journey take?

b Did the cyclist travel at a constant speed at any time during the journey? Justify your answer.

c What was the cyclist's acceleration for the first 15 seconds?

d What happened from $t = 80$ to the end of the journey?

e What was the length of the journey?

f What was the cyclist's average speed for the whole journey?

2 The table shows the velocity of a car for the first minute after starting from rest.

Time (seconds)	5	10	15	20	25	30	35	40	45	50	60
Velocity (m/s)	14	19	20	20	25	32	33	32	29	27	26

Plot these points on a graph, using $2\,\text{cm} \equiv 10\,\text{units}$ on both axes, and draw a smooth curve to pass through them.

a Use your graph to estimate

    i the acceleration of the car after 20 seconds

    ii the acceleration of the car after 25 seconds

b During which period of 10 seconds is the car's speed

    i increasing most   **ii** almost constant.

3 The graph shows the velocity of a car on a road between two sets of traffic lights.

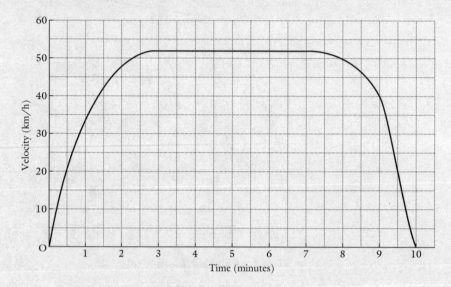

a For how long did the car maintain a steady speed?

b Estimate the acceleration of the car 2 minutes after leaving the first set of lights.

4 The graph shows how a motor cyclist's speed, measured in metres per second, varies in the first 5 seconds after starting from rest.

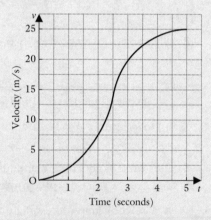

a Copy the graph and draw the tangent to the curve at the point where $t = 3$.

b Find the gradient of this tangent.

c What does this gradient represent?

FUNCTIONS

A rule that maps one number into another is called a **function**.

The result of applying a function f to x is written $f(x)$.
Hence, when f is the function 'square' we write $f(x) = x^2$.
Then if $x = 2$, $f(2) = 2^2 = 4$

To show how $f(x)$ changes as x changes, plot the graph of $y = f(x)$.
For example, if $f(x) = 5x + 2$ plot the graph of $y = 5x + 2$.

Since $y = f(x)$ we can label the vertical axis y or $f(x)$.

This is the graph of $y = 5x + 2$.

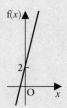

TRANSFORMATION
OF GRAPHS

The curve $y = -f(x)$ is a reflection
in the x-axis of the curve $y = f(x)$.

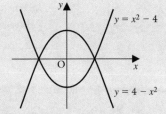

The graph of $y = f(-x)$
is a reflection of $y = f(x)$
in the y-axis.

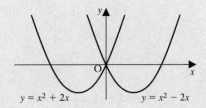

The graph of $y = f(x) + k$
is a translation, parallel to the
y-axis, of the graph of $y = f(x)$,
k units upwards if k is positive
or k units downwards if k is negative.
For example, $f(x) = x^3$ and
$f(x) = x^3 + 3$ are shown opposite.

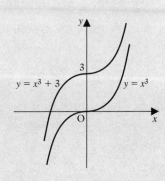

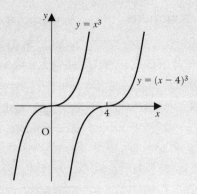

The curve $y = f(x - a)$ is given by translating $y = f(x)$ by a distance a in the positive direction of the x-axis. If a is negative the graph moves to the left.

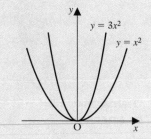

The curve $y = af(x)$ is sketched by stretching the graph of $y = f(x)$ by a factor of a in the y-direction.
(If $a < 1$, $y = f(x)$ is reduced by a factor a in the y-direction.)

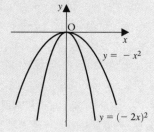

To sketch the graph of $y = f(ax)$ start with $y = f(x)$ and reduce it to $\dfrac{1}{a}$ of its original width.
(If $a < 1$, the graph is stretched by a factor $\dfrac{1}{a}$ in the x-direction.)

TRIGONOMETRIC FUNCTIONS

The functions $f(x) = \sin x°$ and $f(x) = \cos x°$ are **trigonometric functions**. Their graphs are shown below.

Graph of $f(x) = \sin x°$

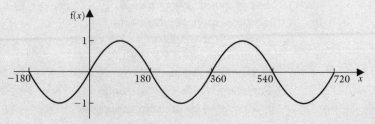

Graph of $f(x) = \cos x°$

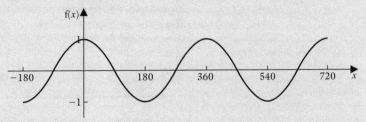

Note that if the graph for $f(x) = \sin x°$ is moved 90 units to the left it gives the graph for $f(x) = \cos x°$. For both graphs the maximum value of the function is 1 and the minimum value is -1 and the shape repeats every interval of 360.

The graph of $f(x) = \tan x°$ for $-90 \leqslant x \leqslant 450$ is shown below.

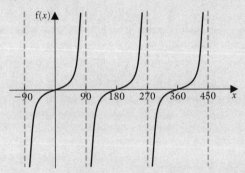

Its properties are different from those of the sine and cosine functions, namely

$\tan x°$ has no greatest or least value
the graph of $y = \tan x°$ does have a repeating pattern,
but it repeats at intervals of 180
the graph has breaks in it: these occur at 90 and every 180 step from
from there along the x-axis.

To solve the equation $\sin x° = -0.3$ for $0 \leqslant x \leqslant 360°$ graphically, draw the graph of $y = \sin x°$ for values of x between 0 and 360 and the line $y = -0.3$. These graphs intersect where $x = 197$ and 343. Hence if $\sin x° = -0.3$, $x = 197$ or 343.

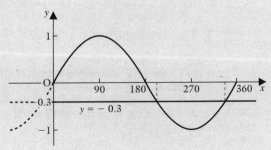

Sketch graphs are useful to make sure that no solution to an equation is overlooked. For example, to solve the equation $\cos x° = 0.5678$ for $0 \leqslant x \leqslant 360$ sketch the graphs of $y = \cos x°$ and $y = 0.5678$.

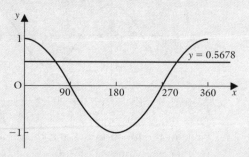

These show that there are two solutions within the required range and that they lie equally spaced before and after 180. Using a calculator, an angle whose cosine is 0.5678 is 55.4°. This is 124.6° before 180° so another value is $180° + 124.6°$ i.e. 304.6°.

The solutions of the equation $\cos x° = 0.5678$ for $0 \leqslant x \leqslant 360$ are therefore $x = 55.4$ and $x = 304.6$.

Check these values by finding the cosine of each angle using your calculator.

EXERCISE 2.8

1

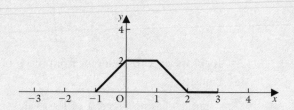

This is a sketch of the function $f(x)$ for $-1 \leqslant x < 3$. Copy the diagram and on the same axes sketch the graph of

  a $f(-x)$ **b** $f(x-1)$ **c** $f(2x)$

2 a Find the value of k if $f(x) = kx^2 + 4$ and $f(2) = 6$.

  b On the same set of axes draw a sketch of the graph of
    i $y = x^2$ **ii** $y = x^2 - 3$ **iii** $y = (x - 3)^2$

3 The sketch shows the graph of $y = f(x)$. On separate diagrams draw sketches of

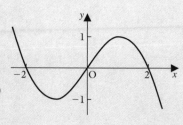

  a $y = -f(x)$ **c** $y = f(x+2)$

  b $y = f(-x)$ **d** $y = f(2x)$

4 Each curve is a transformation of the curve $y = x^2$. Describe the transformation and give the equation of the transformed curve.

a

b

c

d

5 This is a sketch of the graph of $f(x) = (1 - x)(x + 2)(x + 3)$

    a Find $f(0)$.

    b Sketch the graph of $y = f(-x)$ showing clearly where the graph crosses the x-axis.

    c Describe the single geometric transformation that maps the graph of $y = f(x)$ onto the graph of $y = f(-x)$.

6 This is a sketch of the curve $y = f(x)$ for $-5 \leqslant x \leqslant 10$. The point A$(5, -1)$ is the minimum value of y. Write down the coordinates of the minimum value of y on the curves

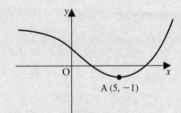

A $(5, -1)$

    a $y = f(x - 2)$ **c** $y = 2f(x)$

    b $y = f(x) - 2$ **d** $y = f(2x)$

7 This is the graph of $y = \cos x°$ for $-360 \leqslant x \leqslant 360$

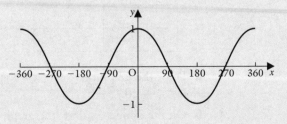

Sketch the graphs of

    a $y = 90 + \cos x°$ **b** $y = \cos(90° + x°)$

8 $f(x) = \sin x$

  a Sketch the graph of $y = f(x)$ for $0 \leqslant x \leqslant 180°$.

  b On the same diagram sketch the graph of $y = f(4x)$.

9 a Draw a sketch of the curve $f(x) = \sin x°$ for $0 \leqslant x \leqslant 360$.

  b For what values of x is $f(x) = 0$?

  c On the same axes draw lines to show how the values of x can be
found for which   **i** $f(x) = 0.5$ **ii** $f(x) = -0.4$

  d On separate diagrams, for $0 \leqslant x \leqslant 360$, sketch the graph of

    i $y = 1 + \sin x°$ **ii** $y = \sin \frac{1}{2}x°$

10 Sketch the graphs of $y = \sin x°$ and $y = 1 - \dfrac{x}{45}$ for

  $0 \leqslant x \leqslant 360$.

  Hence show that there is one solution of the equation

$$45 - x = 45 \sin x°$$

  within the given range and estimate its value correct to the
nearest 10.

11 a Draw a graph of $y = \sin x° + 3$ for $0 \leqslant x \leqslant 180$.

  b Use your graph to solve the equation

$$\sin x° + 3 = 3.75$$

  for values of x between 0 and 180°.

12 The table shows the value of a function $f(x) = (x+1)^2 + 3$ for
various values of x.

x	-4	-3	-2	-1	0	1	2	3
$f(x)$	12	7	4	3	4	7	12	19

  a Draw the graph of $y = f(x)$ for $-4 \leqslant x \leqslant 3$ and
$0 \leqslant y \leqslant 20$.

  b On the same axes draw the graph of $y = x^2$.

  c Describe how the graph of $y = (x+1)^2 + 3$ can be obtained
from the graph of $y = x^2$ by a transformation.

13 The height, h metres, of the tide at Limpster is given relative to a particular rung marked A, on a ladder fixed to the sea wall.
On 1 June the height was given by the equation $h = 6 \cos(30t)°$, where t is the time in hours after 6 a.m.

a Draw a graph of $h = 6 \cos(30t)°$ for values of t from 0 to 12 plotting points at 1 hour intervals.
Take $2 \text{ cm} \equiv 1 \text{ hour}$ on the t-axis and $1 \text{ cm} \equiv 1 \text{ m}$ on the h-axis.

b Use your graph to estimate

i when the height of the tide will be 4 m above A
ii the height of the tide above A at 10 a.m.

MIXED EXERCISE 2.9

1 a Rearrange the equation $3x + 4y = 12$ to express y in terms of x.

b **i** Factorise completely $14n - 4n^2$.

ii Find the integer values of n for which $14n - 4n^2 > 0$ (OCR)

2 This question is about finding a solution to the equation $x^3 + 2x = 7$ by a 'trial and improvement' method.

a Copy and complete the entries in the table.

b Between which two consecutive values of x used in the table in part **a** does the solution of the equation lie?

c Using a value of x with 2 decimal places, determine which of your two values in part **b** is the solution of the equation correct to 1 decimal place.

x	$x^3 + 2x$
1.4	5.544
1.5	
1.6	
1.7	8.313

(WJEC)

3 a **i** Rearrange the inequality $4n - 7 > 26 + n$ into the form $n >$ a number

ii Write down the least whole number value of n that satisfies this inequality.

b Vincent and Rowena both start to rent television sets at the same time. Vincent pays £14 a month for the rental of his television.
Rowena uses a different scheme in which she pays one amount of £50 and then a rental of £8 a month. Let x be the number of months that both Vincent and Rowena have been renting their televisions.

i Write down the inequality satisfied by x for the number of months that the total amount Vincent has paid is less than the total amount Rowena has paid.

ii Solve your inequality. Explain what your solution tells you. (WJEC)

4 The heat setting of a gas oven is called its Gas Mark.
A Gas mark, G, may be converted to a temperature, F, in degrees Fahrenheit, using the formula

$$F = 25G + 250$$

a Factorise completely $25G + 250$.

A Gas Mark, G, may be converted to a temperature, C, in degrees Celsius, using the formula

$$C = 14G + 121$$

b Make G the subject of the formula $C = 14G + 121$. (Edexcel)

5 The cost, W pounds, of a chest of drawers may be calculated using the formula

$$W = k + md$$

k and m are constants and d is the number of drawers.
The cost of a chest of drawers with 4 drawers is £117.
The cost of a chest of drawers with 6 drawers is £149.

a Use the information to write down two equations in k and m.

b Solve the equations to find the value of m. (Edexcel)

6 a Solve the inequality $7x + 3 > 2x - 15$

b Solve the simultaneous equations

$$x + y = 50$$
$$2x + 5y = 145$$

(AQA)

7 The percentage profit, p, on the sale of an item is given by the formula

$$p = \frac{100(s - c)}{c}$$

where s is the selling price and c is the cost price.
Express c in terms of s and p. (OCR)

8 The number of coins, N, with diameter d cm and with a fixed thickness that can be made from a given volume of metal can be found by using the formula

$$N = \frac{k}{d^2}$$

where k is a constant.

a Given that 5000 coins of diameter 2.5 cm can be made from the given volume of metal, calculate how many coins of diameter 2 cm can be made from an equal volume of metal.

b Rearrange the formula $N = \frac{k}{d^2}$ to make d the subject.

c 2000 coins are to be made using an equal volume of metal. Calculate the diameter of these coins. (OCR)

9 Greg sold 40 tickets for a concert.
He sold x tickets at £2.00 each and y tickets at £3.50 each. He collected £92.00.
Write down two equations connecting x and y.
Solve these simultaneous equations to find how many of each kind of ticket he sold.
(AQA)

10 n is a non-zero integer.

$$\frac{42}{n^2} > 5$$

List the possible values of n. (AQA)

11 Solve the equations **i** $4y^2 - 81 = 0$ **ii** $\dfrac{1}{x+2} + \dfrac{1}{3} = -1$ (Edexcel)

12 Solve the equation $x^2 + 5x = 3$ giving your solutions to 1 decimal place. (AQA)

13 Factorise fully $3x^2 - 6x$. (AQA)

14 There is a relationship between the terms in rows A, B and C.

Row A	1	2	3	4	5
Row B	1	4	9	16	25
Row C	2	6	12	20	30

a What is the formula for the nth term in row B?
b What is the formula for the nth term in row C? (AQA)

15 The frequency of radio waves (F khz) is inversely proportional to the wavelength
(L metres).
The formula which connects wavelength and frequency is

$$L = \frac{C}{F},$$

where C is a constant.

The table below shows some radio stations with their frequencies and wavelengths.

Radio Station	Frequency (F khz)	Wavelength (L metres)
Radio Atlantic	252	1179
Virgin Radio	1215	
BBC World Service		458.5

a Express C in terms of L and F.
b Calculate the wavelength of Virgin Radio, to the nearest metre.
c Calculate the frequency of the BBC World Service. (OCR)

16 British Airways flies Boeing 747s and Concordes from London to New York. For this journey the average speed of Concorde is 1.8 times the average speed of a Boeing 747.

a If the average speed of a Boeing 747 is x km/h write down, in terms of x, the average speed of Concorde.

The distance from London to New York is 6660 km.

b Write down expressions, in terms of x for the time in hours taken for the journey for
  i a Boeing 747   **ii** Concorde.

A Boeing 747 takes 3 hours 20 minutes longer for the journey than Concorde.

c Use your answers to part **b** to write down an equation in x and solve it to find the average speed of a Boeing 747. (OCR)

17 The diagram represents a greenhouse.
The volume of the greenhouse is given by the formula

$$V = \tfrac{1}{2}LW(E+R).$$

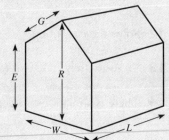

a Make L the subject of the formula, giving your answer as simply as possible.

The surface area A, of the greenhouse is given by the formula

$$A = 2GL + 2EL + W(E+R)$$

$V = 500$, $A = 300$, $E = 6$ and $G = 4$.

b By substituting these values into the equations for V and A show that L satisfies the equation.

$$L^2 - 15L + 50 = 0.$$

Make the steps in your working clear.

c Solve the equation $L^2 - 15L + 50 = 0$. (Edexcel)

18 The height, h metres, of a sky rocket t seconds after being launched is given by the formula

$$h = at^2 + bt + 2$$

where a and b are constants. The heights of the rocket above the ground at two different times are given in the table below.

t seconds	1	2
h metres	37	62

a At what height above the ground is the rocket launched?

b **i** Use the table of values to show that

$$a + b = 35$$
$$\text{and} \quad 4a + 2b = 60.$$

  ii Solve these simultaneous equations to find the value of a and the value of b.

c What was the height of the sky rocket $7\tfrac{1}{2}$ seconds after it was launched? (OCR)

19 Solve $3x + y = 4$

$y = x + 2$ (AQA)

20 Solve the inequalities **a** $4x + 3 \geqslant 17$ **b** $3x^2 \leqslant 12$ (OCR)

21 a Write the following expression as a single fraction in its simplest form.

$$\frac{3}{x + 2} + \frac{5}{x - 4}$$

b Hence show that $\dfrac{3}{x + 2} + \dfrac{5}{x - 4} = 1$ can be rearranged into

the form $x^2 - 10x - 6 = 0$.

c **i** Express $x^2 - 10x - 6$ in the form $(x - a)^2 - b$ where a and b are numbers to be found.

ii Hence, or otherwise, solve the equation $x^2 - 10x - 6 = 0$ (OCR)

22 a Simplify   **i** $\dfrac{12x^5 \times 3y^3}{9x^2 y}$ **ii** $(4p^2 q^3)^2$

b Factorise completely   **i** $9a^2 b^3 + 15a^3 b^2$ **ii** $x^2 + 7x - 60$

c On a number line show the solution to these inequalities.

$$-7 \leqslant 2x - 3 < 3$$ (Edexcel)

23 Solve the simultaneous equations $3p + 2q = 6$

$2p + 5q = -7$ (Edexcel)

24 Solve the equation $2x^2 - 5x - 4 = 0$ (Edexcel)

25 a Simplify the following expression. $\dfrac{x^2 + x - 2}{x^2 - 4}$

b Solve $\dfrac{2x + 1}{x - 1} = \dfrac{7x + 3}{4x - 3}$

c Solve $(x - 5)(x + 1) = 0$ (OCR)

26 a Simplify $2a^3 \times 4a^2$.

b Solve the equation $x^2 - 8x + 10 = 0$.

c Make y the subject of this formula.

$$x(2y - 3) = 5(y - 2)$$

d Solve algebraically these simultaneous equations. Show your method clearly.

$$x + y = 5$$
$$x^2 + 3y^2 = 49$$ (Edexcel)

27 Solve the equation $\dfrac{2(x - 2)}{x^2 - 4} + \dfrac{3}{2x - 1} = 1$ (Edexcel)

28 a Expand $x(3x^2 + 4)$.

b Factorise completely $ax + ay - bx - by$.

c Expand $(2x^2y)^3$.

d Solve the simultaneous equations $4x + 3y = -8$
$$6x - 2y = 27$$ (Edexcel)

29 Solve the simultaneous equations $y = 2x - 7$
$$x^2 + y^2 = 61$$ (Edexcel)

30 a List the values of n, where n is an integer, such that
$$3 < 3n < 18.$$

b Solve the simultaneous equations $2x + y = 9$
$$x - 2y = 7$$

c Factorise the expression $n^2 - n$

d Simplify $3a^2b \times 2a^3b$ (AQA)

31 Solve the equation $\dfrac{1}{x} + \dfrac{3x}{x-1} = 3$ (AQA)

32 a Factorise $x^2 - 9$.

b Simplify fully the expression $\dfrac{x^2 - 9}{2x^2 + 5x - 3}$ (AQA)

33 a The expression $x^2 - 10x + a$ can be written in the form $(x + b)^2$.
Find the values of a and b.

b Solve the equation $x^2 - 10x + 20 = 0$.
Give your answers to two decimal places.

c State the minimum value of y if $y = x^2 - 10x + 20$. (AQA)

34 Kath discovers the following results when doing some fraction problems:

$$\frac{2}{1} - \frac{1}{2} = \frac{3}{2}$$

$$\frac{3}{2} - \frac{2}{3} = \frac{5}{6}$$

$$\frac{4}{3} - \frac{3}{4} = \frac{7}{12}$$

$$\frac{5}{4} - \frac{4}{5} = \frac{9}{20}$$

The nth line of this sequence is

$$\frac{(n+1)}{n} - \frac{n}{(n+1)} = \frac{2n+1}{n(n+1)}$$

Prove that the left-hand side of this expression is equal to the right-hand side.
Show all your working clearly. (AQA)

35 a Simplify to a single fraction in its lowest terms $\dfrac{4a}{5} + \dfrac{a}{15} - \dfrac{3a}{10}$

b Simplify this expression $\dfrac{2x - 5}{6x - 15}$

c Solve the simultaneous equations $x^2 + y^2 = 100$
$$x - y = 2$$
(AQA)

36 a A rectangle has a perimeter of 12 cm
and an area of 7 cm^2.
The length of the rectangle is x centimetres.
Show that $x^2 - 6x + 7 = 0$.

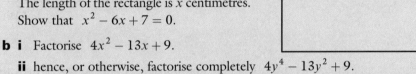

b i Factorise $4x^2 - 13x + 9$.

ii hence, or otherwise, factorise completely $4y^4 - 13y^2 + 9$.

37

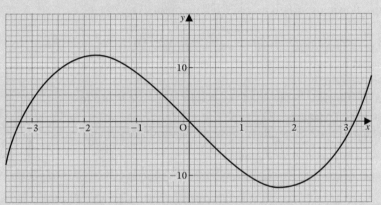

The graph of $y = x^3 - 10x$ has been drawn on the grid above.

a On a copy of the same grid, draw the graph of $y = 2x^2$.

b Use the graphs to find two solutions to the equation $x^3 - 2x^2 - 10x = 0$ (Edexcel)

38 Here is a velocity time graph of a car travelling between two sets of traffic lights.

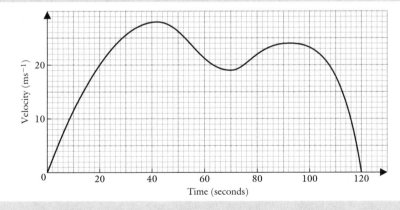

Calculate an estimate for the acceleration of the car when the time is equal to 20 seconds.
(Edexcel)

39 a Complete the table opposite to show the values
of $y = 4 \sin x$ for values of x from $0°$ to $180°$
in steps of $30°$.

b Draw the graph of $y = 4 \sin x$ for values of
x from $0°$ to $180°$. Use your graph to find all the
solutions of the equation $4 \sin x = 1.5$ between
$x = 0°$ and $x = 180°$, showing how you have
arrived at your solutions.

x	$4 \sin x$
$0°$	
$30°$	
$60°$	
$90°$	
$120°$	
$150°$	
$180°$	

(WJEC)

40 In an electrical appliance, the power, P watts, is proportional to the square of the current,
I amps, flowing through it. Sketch a graph of P against I. (AQA)

41 a Liquid is poured at a steady rate into the bottle shown in
the diagram.
As the bottle is being filled the height, h, of the liquid in
the bottle changes.
Which of the five graphs shown below shows this change ?

Give a reason for your choice.

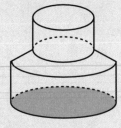

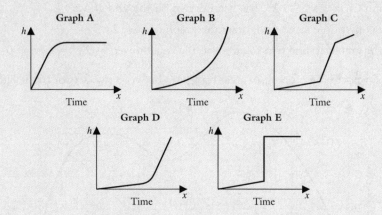

b Liquid is poured at a steady rate into another
container. The graph on the right shows how the
height, h, of the liquid in this container changes.
Sketch a picture of the container. (AQA)

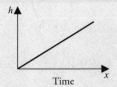

42 John places a cake in his freezer. The temperature, $T\,°C$, of the cake after t minutes is given by the formula.

$$T = 32(2^{-t}) - 18.$$

a Complete the table below.

t (minutes)	0	1	2	3	4
$T(°C)$					

b Draw the graph of T against t.

c John knows that the cake's temperature is $14\,°C$ when he places it in the freezer. He does not know the formula for its temperature after t minutes.
He estimates that its temperature will fall by $10\,°C$ every minute.
On the grid, draw the graph showing how John thinks the temperature will vary during the first three minutes.

d Use your graph to find the time when the estimated temperature is the same as the true temperature of the cake. (OCR)

43 a Sketch the graph of $y = \cos x°$ from $x = -180$ to $x = 540$.

b This is the graph of $y = f(x)$.

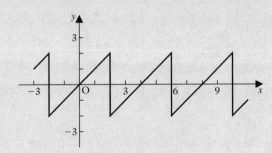

Sketch the graphs of the following.
Make clear the scales on your axes and compare each graph with $y = f(x)$.

i $y = 3f(x)$ **ii** $y = f(2x)$ **iii** $y = f(x-1)$ **iv** $y = 3 + f(x)$.   (OCR)

44 The graph shows the speed of a Grand Prix car over part of a racing circuit.

a Explain what is happening at
i A and **ii** B.

b Find the acceleration at time 4 seconds.

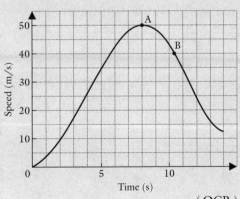

(OCR)

45 This is a sketch of the graph of $y = f(x)$ where

$$f(x) = (x+3)(x-2)(x-4)$$

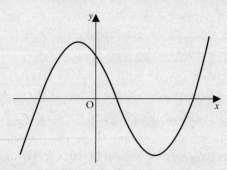

a Calculate the value of $f(0)$.

b On a copy of the axes above sketch the graph of $y = f(-x)$.

c Describe fully the single geometric transformation which maps the graph of $y = f(x)$ onto the graph of $y = f(-x)$.

The equation $f(x) = f(-x)$ has a solution $x = 0$.
It also has a positive solution x such that

$$n < x < n+1$$

where n is a positive integer.

d Write down the value of n. (Edexcel)

46 a Draw the graph of $y = x^2 - 3x$, for values of x from -1 to 4.

b By drawing a suitable straight line on the same diagram, estimate, correct to one decimal place, the solutions to the equation $x^2 - 2x - 1 = 0$.

c By drawing another straight line on the same diagram, solve the inequality $x^2 - 3x \geqslant 1$. (AQA)

47 The diagram shows a sketch of $y = x^4$.

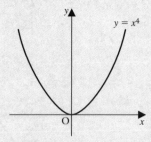

Sketch the following curves. Mark clearly the coordinates of any points where your curves meet the axes.

a $y = -x^4$ **b** $y = 2 + x^4$ **c** $y = (x+3)^4$ (WJEC)

48 The region R in the diagram can be described as the intersection of three inequalities:

$$x \geqslant 1$$
$$x + y \leqslant 6$$
$$y \geqslant x + 2$$

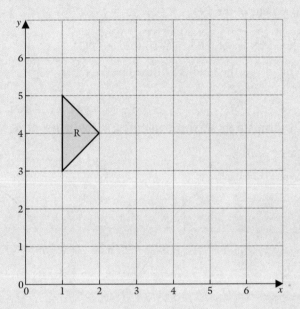

The region Q is the reflection of R in the line $y = x$.
Q can be described as the intersection of three inequalities.

What are these inequalities: (AQA)

49 a Write down the equation of the line which is parallel to $y = 2x$ and which passes through the point (0, 8).

This line crosses the x-axis at the point P.

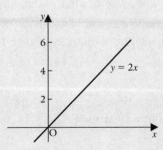

b Calculate the coordinates of P. (Edexcel)

50 P is the point (1, 8) and Q is the point (11, 3).
M is the midpoint of the line PQ.

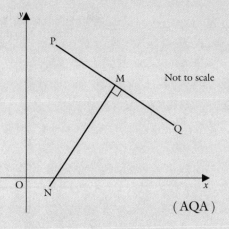

Not to scale

a Find the coordinates of the point M.

The gradient of the PQ is $-\dfrac{1}{2}$.

The line MN is perpendicular to the line PQ.

b Write down the gradient of the line MN.

(AQA)

51 a The diagram shows the graph of $y = \cos x$ for $0° \leqslant x \leqslant 360°$.

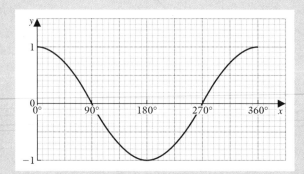

i On the diagram show the location of the two solutions of the equation $\cos x = -0.5$.

ii The angle x is between $0°$ and $360°$.

Work out accurately the two solutions of the equation $\cos x = -0.5$.

b On a particular day the height, h metres, of the tide at Weymouth, relative to a certain point, can be modelled by the equation

$$h = 5 \sin (30t)°,$$

where t is the time in hours after midnight.

i Sketch the graph h against t for $0 \leqslant t \leqslant 12$.

ii Estimate the height of the tide, relative to the same point, at 2 p.m. that day.

(AQA)

52 The graph of the equation $\qquad y = ax + b$
intersects the graph of the equation $y = cx + d$
at point P.

Show that the x-coordinate of P is $x = \dfrac{d - b}{a - c}$.

(Edexcel)

53

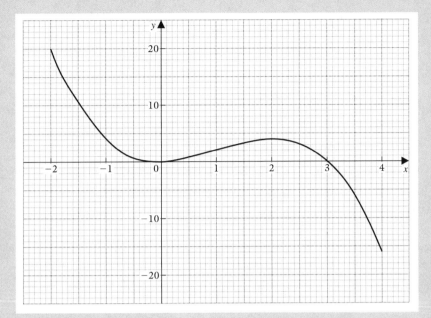

The diagram shows the graph of $y = 3x^2 - x^3$ for values of x from -2 to 4.

a Use the graph to find three values of x which satisfy the equation $3x^2 - x^3 = 2$.

b The equation $3x^2 - x^3 = k$ is satisfied by only one value of x between -2 and 4. What can be said about the number k?

c **i** On a copy of the diagram, draw the reflection of the graph in the x-axis.
ii Write down the equation of this reflection. (OCR)

54 The lines shown have equations

i $3x - y = 1$ **ii** $y = x$ **iii** $2x + y = 5$ **iv** $2y - x = 6$

a State which of the lines **A**, **B**, **C** or **D** fits each of these equations.

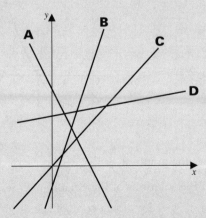

b Show clearly on a copy of the diagram the region which satisfies all the following conditions: $3x - y \geqslant 1$ $2x + y \geqslant 5$ $y \geqslant x$ $2y - x \leqslant 6$ (OCR)

55 a On a copy of the axes below, draw the graph of $y = \sin x$.

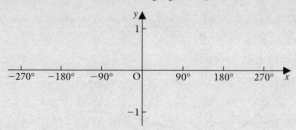

b Given that $-270° \leqslant x \leqslant 270°$; find the three values of x which satisfy the equation

$$\sin x = 0.876.$$

Write your answers correct to one decimal place. (OCR)

56 The number of bacteria in a colony multiplies by a factor of 3 every hour. Initially there are 20 bacteria.

a Complete the table.

Time (hours)	0	1	2	3	4	5
Number of bacteria	20	60				

b Write down the formula for the number of bacteria, n, after t hours.

c Using the table of values in part **a** draw the curve showing how the number of bacteria changes in the 5 hours.

d i By drawing a tangent to the curve in part **c**, calculate the gradient of the curve at the point where $t = 3.5$.

ii What information does this gradient value represent? (OCR)

57

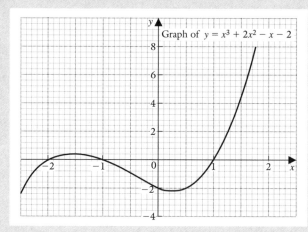

The diagram above shows the graph of $y = x^3 + 2x^2 - x - 2$.

a Use the graph to find the solutions of the equation $x^3 + 2x^2 - x - 2 = 0$

b By drawing the graph of $y = 2x^2$ on a copy of the diagram, find the solution of the equation $x^3 - x - 2 = 0$.

c Use the graph to find solutions of the equation $x^3 + 2x^2 - x - 1 = 0$. (OCR)

58

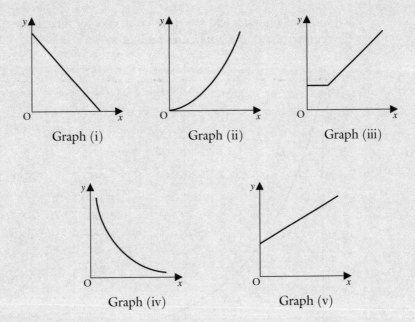

Graph (i)   Graph (ii)   Graph (iii)

Graph (iv)   Graph (v)

Select from the five graphs one which illustrates each of the following statements.

a The time (y) taken for a journey is inversely proportional to the average speed (x).

b The surface area (y) of a sphere is proportional to the square of the radius (x).

c The cost (y) of an electricity bill consists of a fixed charge plus an amount proportional to the number of units of electricity used (x). (OCR)

59 a Sketch the graph of $y = \tan x°$ on a copy of the given axes, for $-90 \leqslant x \leqslant 450$.

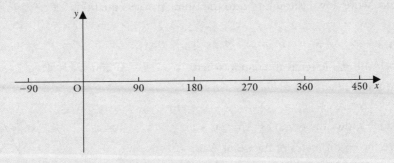

b Use your calculator to find the value of x between 0 and 90 for which $\tan x° = 2$.

c Use your graph and the answer to part **b** to solve the equations

   i $\tan x° = 2$ for $180 \leqslant x \leqslant 360$

   ii $\tan x° = -2$ for $180 \leqslant x \leqslant 360$. (OCR)

60 A glass was filled with boiling water and was then left to cool for three hours. The graph below shows the temperature of the water after t minutes.

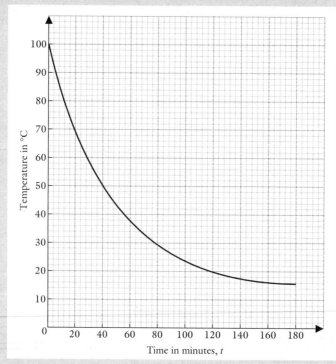

a On a copy of the graph draw the tangent to the curve at the point where $t = 80$.

b Find the gradient of the curve at the point where $t = 80$ and state its units.

c What does the gradient of this graph measure? (OCR)

61 a For $0 \leqslant x \leqslant 10$ and $-3 \leqslant x \leqslant 8$, draw the graphs of
$$y = 2x - 3 \quad \text{and} \quad 3x + 4y = 24.$$

b Hence write down the solution to the simultaneous equations
$$y = 2x - 3$$
$$3x + 4y = 24$$

c Label with the letter R the region where $y \geqslant 2x - 3$ and $3x + 4y \leqslant 24$.

d Find the gradient of the line $3x + 4y = 24$. (AQA)

62 The sketch shows the graph of $y = 2x + 1$.

i Give the co-ordinates of the point A.

ii On a copy of the diagram, sketch the graph of $y = 2x - 1$.

iii From your graphs explain why the equations $y = 2x + 1$ and $y = 2x - 1$ cannot be solved simultaneously. (AQA)

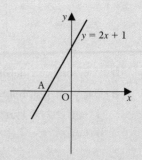

63 Two people start a rumour that goes round a school.
The following table shows the number of people, n, who have heard the rumour after t hours.

Time, t, in hours	0	1	2	3	4	5
Number of people, n	2	6	18	54	162	486

a Write down a formula for n in terms of t.

The graph below shows the number of people, n, who have heard the rumour after t hours.

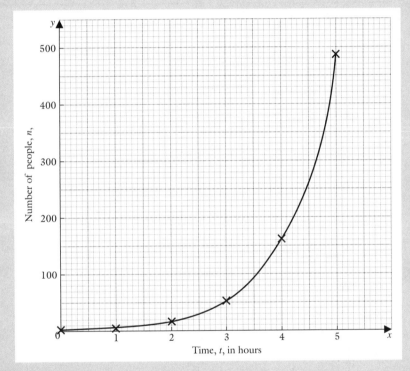

b i Find the gradient of the curve when $t = 3$.
  ii What does the gradient represent?   (OCR)

64 a Complete the table of values for the graphs of
$$y = x^3 - 2 \quad \text{and} \quad y = 3x^2 + 3x - 6.$$

x	−2	−1	0	1	2	3	4
$y = x^3 - 2$		−3	−2	−1		25	62
$y = 3x^2 + 3x - 6$	0		−6	0	12	30	

b i On graph paper draw the graphs of
$$y = x^3 - 2 \quad \text{and} \quad y = 3x^2 + 3x - 6$$
  ii Use your graphs to solve the equation $x^3 - 3x^2 - 3x + 4 = 0$.   (Edexcel)

65 A doctor injects a patient with a pain killer. The amount, y mg, remaining in the bloodstream after t hours is given by the equation

$$y = 5 \times 3^{-0.5t}.$$

a Write down the amount of pain killer initially injected by the doctor.

b Complete the following table of values. Give your answers to 2 decimal places.

Time after injection, t hours	0	1	2	3	4	5
Amount of pain killer remaining, y mg						0.32

c On a copy of the grid draw the graph of y against t.

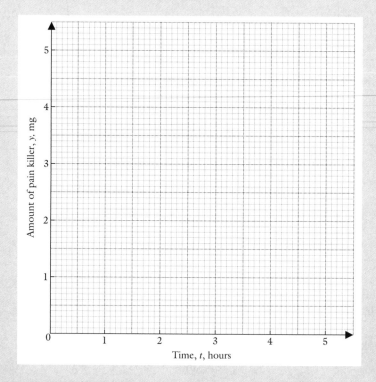

d The pain killer is only effective whilst there is more than 0.75 mg remaining in the bloodstream. Use your graph to determine for how long it is effective.

e The amount of a different pain killer remaining in the bloodstream is given by the equation $y = 5 \times 3^{-0.7t}$.
Does this pain killer lose its effectiveness more or less rapidly than the first? Give a reason for your answer.

(OCR)

66 a i Factorise $x^2 - 4x - 12$.   **ii** Solve $x^2 - 4x - 12 = 0$.

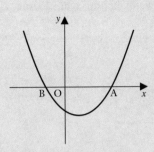

The diagram shows a sketch of the graph of $y = x^2 - 4x - 12$.
The curve cuts the x-axis at the points A and B.

b Write down the coordinates of A and B.

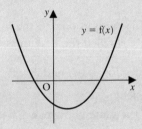

$f(x) = x^2 - 4x - 12$

c Sketch on a copy of the axes above the graph of $y = f(x - 2)$.   (Edexcel)

67 The diagram shows the graph of $y = x^3$. By drawing the graph $y = x^2 + 1$ for
$0 \leqslant x \leqslant 2$, estimate the solution of the equation $x^3 = x^2 + 1$.

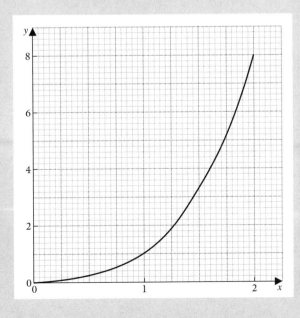

(AQA)

68 The cross-section of a swimming pool is shown. It is filled, from empty, at a uniform rate. Sketch a graph of the water height against time, t, as the pool fills.

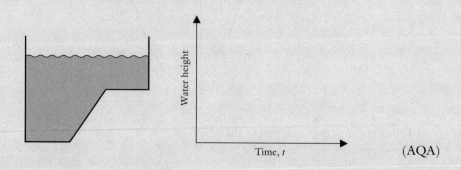

(AQA)

69 A man's weight, W kilograms, is proportional to the square of his height, H metres. The constant of proportionality is called the body mass index, B.

a Write down a formula connecting W, H and B.

b Ali is 1.8 m tall and weighs 81 kg.
Stephen is 2 m tall and has the same body mass index as Ali.
What is Stephen's weight?

c The graph of W against H for people with a body mass index of 16 is drawn.

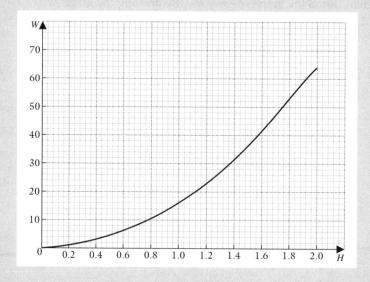

   i Find the gradient of the curve when $H = 1.7$ m.
   ii What are the units of the gradient?

(AQA)

70 Two trains A and B start from rest on parallel tracks.
On the grid below are two graphs which show how the speed of each train varies with time over the first twenty five seconds of their journey.

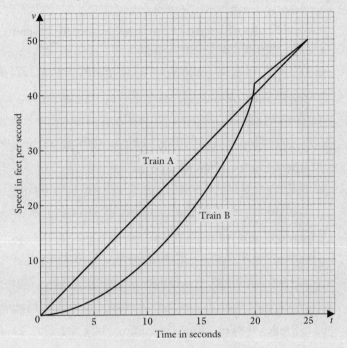

a What is the equation of the graph relating to train A?

b What is the value of t when the trains are travelling at the same speed?

c **i** After how many seconds do the trains have the same acceleration?

  ii Calculate the size of this acceleration. (AQA)

71 a A sketch of the graph of $y = \sin x$ for $0° \leqslant x \leqslant 720°$ is shown.

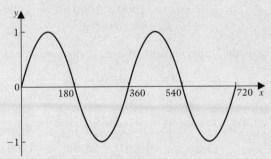

  Find all the values of x in the interval $0° \leqslant x \leqslant 720°$ that satisfy $\sin x = \frac{1}{2}$

b Find all the values of x which satisfy the equation $\cos x = -\cos 30°$
in the interval $0° \leqslant x \leqslant 720°$. (AQA)

3 Shape, Space and Measure

Metric units of length in common use are the kilometre, metre, centimetre and millimetre, where

$$1\,\text{km} = 1000\,\text{m}, \quad 1\,\text{m} = 100\,\text{cm}, \quad 1\,\text{cm} = 10\,\text{mm}$$

Metric units of mass are the tonne (t), kilogram (kg), gram (g) and milligram (mg), where

$$1\,\text{tonne} = 1000\,\text{kg}, \quad 1\,\text{kg} = 1000\,\text{g}, \quad 1\,\text{g} = 1000\,\text{mg}$$

Imperial units of length in common use are the mile, yard (yd), foot (ft) and inch (in), where

$$1\,\text{mile} = 1760\,\text{yards}, \quad 1\,\text{yard} = 3\,\text{feet}, \quad 1\,\text{foot} = 12\,\text{inches}$$

Imperial units of mass still in common use are the ton, hundredweight (cwt), stone, pound (lb) and ounce (oz), where

$$1\,\text{ton} = 20\,\text{cwt} = 2240\,\text{lb}, \quad 1\,\text{stone} = 14\,\text{lb}, \quad 1\,\text{lb} = 16\,\text{ounces}$$

For an approximate conversion between metric and Imperial units, use

$$5\,\text{miles} \approx 8\,\text{km}, \quad 1\,\text{inch} \approx 2.5\,\text{cm}, \quad 1\,\text{kg} \approx 2.2\,\text{lb}, \quad 1\,\text{tonne} \approx 1\,\text{ton}$$

For a quick but very rough conversion use

$$1\,\text{km} \approx \tfrac{1}{2}\,\text{mile}, \quad 1\,\text{yard} \approx 1\,\text{m}, \quad 1\,\text{kg} \approx 2\,\text{lb}$$

Area is measured by standard-sized squares.

$$1\,\text{cm}^2 = 10 \times 10\,\text{mm}^2 = 100\,\text{mm}^2$$
$$1\,\text{m}^2 = 100 \times 100\,\text{cm}^2 = 10\,000\,\text{cm}^2$$
$$1\,\text{km}^2 = 1000 \times 1000\,\text{m}^2 = 1\,000\,000\,\text{m}^2$$
$$1\,\text{hectare} = 10\,000\,\text{m}^2$$

The main **Imperial unit of area** is the acre where

$$1\,\text{acre} = 4840\,\text{sq yd}$$
and $$1\,\text{hectare} = 2.5\,\text{acres}, \text{ correct to 1 d.p.}$$

Volume is measured by standard-sized cubes.

$$1\,\text{cm}^3 = 10 \times 10 \times 10\,\text{mm}^3 = 1000\,\text{mm}^3$$
$$1\,\text{m}^3 = 100 \times 100 \times 100\,\text{cm}^3 = 1\,000\,000\,\text{cm}^3$$

The **capacity** of a container is the volume of liquid it could hold.
The main **metric units of capacity** are the litre and the millilitre (ml), where

$$1\,\text{litre} = 1000\,\text{ml} \quad \text{and} \quad 1\,\text{litre} = 1000\,\text{cm}^3 \quad \text{so} \quad 1\,\text{ml} = 1\,\text{cm}^3$$

The main **Imperial units of capacity** are the gallon and the pint, where

$$1 \text{ gallon } = 8 \text{ pints}$$

Rough conversions between metric and Imperial units of capacity are given by

$$1 \text{ litre } \approx 1.75 \text{ pints} \quad \text{and} \quad 1 \text{ gallon } \approx 4.5 \text{ litres}$$

Changing units

To change a quantity from a large unit to a smaller unit *multiply* the given quantity by the number of smaller units in 1 large unit,

e.g.   since 1 tonne $= 1000 \text{ kg}$
$$4.5 \text{ tonnes } = 4.5 \times 1000 \text{ kg} = 4500 \text{ kg}$$

To change a quantity from a small unit to a larger unit *divide* the quantity by the number of smaller units in 1 large unit,

e.g.   since 1 foot $= 12 \text{ inches}$
$$108 \text{ inches } = 108 \div 12 \text{ feet} = 9 \text{ feet}$$

Compound units

If a body covers a miles in t hours

the speed of the body is $\dfrac{a}{t}$ miles per hour (mph).

The corresponding unit in the metric system
is kilometres per hour (km/h).
To convert between mph and km/h use $5 \text{ mph} \approx 8 \text{ km/h}$
To convert from km/h to m/s multiply by 1000 ($1 \text{ km} = 1000 \text{ m}$)
and divide by 60×60 ($1 \text{ h} = 60 \times 60 \text{ s}$)

e.g. $36 \text{ km/h} = 36 \times \frac{1000}{3600} \text{ m/s} = 36 \times \frac{5}{18} \text{ m/s} = 10 \text{ m/s}$

EXERCISE 3.1

In questions **1** to **4** express each given quantity in terms of the units in brackets.

1 **a** 5240 m (km)   **e** 500 g (kg)   **i** 33 cm (mm)
     b 240 mm (cm)   **f** 67 cm (m)   **j** 42 cm (m)
     c 0.06 kg (g)   **g** 1.39 m (cm)   **k** 5.2 g (mg)
     d 7500 mg (g)   **h** 2450 mm (m)   **l** 0.007 t (kg)

2 **a** 0.0826 m (mm)   **e** 0.06 kg (mg)   **i** 8 yd (ft)
     b 2.4 t (kg)   **f** 2 miles (yd)   **j** 6 ft (in)
     c 72 in (ft)   **g** 85 in (ft and in)   **k** $3\frac{1}{2}$ miles (ft)
     d 140 ft (yd and ft)   **h** 4 lb 3 oz (oz)   **l** 5 yd (in)

3　**a** 11 st 3 lb (lb)
　　b 2 tons 14 cwt (cwt)
　　c 2 ft 8 in (in)
　　d 3 m 35 cm (cm)
　　e 176 lb (stones and lb)
　　f 70 oz (lb and oz)

　　g 3 yd 2 ft (ft)
　　h 2 yd 8 in (in)
　　i 0.4 hectares (m^2)
　　j 250 000 m^2 (hectares)
　　k 72 600 yd^2 (acres)
　　l 0.55 acres (yd^2)

4　**a** 5.6 litres (ml)
　　b 40 gallons (pints)
　　c 420 000 cm^3 (litres)
　　d 0.077 litres (ml)
　　e 85 pints (gall and pints)
　　f 8206 cm^3 (litres)

　　g 12 000 ml (litres)
　　h 0.42 litres (cm^3)
　　i 120 pints (gallons)
　　j 76 000 ml (litres)
　　k 645 ml (litres)
　　l 0.035 litres (ml)

5　Express

　　a 56 mm in cm
　　b 0.072 km in m
　　c 36 inches in feet
　　d 48 feet in yards
　　e 15 cm^2 in mm^2
　　f 8000 cm^2 in mm^2

　　g 0.85 m^2 in cm^2
　　h 0.007 m^3 in cm^3
　　i 10 000 mm^3 in cm^3
　　j 5.3 cm^3 in mm^3
　　k 1.5 sq ft in sq inches
　　l 288 sq inches in sq ft

6　Find, giving your answer in the unit in brackets

　　a 4 m + 50 cm (m)
　　b 3 kg + 500 g (kg)
　　c 5 yd − 2 ft (ft)
　　d 2 tons − 6 cwt (cwt)
　　e 4 yd − 4 ft + 32 in (in)
　　f 1.8 t − 890 kg (kg)

　　g 3 m + 64 cm + 112 mm (cm)
　　h 3 tons − 13 cwt − 300 lb (lb)
　　i 0.7 m − 466 mm (cm)
　　j 124 g + 0.03 kg + 1940 mg (g)
　　k 4.5 cm^2 + 560 mm^2 (mm^2)
　　l 3 sq ft + 92 sq in (sq in)

7　In this question write the given quantity, approximately, in terms of the unit in brackets.

　　a 5 kg (lb)
　　b 50 miles (km)
　　c 10 ft (m)
　　d 2 m (in)
　　e 15 m (yd)

　　f 9 lb (kg)
　　g 50 km (miles)
　　h 250 g (oz)
　　i 2 in (cm)
　　j 4 t (tons)

　　k 5 gallons (litres)
　　l 18 litres (gallons)
　　m 120 hectares (acres)
　　n 100 acres (hectares)
　　p 40 cm (in)

8　Convert the given quantity to the unit given in brackets.

　　a 72 km/h (m/s)
　　b 500 mph (km/h)
　　c 600 kg/m^3 (g/cm^3)

　　d 88 km/h (mph)
　　e 0.2 g/cm^3 (kg/m^3)
　　f 35 m/s (km/h)

**LENGTH, AREA,
VOLUME AND
DENSITY**

The **perimeter** is the distance all round the edge of a shape.

The **area of a square** = (length of a side)^2,
The **area of a rectangle** = length × breadth.

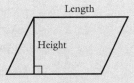

The **area of a parallelogram** is given by
A = length × height

The **area of a triangle** is given by
$A = \frac{1}{2}$ base × height

When we talk about the height of a triangle or of a parallelogram, we mean the perpendicular height.

The **area of a trapezium** is equal to
$\frac{1}{2}$(sum of the parallel sides) × (distance between them)

$$= \tfrac{1}{2}(a+b) \times h$$

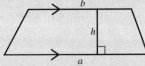

The **diameter of a circle** is twice the **radius**.

The **circumference** is given by $C = 2\pi r$,
where r units is the radius of the circle
and $\pi = 3.14159\ldots$

The **area of a circle** is given by $A = \pi r^2$

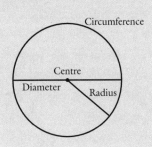

Part of the circumference of a circle is
called an **arc**.

Length of the arc AB is $\dfrac{\widehat{AOB}}{360°} \times 2\pi r$

The slice of the circle enclosed by the arc
and the two radii is called a **sector**.

The **area of the sector AOB** is $\dfrac{\widehat{AOB}}{360°} \times \pi r^2$

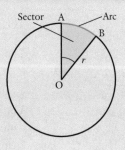

The **volume of a cuboid** = length × breadth × height

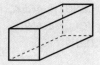

A solid with a constant cross-section is called
a **prism**.

The **volume of a prism** is given by

    area of cross-section × length

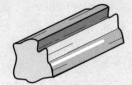

The **volume of a pyramid** is given by
$\frac{1}{3}$ area of base × perpendicular height

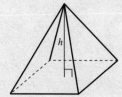

The **volume of a cylinder** is given by $V = \pi r^2 h$

The **curved surface area of a cylinder**
is given by $A = 2\pi rh$

The **volume of a cone** is given by $V = \frac{1}{3}\pi r^2 h$

The **curved surface area of a cone** is given
by $A = \pi rl$

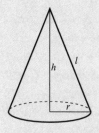

The **volume of a sphere** is given by $V = \frac{4}{3}\pi r^3$

The **surface area of a sphere** is given $A = 4\pi r^2$

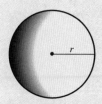

The **density** of a material is the mass of one unit of volume of the
material, for example, the density of silver is $10.5\,\text{g/cm}^3$. This means that
$1\,\text{cm}^3$ of silver weighs $10.5\,\text{g}$.

Compound shapes

Compound areas and solids can often be divided into simple familiar shapes.

For example, this area can be divided into a square, a rectangle and a semicircle,

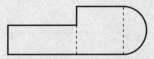

while this solid can be divided into a cylinder and a cone.

Dimensions of a formula

Checking the dimensions of a formula can prove very useful.
For example, if r is a number of length units and a is a number of area units, then

$$2r \text{ is one-dimensional} \qquad (\text{length})$$
$$3r^2 \text{ is two-dimensional} \qquad (\text{area})$$
$$2\pi a \text{ is two-dimensional} \qquad (\text{area})$$
$$\pi r^3 \text{ is three-dimensional} \qquad (\text{volume})$$
$$ar \text{ is three-dimensional} \qquad (\text{volume})$$

Numbers that are not a number of length, area or volume units do not affect the dimensions of a formula. For example, $c = 2\pi r$ where r units is the radius, and hence a length, is one-dimensional.

Checking dimensions and units helps to identify whether a quantity represents length, area or volume,
e.g. a quantity given as $x\,\text{cm}^2$ must be an area.

Applying the same check helps to spot an incorrect formula. Suppose, for example, that a formula for the volume, V, of an object is given as $V = 4\pi xy$ where x and y are numbers of length units.
Volume is three-dimensional, whereas $4\pi xy$ is only two-dimensional, so the formula cannot be correct.

EXERCISE 3.2

1 For each shape find **i** the perimeter **ii** the area.

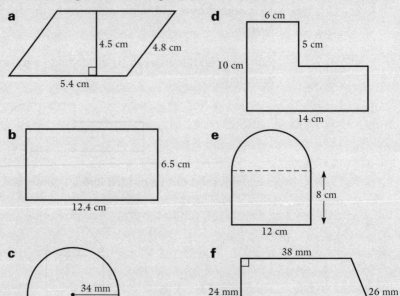

a

4.5 cm 4.8 cm

5.4 cm

b

6.5 cm

12.4 cm

c

34 mm

d

6 cm

5 cm

10 cm

14 cm

e

8 cm

12 cm

f

38 mm

24 mm 26 mm

48 mm

2 For the shaded part of each diagram find
  i the perimeter   **ii** the area.

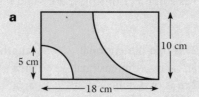

a

10 cm

5 cm

18 cm

b

5 cm ←—18 cm—→ 5 cm

12 cm 12 cm

21 cm

3 For each of the following shapes the area is given, together with one
dimension. Find the measurement marked with a letter.

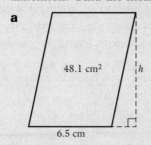

a

48.1 cm² h

6.5 cm

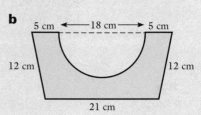

c

a

8.3 cm

15 cm

Area = 99.6 cm²

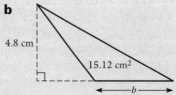

b

4.8 cm

15.12 cm²

b

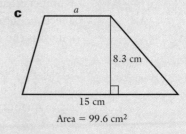

d

l

24 m

Area = 1484 m²

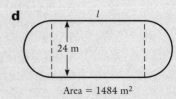

4 For each solid find
   i the volume   **ii** the surface area.

a

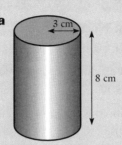

3 cm

8 cm

c

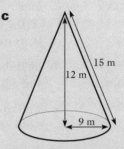

15 m

12 m

9 m

b

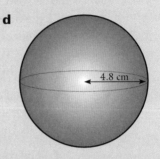

10 mm

10 mm

45 mm

10 mm

d

4.8 cm

5 Find the volume of each solid.

a

36 mm

25 mm

25 mm

25 mm

b

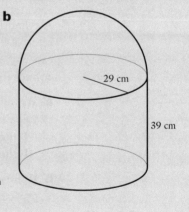

29 cm

39 cm

6 A circular sheet of metal of radius 20 cm
is to be used to make an open right
circular cone by using the major
sector AOB.
If AÔB = 130° find

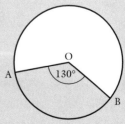

O

A

130°

B

  a the length of the minor arc AB

  b the length of the major arc AB

  c the area of metal used to make the cone

  d the radius of the circular base of the cone.

7 a A cubical wooden block measures 3.6 cm by 4.5 cm by 12 cm. It is made from oak that has a density of 0.9 g/cm³.
Find
  i the volume of the block
  ii its mass in kg.

b Find the capacity of a cylindrical can which has a diameter of 8 cm and a height of 9.5 cm.

c The volume of a cuboid is 84 cm³. It is 6 cm long and 3.5 cm wide. Find its height.

d A solid metal sphere has a radius of 5.5 cm. It is made from metal, the density of which is 6.8 g/cm³.
Find   **i** the volume of the sphere
        ii its surface area
        iii its mass in kilograms.

e A solid copper cylinder has a mass of 4.5 kg. If its diameter is 8 cm and it is made from metal that has a density of 8.8 g/cm³ find
  i its volume   **ii** its height.

f The volume of a pyramid is 260 cm³. The height of the pyramid is 39 cm. Find the area of its base.

8 A cylindrical can of orange juice has an internal diameter of 6.3 cm and has a capacity of 330 ml. Each measurement is correct to 2 significant figures.

a Write down the smallest possible value for
  i the radius of the base
  ii the area of the base.

b Write down the greatest possible capacity of the can.

c Calculate the largest possible height of the can.

9 Rainwater is collected in a rain gauge which has a circular top of radius 5 cm. The water is drained into a cylinder of diameter 3 cm. By how much will the water level in the cylinder rise as the result of a storm in which the rainfall is 9 mm?

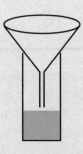

10 The radius of the inside of a circular running track is 63 m and the track is 10 m wide.

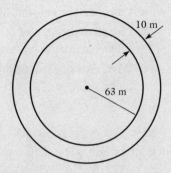

Find

a how much further it is around the outside of the track than around the inside

b the total area of the track

c the volume of material used to lay the track if it is 12 cm deep.

11 This haystack is a cylinder with a cone on top.

From the list given below, which formula could give

a the total length of the edges

b the surface area of the haystack excluding the base

c the volume of the haystack?

$$P = \sqrt{2}\pi r^2 + 2\pi rh \qquad Q = \sqrt{2}\pi r^2 + 2\pi r^2 h$$
$$R = 4\pi r \qquad\qquad S = \pi r^2 h + \pi r^2$$
$$T = 2\pi r + 2rh \qquad U = \tfrac{1}{3}\pi r^3 + \pi r^2 h$$

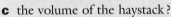

12 These railings are to be painted black except for the spheres on the tops of the uprights. These are to be painted gold. There are 150 of these spheres and each one has a diameter of 10 cm. One tin of gold paint covers 5 square metres. How many tins of gold paint are needed?

13 A globe-puzzle consists of a metal globe of diameter 30 cm over which a world map is assembled using cardboard pieces 1 mm thick.

  a Calculate the amount of space occupied by the metal globe.

  b Calculate the area of card, 1 mm thick, needed for the puzzle. What volume, in cm³, does this card occupy?

  c What area of card, of the same thickness, is needed to cover a globe of radius 7.5 cm?

STRAIGHT LINE GEOMETRY

An **acute angle** is less than one right angle.

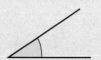

An **obtuse angle** is larger than 1 right angle but less than 2 right angles.

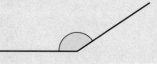

A **reflex angle** is larger than two right angles.

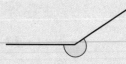

Vertically opposite angles are equal.

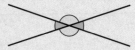

Angles on a straight line add up to 180°.

Two angles that add up to 90° are called **complementary** angles.

$$a° + b° = 90°$$

Two angles that add up to 180° are called **supplementary** angles.

$$c° + d° = 180°$$

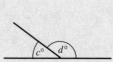

Angles at a point add up to 360°.

$$e° + f° + g° + h° = 360°$$

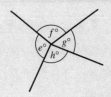

When two **parallel lines** are cut by a transversal,

the **corresponding angles** are equal,

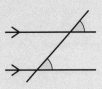

the **alternate angles** are equal,

the **interior angles** add up to 180°.

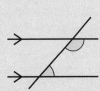

The **three angles in a triangle** add up to 180°.

$$a° + b° + c° = 180°$$

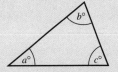

An **equilateral triangle** has all three sides equal and each angle is 60°.

An **isosceles triangle** has two equal sides and the angles at the base of these sides are equal.

A **quadrilateral** has four sides.
The four angles in a quadrilateral add up to 360°.

$$a° + b° + c° + d° = 360°.$$

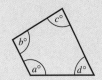

Special quadrilaterals

In a **square**

- all four sides are the same length
- all four angles are right angles
- both pairs of opposite sides are parallel
- the diagonals are equal and bisect each other at right angles.

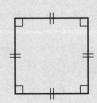
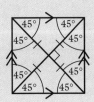

In a **rectangle**

- both pairs of opposite sides are
  the same length
- all four angles are right
  angles
- both pairs of opposite
  sides are parallel
- the diagonals bisect each other
- the diagonals are equal.

In a **rhombus**

- all four sides are the same length
- the opposite angles are equal
- both pairs of opposite sides
  are parallel
- the diagonals bisect each
  other at right angles.

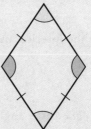

In a **parallelogram**

- the opposite sides are the same length
- the opposite sides are parallel
- the opposite angles are equal
- the diagonals bisect each other.

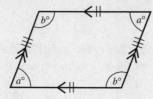
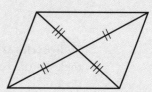

In a **trapezium**

- just one pair of opposite sides are parallel.

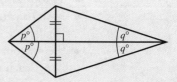

In a **kite**

- one pair of opposite angles are equal
- two pairs of adjacent sides are equal
- one diagonal is a line of symmetry and this diagonal bisects the other
  diagonal at right angles.

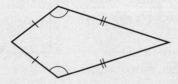
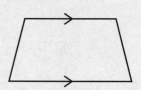

A **polygon** is a plane figure bounded
by straight lines, e.g.

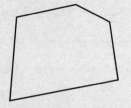

A **regular polygon** has all angles equal and all sides the same length.

This is a regular hexagon.

The **sum of the exterior angles** of any
polygon is 360°.

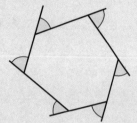

The **sum of the interior angles** of a polygon
depends on the number of sides.

For a polygon with n sides, this sum is
$(180n - 360)°$ or $(2n - 4)$ right angles.

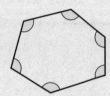

EXERCISE 3.3

1 Find the size of each angle marked with a letter.

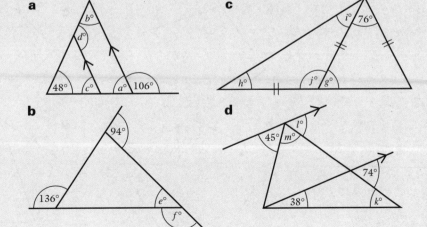

a

c

b

d

2 Find the size of each marked angle.

a

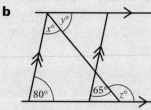

c

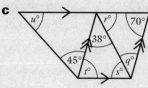

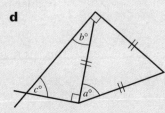

b

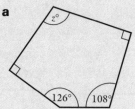

d

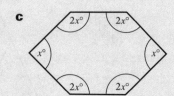

3 **a** Find the size of each exterior angle of a regular polygon with
  i 12 sides **ii** 18 sides.

  b Find the size of each interior angle of a regular polygon with
  i 10 sides **ii** 15 sides.

  c How many sides has a regular polygon
  i if each exterior angle is 18° **ii** if each interior angle is 165°?

  d Is it possible for each exterior angle of a regular polygon to be
  i 45° **ii** 50°?
  If it is, give the number of sides.

  e Is it possible for each interior angle of a regular polygon to be
  i 168° **ii** 154°?
  If it is, give the number of sides.

4 Give the size of each marked angle.

a

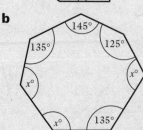

c

b

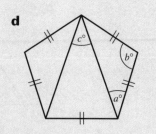

d

5 The interior angles of a hexagon are
$2x°, (3x - 15)°, (3x - 30)°, (x + 65)°, (2x - 45)°$ and $(x + 85)°$.
Find them.

6 The diagram shows an ironing board.
Find the size of each marked angle.

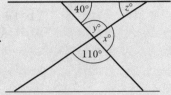

7 ABCDE is a pentagon with all its sides equal.
$AC = CE$, $A\widehat{C}E = 36°$ and $B\widehat{A}C = 44°$.

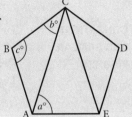

a Find each angle marked with a letter.

b Is the pentagon regular?
Give a reason for your answer.

8 Use squared paper and draw axes for x and y in the range
$-6 \leqslant x \leqslant 6$, $-6 \leqslant y \leqslant 6$ using 1 square to 1 unit. Draw the figure
and write down the coordinates of the point that is not given. Find
the area of each shape in square units.

a Square ABCD with $A(-1, -3)$, $B(-1, 4)$ and $C(6, 4)$.

b Kite ABCD with $A(3, -5)$, $B(1, 3)$ and $C(3, 5)$ in which
AC is the axis of symmetry.

c Parallelogram ABCD with $A(2, -5)$, $B(2, 2)$ and $C(5, 5)$.

d Rectangle ABCD with $A(3, 5)$, $B(3, -5)$ and $C(-2, -5)$.

9 Use squared paper and draw axes for x and y in the range
$-6 \leqslant x \leqslant 6$, $-6 \leqslant y \leqslant 6$ using 1 square to 1 unit.

a ABCD is a rectangle with $A(-3, 1)$ and $B(-3, 5)$. The
diagonals intersect at E which has coordinates $(1, 3)$.
Plot the points A, B, E, C and D. Write down the coordinates of
C and D.

b ABCD is a parallelogram, whose diagonals cut at E, with
$A(-1, 4)$, $B(6, 3)$ and $E(2, 1)$.
Find the coordinates of C and D.

c ABCD is a kite whose diagonals intersect at E with $A(-4, 3)$,
$B(0, 0)$ and $C(6, 3)$. Draw the figure and write down the
coordinates of D and E. What is the area of the kite?

d The diagonals of a square ABCD cross at E with $A(1, 6)$,
$C(3, 0)$ and $E(2, 3)$. Draw the square and write down the
coordinates of the ends of the diagonal BD.

10 The diagram shows a pattern
made with four regular polygons.

a What is the name of the shape
of the gap ABCD?

b Find the sizes of the angles
i AD̂C   **ii** DÂB

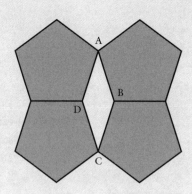

**CONGRUENT
AND SIMILAR
FIGURES**

Two figures are **congruent** when they are exactly the same shape and size.

Congruent triangles
Two triangles can be proved to be congruent if it can be shown that

either   the three sides of one triangle are equal to the three sides of the
other triangle,

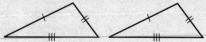

or   two angles and a side of one triangle are equal to two angles and
the corresponding side of the other triangle,

or   two sides and the included angle of one triangle are equal to two
sides and the included angle of the other triangle,

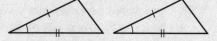

or   two triangles each have a right-angle and the hypotenuse and a
side of one triangle are equal to the hypotenuse and a side of the
other triangle.

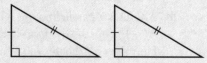

Similar figures

Two figures are **similar** if they are the same shape but different in size,
that is one figure is an enlargement of the other. (One figure may be
turned over or round with respect to the other.) It follows that the
lengths of corresponding sides are all in the same ratio.

The **ratio of the areas of similar figures** is equal to the square of the ratio of their sides.

The **ratio of the volumes of similar objects** is equal to the cube of the ratio of their edges.

Similar triangles

Two triangles can be proved to be similar if it can be shown that either the three angles of one triangle are equal to the three angles of the other. (In practice only two pairs of angles need to be shown to be equal because, since the sum of the three angles in any triangle is 180°, it follows that the third pair must be equal.)

or the three pairs of corresponding sides are in the same ratio. (Two pairs of sides in the same ratio is not enough to prove that the triangles are similar.)

EXERCISE 3.4

1

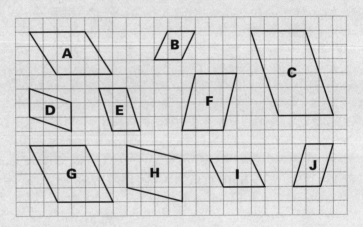

Which of these shapes are

a congruent

b similar but not congruent?

2 The following diagrams are not drawn to scale. Determine whether the two figures are similar. Give reasons for your decision.

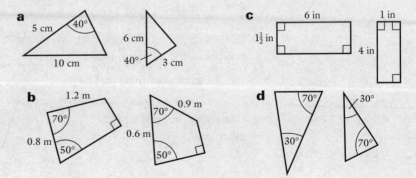

3

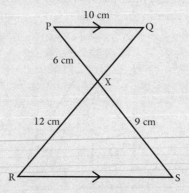

a Which two triangles are congruent and why?

b Which two triangles are similar and why?

4 PQ and RS are parallel.
PXS and QXR are straight lines.

a Show that triangles PXQ and SXR are similar.

b If PQ = 10 cm, PX = 6 cm, XS = 9 cm and RX = 12 cm,
find **i** RS **ii** QX.

5 Two triangles are similar. The length of a side in one triangle is 9 cm
and the length of the corresponding side in the other triangle is 15 cm.

a What is the ratio of the lengths of corresponding sides?

b The length of another side in the smaller triangle is 12 cm. What is
the length of the corresponding side in the larger triangle?

c What is the ratio of the area of the smaller triangle to the area of
the larger triangle?

6 The ratio of the heights of two similar cones is 3 : 5.
The volume of the larger cone is 112.5 cm³.

a What is the ratio of their surface
areas?

b Calculate the volume of the
smaller cone.

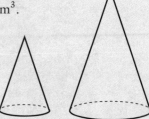

7 The ratio of the surface areas of two spheres is $9:16$. The volume of the smaller sphere is $135\,\text{cm}^3$. Find

a the ratio of the radius of the larger sphere to the radius of the smaller sphere.

b the volume of the larger sphere.

8

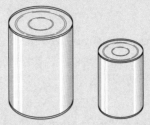

Two cylindrical cans are mathematically similar. The larger can has a capacity of $454\,\text{cm}^3$ and the smaller can has a capacity of $420\,\text{cm}^3$.

a Write, as a ratio in the form $1:n$, where n is correct to 3 significant figures,

  i the height of the smaller can to the height of the larger can

  ii the surface area of the smaller can to the surface area of the larger can.

b The diameter of the larger can is $80\,\text{mm}$. Calculate, correct to the nearest mm,

  i the height of the larger can

  ii the diameter of the smaller can.

c The labels go round each can with an overlap of $1\,\text{cm}$ and are as wide as the can is high. Calculate

  i the length of the label, including the overlap, that fits around the larger can

  ii the area of the label needed for the smaller can.

9 ABCD is a parallelogram.
Prove that triangles ABC and ADC are congruent.

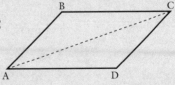

10 The line DE is parallel to the line AC.
Prove that the triangles ABC and DBE are similar.

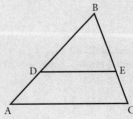

**CIRCLE
GEOMETRY**

Angles standing on the same arc of a circle and in the same segment are equal.

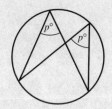

The **angle subtended at the centre of a circle** by an arc is twice the angle subtended at the circumference by the same arc.

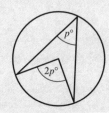

The **angle in a semicircle** is a right angle.

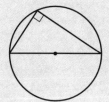

The **opposite angles of a cyclic quadrilateral** are supplementary. An **exterior angle of a cyclic quadrilateral** is equal to the interior opposite angle.

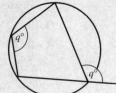

A straight line which cuts a circle in two points is called a **secant**. The section of the line inside the circle is a **chord**. A line that touches a circle is called a **tangent**.

A **segment** of a circle is the part of a circle cut off by a chord.

A **tangent to a circle is perpendicular to the radius** drawn through the point of contact.

Two tangents drawn from an external point to a circle are the same length.

The angle between a tangent and a chord drawn through the point of contact is equal to the angle in the alternate segment. This result is called the **alternate segment theorem**.

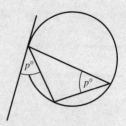

EXERCISE 3.5

1 Find the marked angles.

a

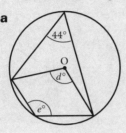

c

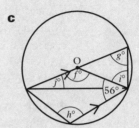

b

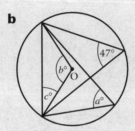

d

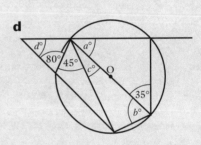

2 Find each angle marked with a letter.

a

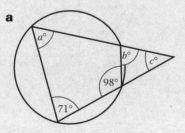

c

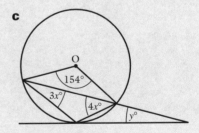

b

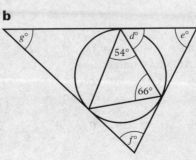

d

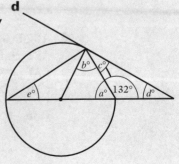

3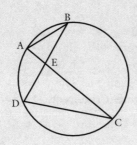

A, B, C and D are four points on the circumference of a circle.

Prove that △ABE is similar to △DCE.

4 TA and TB are tangents from an external point T to a circle. TB is produced to D. $\hat{BAT} = 57°$ and $\hat{CBD} = 63°$. Find

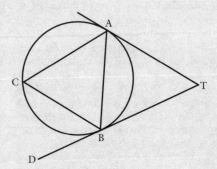

a $A\hat{B}T$ **c** $A\hat{C}B$

b $A\hat{T}B$ **d** $C\hat{A}B$

5 ATB is a tangent to the cirlce and ST is a diameter.

$T\hat{U}V = x°$, $S\hat{T}U = y°$

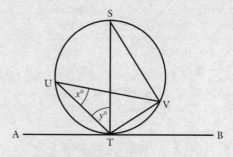

Find, in terms of x or y, the size of

a $B\hat{T}V$ **b** $S\hat{T}V$ **c** $U\hat{V}T$

6 ABCD is a cyclic quadrilateral. ABE and ADF are straight lines. $B\hat{A}C = 35°$, $A\hat{C}D = 40°$ and $C\hat{D}F = 73°$.

Find   **a** $C\hat{A}D$ **b** $A\hat{B}C$ **c** $A\hat{C}B$

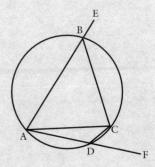

7

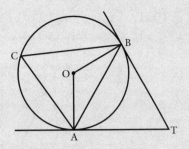

ABC is a triangle inscribed in a circle centre O. TA and TB are tangents to the circle. If $A\widehat{C}B = 53°$ find

a $A\widehat{O}B$ **b** $B\widehat{T}A$ **c** $A\widehat{B}T$

8 A, B, C and D are points on a circle. TCX is a tangent to the circle and TDB is a straight line.

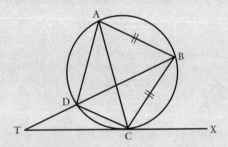

If $AB = BC$, $B\widehat{T}C = 32°$ and $D\widehat{A}C = 43°$ find

a $T\widehat{D}C$ **b** $C\widehat{A}B$ **c** $A\widehat{B}C$ **d** $B\widehat{C}X$

9 AED and BCD are straight lines. AB is a diameter.

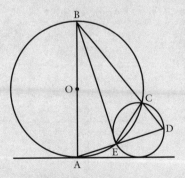

If $A\widehat{B}E = 2x°$ and $E\widehat{B}D = 3x°$ find, in terms of $x°$, the size of the angle

a $B\widehat{A}E$ **b** $B\widehat{E}C$ **c** $E\widehat{C}D$ **d** $C\widehat{E}D$ **e** $C\widehat{D}E$

10 In the diagram, TA is a tangent to the circle and TB is a diameter. DF is parallel to BT.

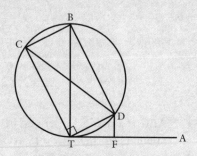

a Show that △CDT is similar to △TDF.

b Show that triangles BTD and CTD are congruent.

CONSTRUCTIONS AND LOCI

The diagrams show 'ruler and compasses only' constructions.

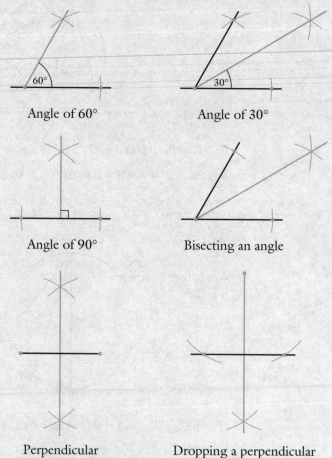

Angle of 60°

Angle of 30°

Angle of 90°

Bisecting an angle

Perpendicular bisector of a line

Dropping a perpendicular from a point to a line

A **locus** is the shape formed by the positions of all the points that satisfy a given rule.

The locus of a point that moves so that it is at a fixed distance from a given point is the circumference of a circle.

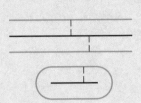

The locus of a point that moves at a constant distance from a fixed straight line is a pair of parallel lines. In the case of a line segment, semicircular ends join the lines.

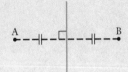

The locus of a point that moves so that it is equidistant from two fixed points, A and B, is the perpendicular bisector of the line joining AB.

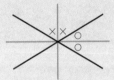

The locus of a point that moves so that it is equidistant from two intersecting straight lines is the pair of bisectors of the angles between the lines.

EXERCISE 3.6

1

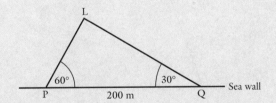

Jane wants to find the distance of a lighthouse, L, from the sea wall. She marks a straight line, 200 m long, from a point P on the wall to another point Q on the wall. Next she measures the angles at P and Q between the line PQ and the lines PL and QL. These angles are respectively 60° and 30°.

a Using ruler and compasses only show this information on a scale drawing. Use a scale of 1 cm ≡ 20 m and leave all construction lines clearly visible.

b Construct the perpendicular from L to PQ. How far is the lighthouse from the sea wall?

2 Construct a quadrilateral PQRS in which $\widehat{P} = 60°$, $\widehat{Q} = 150°$, PQ = 7.5 cm, PS = 12 cm and QR = 5 cm.

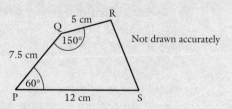

Not drawn accurately

Measure and record

a **i** $\widehat{R}$
  ii $\widehat{S}$

b the length of **i** RS
              ii PR.

3

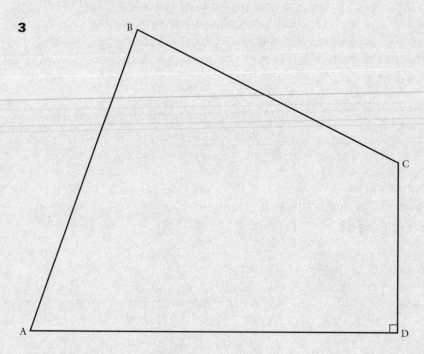

The diagram shows a field. It is drawn using a scale of 1 : 2000. AD = 10 cm and $A\widehat{D}C = 90°$.

a How far is it from A to B in the field?

b Copy the diagram as accurately as you can. Use a protractor to measure the angles.

c A water trough is to be placed in the field so that it is equidistant from B and C, and 60 m from the hedge AD. Find the position of the trough and mark it T. How far is the trough from A?

4 **a** The diagram shows a flower bed which is one third of a circle of radius 6 m.
Make a scale drawing of the flower bed using a scale of 1 cm ≡ 1 m.

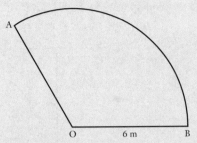

b A gardener wishes to plant a shrub so that it is nearer to O than it is to B, and is at least 6.5 m from A.
On your scale drawing shade the area representing the section of the bed in which the shrub could be planted.

5 The sketch shows the position of Jeff's garage in his garden which is adjacent to the house. Apart from his garage the remainder of the garden is lawn which Jeff cuts regularly using an electric mower. The cable for the mower is 10 m long and the power point from which he operates it is in the corner of the garage at A.

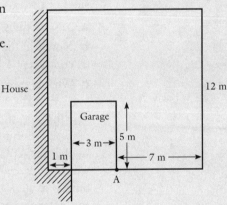

a Make a scale drawing of Jeff's garden and on it shade the area that he cannot mow unless he has an extension cable. Use a scale of 1 cm ≡ 1 m.

b What is the minimum length of cable needed for an extension lead that will enable Jeff to cut all the grass in his garden?

6 A television company wishes to erect a transmitter to give acceptable pictures within a square of side 70 km. The transmitters they install give satisfactory pictures within a radius of 40 km.

a If the company can afford one transmitter only, show, on a scale diagram, the regions that it cannot reach if it is placed at the centre of the square. Use a scale of 1 cm ≡ 10 km.

b Make another scale drawing of the square and show, on your drawing, that it is possible to cover the whole of the square using just two transmitters. Mark, with letters P and Q, possible positions for the two transmitters. Shade the area that is covered by both transmitters.

7 The diagram shows the cross-section through an assembly of a cylindrical shaft, free to rotate inside a box with a square cross-section. It is kept in place in the middle of the box by four cylindrical rollers, one in each corner of the box.
(Only one of these rollers is shown.)

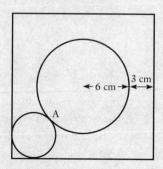

In the diagram, the radius of the circle in the centre is 6 cm and the sides of the box are 18 cm long.

a Make a full size copy of the diagram but leave out the small circle.

b Now add the small circle to the diagram, showing all necessary construction lines.

c What is the diameter of this circle?

8 The diagram shows a section through the tailgate of a car, with the tailgate DE in its fully open position. DE is kept in this position by the stay ABC. The stay consists of two rigid bars, AB and BC which are freely hinged together at B. One end of the stay is hinged to the car body at C and the other end of the stay is hinged to the tailgate at A.

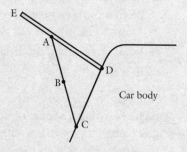

With the tailgate fully open, ABC is a straight line, 50 cm long.
AD and DC are each 30 cm long.
Using a scale of 1 cm to 5 cm make a scale drawing of this section.
The stay is released by pulling the hinge B away from D. By considering the tailgate closing 20° at a time, or otherwise, draw the path traced out by the hinge B as the tailgate closes.

PYTHAGORAS' THEOREM AND TRIGONOMETRY

Pythagoras' theorem states that, in any right-angled triangle with $\widehat{C} = 90°$, $AB^2 = AC^2 + BC^2$.

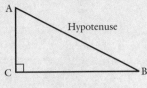

Conversely, if in a triangle PQR, $PR^2 = PQ^2 + QR^2$ then $\widehat{Q} = 90°$

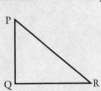

If $PR^2 \neq PQ^2 + QR^2$ then $\widehat{Q} \neq 90°$

If $PR^2 > PQ^2 + QR^2$ then $\widehat{Q} > 90°$

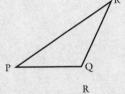

and if $PR^2 < PQ^2 + QR^2$ then $\widehat{Q} < 90°$

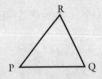

A **three-figure bearing** of a point A from a point B gives the direction of A from B as a clockwise angle measured from the north.

In this diagram, the bearing of A from B is 140°.

If you start by looking straight ahead, the angle that your eyes turn through to look *up* at an object is called the **angle of elevation**, the angle your eyes turn through to look *down* at an object is called the **angle of depression**.

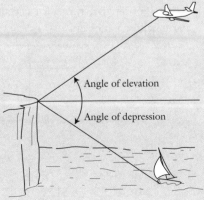

In a right-angled triangle

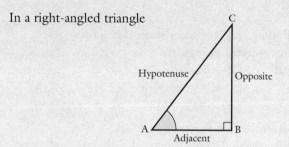

the   **tangent of an angle** $= \dfrac{\text{side opposite the angle}}{\text{side adjacent to the angle}}$

the   **sine of an angle** $= \dfrac{\text{side opposite the angle}}{\text{hypotenuse}}$

the   **cosine of an angle** $= \dfrac{\text{side adjacent to the angle}}{\text{hypotenuse}}$

or more briefly,

$$\tan \widehat{A} = \frac{\text{opp}}{\text{adj}} = \frac{BC}{AB}, \quad \sin \widehat{A} = \frac{\text{opp}}{\text{hyp}} = \frac{BC}{AC}, \quad \cos \widehat{A} = \frac{\text{adj}}{\text{hyp}} = \frac{AB}{AC}$$

The **sine and cosine of an obtuse angle** can be found using

$$\sin x° = \sin(180° - x°) \qquad \cos x° = -\cos(180° - x°)$$

e.g. $\sin 150° = \sin 30° = 0.5$ and $\cos 160° = -\cos 20° = -0.9397$.

In any triangle ABC,

the **sine rule** is $\qquad \dfrac{a}{\sin \widehat{A}} = \dfrac{b}{\sin \widehat{B}} = \dfrac{c}{\sin \widehat{C}}$

the **cosine rule** is $\quad a^2 = b^2 + c^2 - 2bc \cos \widehat{A}$

the **area** is given by Area $= \frac{1}{2}bc \sin \widehat{A}$

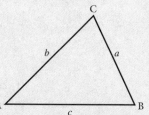

Exact values for certain angles

The trigonometric ratios for certain angles are given in the table, using surds where necessary.

Angle, $x°$	$\sin x°$	$\cos x°$	$\tan x°$
30°	$\dfrac{1}{2}$	$\dfrac{\sqrt{3}}{2}$	$\dfrac{1}{\sqrt{3}}$
45°	$\dfrac{1}{\sqrt{2}}$	$\dfrac{1}{\sqrt{2}}$	1
60°	$\dfrac{\sqrt{3}}{2}$	$\dfrac{1}{2}$	$\sqrt{3}$

EXERCISE 3.7

1 In △ABC, D is the foot of the
perpendicular from B to AC.
AB = 14 cm, AD = 11.2 cm
and DC = 6.3 cm.

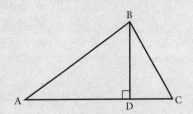

a Find the lengths of
  i BD **ii** BC

b Prove that $A\widehat{B}C = 90°$

2 a Find PQ.

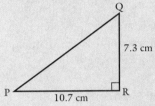

b Find XY.

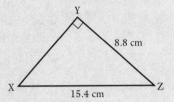

3

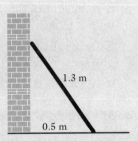

The diagram shows a ladder leaning against a wall.
The ladder is 1.3 m long.
The foot of the ladder is 0.5 m from the base of the wall.
How far up the wall is the top of the ladder?

4

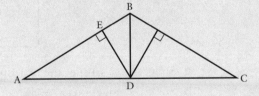

The sketch shows a symmetrical roof truss. AC = 5.8 m and
BD = 1.7 m. E is the foot of the perpendicular from D to AB.

a Find the length of the sloping sides AB and BC.

b If EB = x m use Pythagoras' result in different triangles to find
two expressions for ED^2 in terms of x. By equating these
expressions form an equation in x and solve it. Hence find the
lengths of EB, AE and DE.

5 In △ABC, $\hat{A} = 90°$, AB = 5.2 cm and
BC = 8.3 cm, each side being measured
correct to the nearest mm.

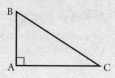

a Write down the upper and lower bounds for the length of
   i AB **ii** BC.

b Calculate upper and lower bounds for the length of the side AC.

c Give the length of AC to an appropriate degree of accuracy.

6 In each part of this question the lengths, in centimetres, of the three
sides of a triangle are given. Sketch each triangle showing clearly
whether the largest angle is less than, equal to or greater than 90°.

a 11.4, 5.9, 9.4   **b** 5.7, 9.5, 7.6   **c** 5.3, 8.7, 7

7 In △ABC, AB = 16 cm,
BÂC = 42° and AB̂C = 90°.
D is the foot of the perpendicular
from B to AC.

Calculate the length of

a BD   **c** BC
b AD   **d** DC

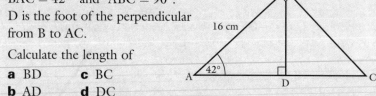

8 ABC is a right-angled triangle.

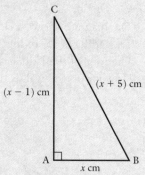

a Show that x satisfies the equation $x^2 - 12x - 24 = 0$

b Solve the equation $x^2 - 12x - 24 = 0$

c Write down the lengths of the sides of the triangle.

9 Calculate

a angle BDA

b the length of BC.

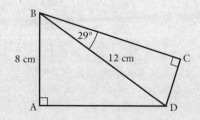

10 Sam is hauling up an 18 metre
flagpole AB into a vertical
position. After one pull AB
makes an angle of 60° with
the ground, and Sam's hands
are 1.8 m above the ground,
i.e. CD = 1.8 m.

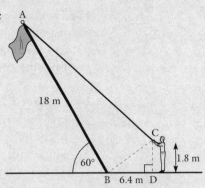

a If BD = 6.4 m calculate the length of the rope from the top of
the flagpole to Sam's hands, i.e. the length of AC.

b When the flagpole is vertical
the end of the rope, C, is
attached to a point E on the
ground. Calculate
  i the angle AEB
  ii the distance BE.

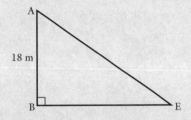

11 From a point A on the ground the angle of elevation of the top, D,
of the Boniface tower is 33.7°. From a point B, 50 m nearer the
base of the tower C, the angle of elevation of D is 36.9°.

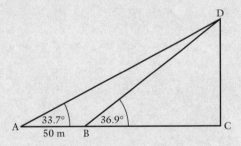

Find

a A$\widehat{B}$D   **c** the length of BD

b A$\widehat{D}$B   **d** the height of the tower.

12 In the △ABC, AB = 25 cm and
BC = 35 cm, each length being
correct to the nearest centimetre.
The angle A is 38°, correct to
the nearest degree.

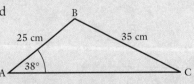

a What is the largest possible value for angle C?

b What is the smallest possible value for angle C?

13 Two tankers, X and Y are 4 km apart and X is due east of Y. The master of a ferry finds that the bearing of X from the ferry is 320° and the bearing of Y from the ferry is 290°. How far is the ferry from the tanker X?

14 Hugo and Ted play the seventh hole at their local golfcourse. The hole is 440 yards from the tee, T, to the hole, H.
Hugo's tee shot reaches position A while Ted's reaches position B. The exact positions of A and B are shown on the diagram.

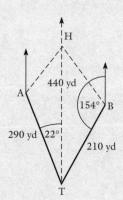

   a Show that $A\widehat{T}B = 48°$

   b How far apart are A and B?

   c Find the distance and direction of the hole from A.

   d Find the distance and direction of the hole from B.

15 In this shape there are two regular octagons. A side of the smaller octagon is 1 unit long. Any two of the shaded triangles can be placed together to form a square.

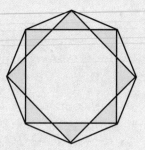

   a Show that the length of a side of one of these squares is $\dfrac{1}{\sqrt{2}}$ units.

   b Find the area of one of the shaded triangles.

   c The smaller octagon can be considered as a square from which four corners have been removed.
   Show that the length of a side of this square is $(1 + \sqrt{2})$ units.

   d Hence show that the area of the smaller octagon is $(2 + 2\sqrt{2})$ sq units.

   e Given that $\cos 45° = \dfrac{1}{\sqrt{2}}$ show that the length of a side of the larger octagon is $\sqrt{\dfrac{\sqrt{2} + 1}{\sqrt{2}}}$.

   f Evaluate this length correct to 3 significant figures.

VECTORS

A **vector** is any quantity that needs to be described by giving both its size (magnitude) and its direction, for example, velocity.

A quantity that needs only size to describe it is called a **scalar**, for example, time.

A vector can be represented by a straight line with an arrow to show direction, e.g.

When vectors are drawn on squared paper they can be described in terms of the number of squares needed to go across to the right and the number of squares needed to go up.

They are written in the form $\begin{pmatrix} a \\ b \end{pmatrix}$ where a is the number of squares across and b is the number of squares up.

In the diagram, $\mathbf{a} = \begin{pmatrix} 6 \\ 3 \end{pmatrix}$ and $\mathbf{b} = \begin{pmatrix} -3 \\ -6 \end{pmatrix}$

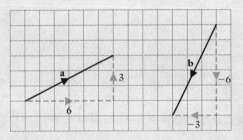

The **magnitude of a vector a**, which is written $|\mathbf{a}|$ or a, is equal to the length of the line representing $\mathbf{a}$.

In the diagram $|\mathbf{a}| = \sqrt{4^2 + 3^2}$
$$= 5$$

Two vectors which have the same magnitude and are in the same direction are *equal*, e.g. $\mathbf{a} = \mathbf{b} = \mathbf{c}$

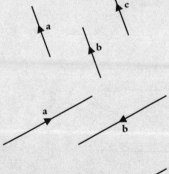

If two vectors, $\mathbf{a}$ and $\mathbf{b}$, have the same magnitude but are in opposite directions then $\mathbf{b} = -\mathbf{a}$.

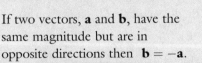

If k is a positive number then $k\mathbf{a}$ is a vector in the same direction as $\mathbf{a}$ and of magnitude $k|\mathbf{a}|$.

Addition of vectors

The triangle law for the addition of two vectors gives

$$\overrightarrow{BA} = \overrightarrow{BC} + \overrightarrow{CA}$$

i.e. $\mathbf{c} = \mathbf{a} + \mathbf{b}$

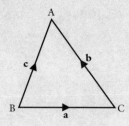

Subtraction of vectors

The triangle law gives $\overrightarrow{AC} = \overrightarrow{AB} + \overrightarrow{BC}$

But $\overrightarrow{BC} = -\overrightarrow{CB}$

$\therefore\ \overrightarrow{AC} = \overrightarrow{AB} + (-\overrightarrow{CB})$

$\Rightarrow\ \overrightarrow{AC} = \overrightarrow{AB} - \overrightarrow{CB}$

i.e. $\mathbf{c} = \mathbf{a} - \mathbf{b}$

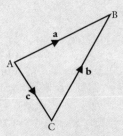

If $\mathbf{a}$ and $\mathbf{b}$ are two non-parallel vectors such that $\mathbf{c} = h\mathbf{a} + k\mathbf{b}$ and $\mathbf{c} = p\mathbf{a} + q\mathbf{b}$ where h, k, p and q are numbers then $h = p$ and $k = q$.

EXERCISE 3.8

1 a Copy the vectors **a** and **b** onto squared paper and on the same unit grid draw line segments to represent the vectors

    i $\mathbf{a} + \mathbf{b}$ **ii** $\mathbf{a} - \mathbf{b}$ **iii** $\mathbf{a} + 2\mathbf{b}$ **iv** $\mathbf{a} - 3\mathbf{b}$

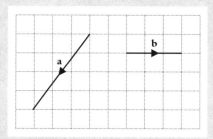

  b **i** If $\mathbf{c} = \mathbf{a} + \mathbf{b}$ and $\mathbf{d} = \mathbf{a} - \mathbf{b}$ find $\mathbf{c}$ and $\mathbf{d}$ in the form $\begin{pmatrix} x \\ y \end{pmatrix}$.

    ii If $\mathbf{e} = \mathbf{a} + 2\mathbf{b}$ how does $\mathbf{e}$ compare with $\mathbf{a}$?

2 a Find the single vector that is equivalent to

    i $\overrightarrow{AB} + \overrightarrow{BD}$

    ii $\overrightarrow{BC} + \overrightarrow{CD} + \overrightarrow{DA}$

    iii $\overrightarrow{DB} - \overrightarrow{CB}$

  b Give an alternative route for

    i $\overrightarrow{AD}$ **ii** $\overrightarrow{CB}$

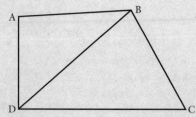

3 ABCD is a quadrilateral.
X is a point inside ABCD.
$\overrightarrow{AX} = \overrightarrow{XC} = \mathbf{a}$
and $\overrightarrow{BX} = \overrightarrow{XD} = \mathbf{b}$

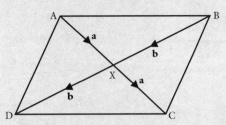

a **i** Is AXC a straight line.
ii Is DXB a straight line?
Justify your answers.

b Find in terms of **a** and **b**
i $\overrightarrow{AB}$ **ii** $\overrightarrow{DC}$ **iii** $\overrightarrow{AD}$ **iv** $\overrightarrow{BC}$

c What special name is given to this quadrilateral?
Justify your answer.

d What special properties for this quadrilateral follow from the results in parts **a** to **c**?

4 D is the mid-point of AB and E is
the mid-point of BC.
$\overrightarrow{BD} = \mathbf{p}$ and $\overrightarrow{BE} = \mathbf{q}$.

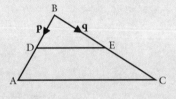

a Write down, in terms of **p** and **q**,
i $\overrightarrow{DB}$ **ii** $\overrightarrow{DE}$ **iii** $\overrightarrow{AC}$

b Hence show that DE is parallel to AC.

5 A bird tries to fly at 40 km/h on a bearing of 165° but is blown off course by a wind blowing at 25 km/h from the west.

a Show this information on a scale drawing and use it to estimate the magnitude and direction of the resultant velocity.

b Calculate the magnitude and direction of the resultant velocity, giving each answer correct to the nearest whole number.

6 M is the mid-point of the base QR of an isosceles triangle in which
PQ = PR. $\overrightarrow{PQ} = \mathbf{a}$, $\overrightarrow{PR} = \mathbf{b}$ and N is the mid-point of PR.

a Write $\overrightarrow{QR}$, $\overrightarrow{QM}$, $\overrightarrow{PM}$ and $\overrightarrow{QN}$
in terms of **a** and **b**.

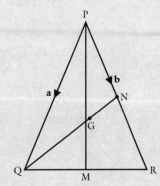

b G is a point on QN such that
QG = 2GN.
Show that $\overrightarrow{QG} = \frac{1}{3}(\mathbf{b} - 2\mathbf{a})$.

c Prove that when PG is produced
it passes through M.

7 In △PQR, $\overrightarrow{PQ} = \mathbf{q}$ and $\overrightarrow{PR} = \mathbf{r}$. S is the point such that $\overrightarrow{PS} = 3\mathbf{r}$ and T is the mid-point of QS.

  a Find, in terms of $\mathbf{q}$ and $\mathbf{r}$, the vectors

    i $\overrightarrow{QS}$ **ii** $\overrightarrow{QT}$ **iii** $\overrightarrow{PT}$.

  b U is the point such that $PU = 2\overrightarrow{PT}$. Find $\overrightarrow{QU}$.

  c What type of quadrilateral is   **i** QUSP   **ii** QURP?

8 D is the midpoint of AC and E is the midpoint of AB.
$\overrightarrow{AC} = \mathbf{p}$ and $\overrightarrow{AB} = \mathbf{q}$.

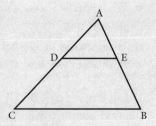

  a Express in terms of $\mathbf{p}$ and $\mathbf{q}$

    i $\overrightarrow{CB}$ **ii** $\overrightarrow{DE}$

  b Hence prove that DE is parallel to CB and equal to half of it.

TRANSFORMATIONS

When an object is **reflected in a mirror line**, the object and its image form a symmetrical shape with the mirror line as the axis of symmetry.

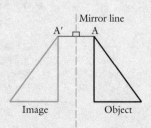

When an object has been reflected, corresponding points on the object and the image are the same distance from the mirror line. Therefore the mirror line can be found by joining a pair of corresponding points on the object and the image, AA′ say, and then finding the line that goes through the midpoint of AA′ and is perpendicular to it.

An object is **translated** when it moves without being turned or reflected to form an image.

A **translation** can be described by the vector which gives the movement of a point on the object to the corresponding point on the image.

In the diagram, the movement from the point A to the corresponding point A′ is 7 units to the left and 4 units up, so the translation is described by the vector $\begin{pmatrix} -7 \\ 4 \end{pmatrix}$.

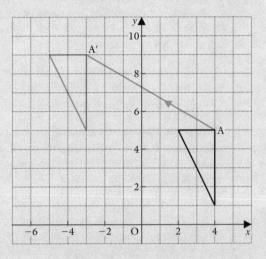

When an object is **rotated about a point** to form an image, the point about which it is rotated is called the **centre of rotation** and the angle it is turned through is called the **angle of rotation**.

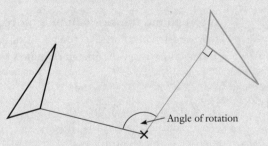

To **find the centre of rotation** in this diagram, draw the perpendicular bisectors of AA′ and either BB′ or CC′.

The point where these lines intersect gives the centre of rotation.

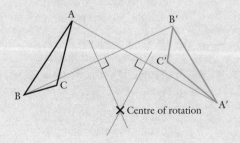

When an object is **enlarged** by a **scale factor** 2, each line on the image is twice the length of the corresponding line on the object.

The diagram shows an enlargement of a triangle, with centre of enlargement X and scale factor 2. The dashed lines are guide-lines.

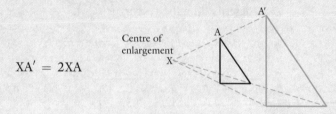

Centre of
enlargement

$XA' = 2XA$

When the scale factor is *less than one*, the image is *smaller* than the object.

When the scale factor is *negative*, the guide-lines are drawn *backwards* through the centre of enlargement, O, so that, if the scale factor is -2, $OA' = 2OA$.

This produces an image, each of whose lines is twice as long as the corresponding line on the object, and is a rotation of the object by $180°$.

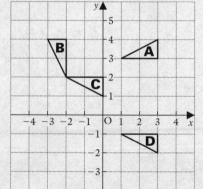

A **compound transformation** is the result of one transformation followed by another, e.g. the result of reflecting an object in the *y*-axis and then rotating the image obtained by $30°$ about the origin.

EXERCISE 3.9

1 a Describe fully the single transformation that maps triangle A onto

    i triangle B

    ii triangle D.

b Describe fully the single transformation that maps triangle B onto triangle C.

c Describe fully the single transformation that maps triangle D onto triangle C.

d Describe a compound transformation that maps triangle A onto triangle C.

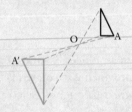

2 Triangles ABC and PQR are shown in the diagram. Copy the diagram and

a enlarge △ABC by a scale factor of 2, centre the origin; label the image $A_1B_1C_1$

b rotate △ABC 90° clockwise about the origin; label this $A_2B_2C_2$

c draw the image of $A_2B_2C_2$ defined by the vector $\begin{pmatrix} -6 \\ -1 \end{pmatrix}$; label the image $A_3B_3C_3$

d Describe the single transformation that maps △$A_3B_3C_3$ onto △PQR.

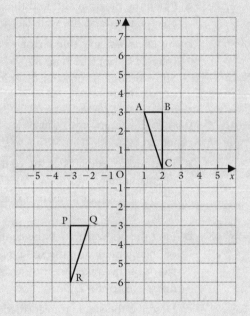

3

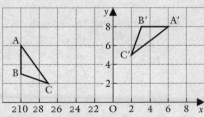

a Describe a compound transformation that maps △ABC onto △A′B′C′.

b Describe the single transformation that maps △ABC onto △A′B′C′.

4

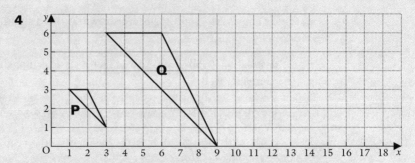

In the diagram triangle Q is an enlargement of triangle P.
The centre of enlargement is X.

a Copy the diagram and find the coordinates of the centre of enlargement and the scale factor.

b An enlargement, scale factor $\frac{2}{3}$ and centre (18, 3) transforms triangle Q onto triangle R.
Show triangle R on your diagram and write down the coordinates of its vertices.

5 Copy the diagram onto squared paper.

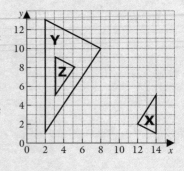

a Find the coordinates of the centre of enlargement and the scale factor of the transformation that maps △X onto △Y.

b Find the coordinates of the centre of enlargement and the scale factor of the transformation that maps △Y onto △Z.

c Describe the single transformation that maps △Z onto △X.

6 a On a map whose scale is 1 : 5000, the area of a field is 4 cm². Find the area of the actual field.

b

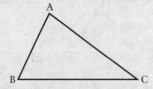

Triangles ABC and DEF are similar and the scale factor for reducing △ABC to △DEF is $\frac{1}{2}$.

  i If AB = 8 cm, find DE.
 ii If BC = 13 cm, find EF.
iii If DF = 4.5 cm, find AC.
 iv If area △ABC = 35.5 cm², find area △DEF.

7 In this diagram, △PQR is an enlargement of △ABC.

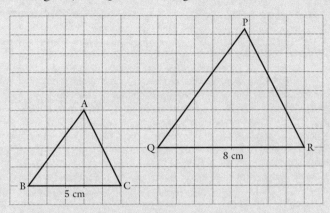

a What is the scale factor of the enlargement?

b If AB = 5 cm, how long is PQ?

c If $\widehat{ABC} = 53°$, what is $\widehat{PQR}$?

d Copy the diagram and draw the image of △PQR under an enlargement with centre Q and a scale factor of $\frac{1}{2}$. Label the image XYZ.

e What is the scale factor of the enlargement which maps △ABC to △XYZ?

8 A rectangle ABCD is 12 cm long and 5 cm wide.
A rectangle PQRS is 8 cm long and $3\frac{1}{3}$ cm wide.

a Explain why the rectangles are similar and give the scale factor that maps ABCD to PQRS.

b Find the scale factor that changes the area of ABCD to the area of PQRS.

9 Triangle A is mapped to triangle B by a reflection in a line. The diagram shows triangle A and one side of triangle B.

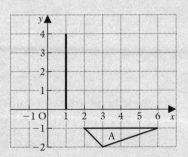

a Copy and complete the diagram to show triangle B.

b Write down the equation of the line.

10

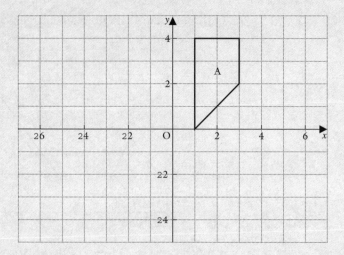

a Draw the reflection of the shape A in the line $y + x = 0$.

b Draw the enlargement of the shape A by a scale factor of $\frac{1}{2}$, with centre $(-7, 0)$.

THREE-DIMENSIONAL OBJECTS

A **right pyramid** has its vertex directly above the centre of its base.

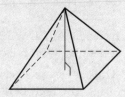

The **angle between a line and a plane** is given by drawing the perpendicular from the top of the line to the plane and finding the angle between the line joining the foot of that perpendicular to the base of the line and the line itself.

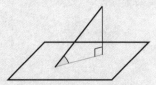

The **angle between two planes** is the angle between two lines, one in each plane, that are perpendicular to the line joining the planes.

EXERCISE 3.10

1 The diagram shows a cuboid measuring 9 cm by 5 cm by 4 cm.

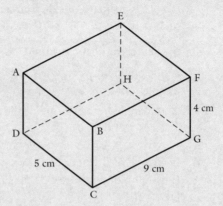

a Find the length of   **i** AC   **ii** BG   **iii** DG   **iv** AG

b Find the angle between the plane ABCD and the line
   i AH   **ii** AG.

c Find the angle between the plane DCGH and the plane ACGE.

d If each dimension is correct to the nearest centimetre find

   i the largest possible volume for the cuboid

   ii the smallest possible value for the area of the base, DCGH, of
   the cuboid.

2 The sketch shows a closed storage box for toys. It is made from sheet
plastic of negligible thickness.

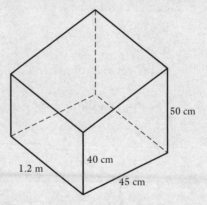

Use the dimensions given with the sketch to calculate

a the capacity of the box in m³

b the angle between the sloping top and the horizontal

c its total surface area in m²

d the length of the longest straight stick that will fit into the empty
box.

3 The diagram shows a square pyramid. The base ABCD is a square of side 5 cm and V is the vertex such that VA = VB = VC = VD = 8 cm. Find

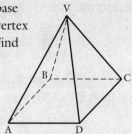

a the height of the pyramid

b the angle between one of the sloping edges and the base

c the angle between one of the sloping faces and the base.

4 Each face of this tetrahedron is an equilateral triangle of side 4 cm. E is the midpoint of BC. AF is the height of the pyramid and DF = 2EF.

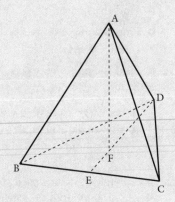

Find, in surd form,

a DE

b AF

c the volume of the tetrahedron

5 The diagram shows a cuboid.

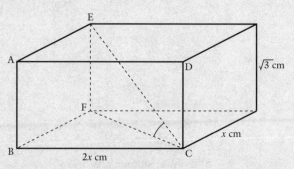

Show that $\cos E\widehat{C}F = \dfrac{x\sqrt{5}}{\sqrt{3 + 5x^2}}$

MIXED EXERCISE 3.11

1 A radio mast is secured by stays as indicated in the diagram. D is the mid-point of CE.

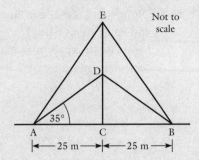

a Find the length of CD.

b Find the angle of elevation of the top of the mast E from A. (AQA)

2 The diagram shows part of a football pitch. The penalty arc is part of a circle, radius 10 yards, with centre P. M is the mid-point of AB and PM = 6 yards.

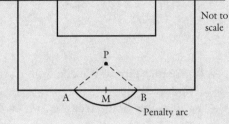

a Calculate angle APB.

b Calculate the length of the penalty arc.

(AQA)

3 A helicopter service links the Channel Islands of Alderney, Guernsey and Jersey.

The pilot sets off from Jersey to Guernsey, a distance of 42 km, on a bearing of 311°.

Alderney is 38 km from Guernsey on a bearing of 035°.

Calculate how long it will take to fly from Alderney to Jersey at an average speed of 180 km/h.

(AQA)

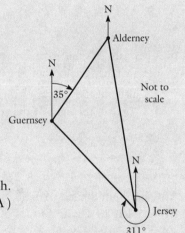

4 Triangle ABC is right-angled at B. The lengths of AB and BC are 21.4 cm and 16.3 cm respectively, correct to the nearest mm.

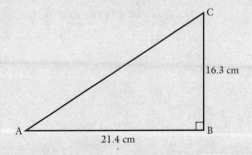

a Calculate the largest possible value of the area of triangle ABC, correct to the nearest tenth of a cm².

b Use Pythagoras' theorem to calculate the range of possible values of the length of AC, correct to the nearest mm.

c Calculate the largest possible value of angle A. (WJEC)

5 a This earring is made from a rectangle
and an isosceles triangle.
Calculate its area showing all your working.

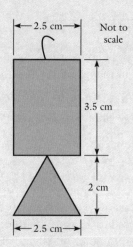

b Another earring is made from a sector of a circle,
radius 3 cm.
The angle of the sector is 70°.
Calculate its area showing all your working.

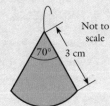

c A pendant of similar shape to the earring is cut from a
larger circle. The ratio of the radii of the two sectors is 2 : 3.

i Calculate the **radius** of the sector used for the pendant.
ii Calculate the **area** of the sector used for the pendant.

(AQA)

6 ABC is a right-angled triangle.
AB is of length 4 m and BC is of length 13 m.

a Calculate the length of AC.
b Calculate the size of angle ABC.

(Edexcel)

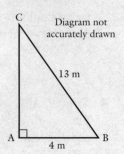

7 Triangle P is mapped to triangle Q by an
enlargement of scale factor −0.5.
If AB is of length 6.4 cm, how long is FD ?

(Edexcel)

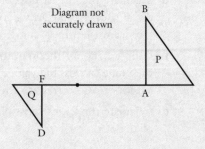

8 An aeroplane's velocity in still air is 400 km per hour on a bearing of 045°.
The wind is blowing at 100 km per hour on a bearing of 070°.
Draw an accurate vector diagram to show the actual velocity of the aeroplane.
(Use a scale of 1 cm to represent 50 km per hour.)   (Edexcel)

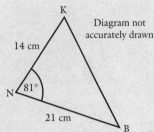

9 a Calculate the length of KB.

   b Calculate the size of the angle NKB.

                                      (Edexcel)

10

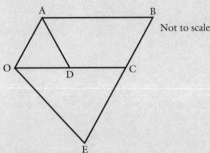

OABC is a parallelogram. $\overrightarrow{OA} = \mathbf{a}$, $\overrightarrow{OC} = \mathbf{c}$.

BCE is a straight line, $\overrightarrow{BE} = 3\overrightarrow{BC}$. D is the mid-point of OC.

   a Write in terms of **a** and **c**   **i** $\overrightarrow{AD}$ **ii** $\overrightarrow{OE}$

   b Deduce the ratio of the lengths of AD and OE.   (OCR)

11 Salima walks from her school, at A, to her home, at
C. She can go along the roads AB and BC or along
a footpath, AC, across the school playing field.
AB = 220 m,   BC = 130 m   and
angle ABC = 115°.
Calculate how much further it is for Salima to walk
from A to C, via B, rather than by the footpath.
                                      (OCR)

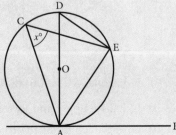

12 In the diagram, O is the centre of the circle,
AD is the diameter and AB is a tangent.
Angle ACE = $x°$.
Find, in terms of x, the size of

   a angle ADE   **c** angle EAB

   b angle DAE   **d** angle AOE.   (OCR)

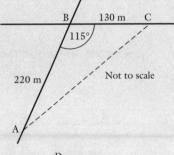

13 Byron and Shelley are two dogs.

Byron's lead is 1 m long. One end of the lead can slide along a rail PQ which is fixed to the wall of the house.

Shelley's lead is 1.5 m long. One end of this lead is attached to a post A, at the corner of his kennel.

The scale diagram below represents the fenced garden PQRS where the dogs live.

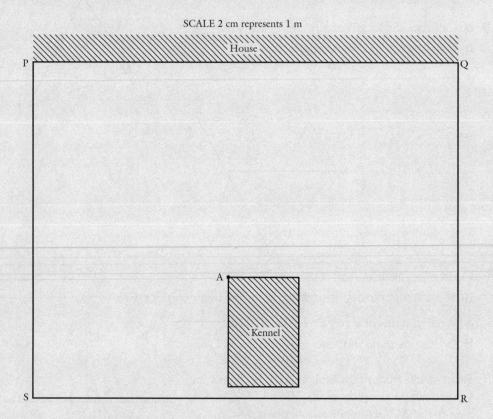

SCALE 2 cm represents 1 m

Show on the diagram all the possible positions of each dog if the leads remain tight.

(OCR)

14

ABCD is a rectangle. The length of AB is 6.3 cm measured to the nearest millimetre.

a Complete the inequality

_____ ⩽ AB < _____

The diagonal AC is 7.6 cm measured to the nearest millimetre.

b Calculate upper and lower bounds for the length of the side BC.

c Give the length of BC to an appropriate degree of accuracy. (OCR)

15 A spherical netball has a volume of
$8200 \, \text{cm}^3$ correct to the nearest $100 \, \text{cm}^3$.
It must be possible for the ball to pass
through a circular hoop.
Calculate the minimum diameter of the
hoop for this to be certain to happen. (OCR)

Not to scale

16

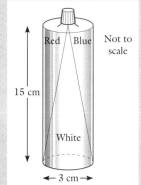

Not to scale

Red Blue

15 cm

White

← 3 cm →

The body of a tube of toothpaste is a cylinder with diameter
3 cm and height 15 cm. It contains three colours of
toothpaste: white, red and blue.

The white toothpaste is in the shape of a cone, as shown in
the diagram; the red and blue share equally the remaining
volume.
a Calculate the volume of each colour of toothpaste.

When the toothpaste is squeezed from the tube, the colours come out in equal
quantities.

b **i** Given that the diameter of the circular opening is 6 mm, calculate the area of the
opening occupied by the blue toothpaste.
ii Given also that the blue toothpaste subtends an angle of $x°$ at the centre of the
opening, as shown in the diagram, find in terms of x an expression for the area
of the blue toothpaste.

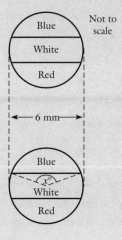

Not to scale

Blue

White

Red

← 6 mm →

Blue

$x°$

White

Red

(OCR)

17 Cheese-wedge & Co. sells cheeses
which are always shaped as right-
angled triangular prisms with height,
width and length in the ratio $3:4:7$.

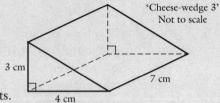

The cheeses are named according to their heights.

A 'Cheese-wedge 3' has height 3 cm, width 4 cm and length 7 cm.

A 'Cheese-wedge 6' has height 6 cm, width 8 cm and length 14 cm.

a Calculate the volume of a 'Cheese-wedge 3'.

b Calculate the total surface area of a 'Cheese-wedge 6'.

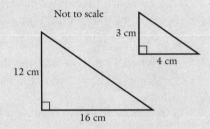

c The diagram above shows the triangular end-faces of a 'Cheese-wedge 12' and a
'Cheese-wedge 3'.

  i How many small triangles can be cut from the large triangle?

  ii How many 'Cheese-wedge 3's's can be made from a 'Cheese-wedge 12'?

(OCR)

18 Chips are sold in containers shaped as a cone.

The cone is made from a sector of a circle with
radius 20 cm and angle $120°$.

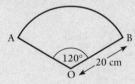

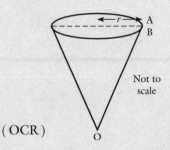

a Calculate the length of the arc AB shown in the above diagram.

The line OA is joined to the line OB to form the cone.

b Calculate the radius, r, of the cone.

The cone holds 64 chips.
A similar smaller cone is made using a sector of a circle
with radius 15 cm and an angle of $120°$.

c Calculate the number of chips which the smaller
cone will hold. Assume all the chips are the same
size and shape.

(OCR)

19 Forces P and Q have magnitude 5 N and 8 N respectively. Find the magnitude of their resultant if the angle between P and Q is

a 0°   **b** 180°   **c** 90°   (OCR)

20 In the diagram O is the centre of the circle and angle ATB is 50°.
TA and TB are tangents.

a Find angle AOB.

b Find angle ACB, giving a reason for your answer.   (OCR)

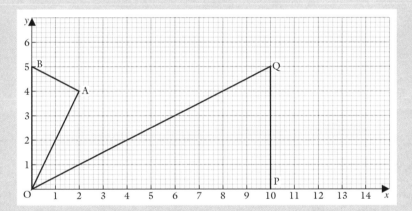

Not to scale

21 The grid shows the triangle OAB with vertices at (0, 0), (2, 4) and (0, 5) and the triangle OPQ with vertices at (0, 0), (10, 0) and (10, 5).

The transformation that maps OAB onto OPQ can be considered as a rotation about the origin followed by a second transformation.

a Measure and write down the angle of rotation.

b Describe fully the second transformation.   (OCR)

22 Jelly beans are sold in tubes which are cylinders of radius 3 cm and height 12 cm.

a Calculate the volume of one of the tubes.

Tubes of jelly beans are packed into a carton in the shape of a cuboid measuring 60 cm by 30 cm by 12 cm.

b **i** How many tubes may be packed into a carton?
   ii Find the volume of empty space in a carton filled with tubes of jelly beans.

(OCR)

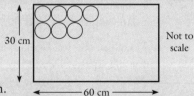

30 cm

60 cm

Not to scale

23 The diagram shows the end view of the framework
for a sports arena stand.

a Calculate the distance AB.

b Calculate the angle x. (AQA)

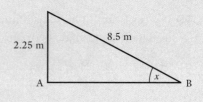

24 The diagram shows a circle with diameters AC and BD.

a Write down the size of angle ADC.
Give a reason for your answer.

b **Prove** that triangle ABD is congruent to triangle DCA.
Remember to give your reasons. (OCR)

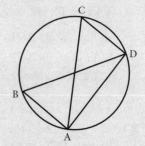

25

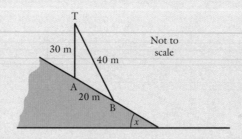

A vertical mast, AT, 30 m high is erected on a straight hillside AB. From the top, T, of
the mast a rope, TB, 40 m long is secured at a point B, 20 m from the foot of the mast A.
Points A, T and B lie in the same vertical plane.

Calculate the angle, x, that the hillside makes with the horizontal. (OCR)

26 The diagram shows a bucket. The base radius is r,
the slant height is $3r$ and the radius of the top is R.

a Which of the following could be the correct expression
for the surface area of the bucket?

$$\pi r^2 + 3\pi r^2 R \qquad 3\pi r^2 R$$

$$4\pi r^2 + 3\pi r R \qquad \frac{16\pi r^3}{3}$$

b Explain how you decided. (OCR)

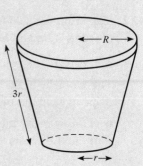

27 The diagram represents a regular pentagon with two of its lines of symmetry shown.

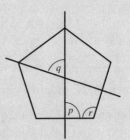

  a Write down the value of angle p.

  b Calculate the value of

   i angle q

   ii angle r (AQA)

28 "Bradley's Soup" is canned by a small family business. Each morning they make 200 litres of soup. This is put in cylindrical tins, each of which is 8.4 cm high and has a diameter of 7.0 cm.

How many of these tins can be filled from the 200 litres of soup?   (AQA)

29 TPK is a tangent to the circle.
TSQ is a straight line.

PQ = QR.

$\widehat{QPK} = 50°$

$\widehat{STP} = 26°$

Calculate the size of

  a angle PQR

  b angle QRS.   (AQA)

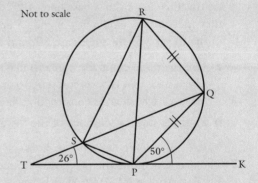

30 A helicopter leaves a heliport H and its measuring instruments show that it flies 3.2 km on a bearing of 128° to a checkpoint C. It then flies 4.7 km on a bearing of 066° to its base B.

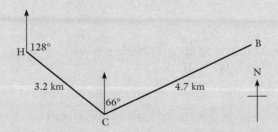

  a Show that angle HCB is 118°.

  b Calculate the direct distance from the heliport H to the base B.   (AQA)

31 a Given that $\overrightarrow{OA} = 3\mathbf{p}$, $\overrightarrow{OB} = 3\mathbf{q}$ and $\overrightarrow{OC} = 4\overrightarrow{OB}$,
show that $\overrightarrow{AC} = 12\mathbf{q} - 3\mathbf{p}$.

b Given that $\overrightarrow{AM} = \frac{2}{3}\overrightarrow{AB}$ and $\overrightarrow{AN} = \frac{1}{3}\overrightarrow{AC}$,
express $\overrightarrow{ON}$ in terms of $\mathbf{p}$ and $\mathbf{q}$.

c Given that $\overrightarrow{OM} = \mathbf{p} + 2\mathbf{q}$
what can you say about the
points O, M and N?   (AQA)

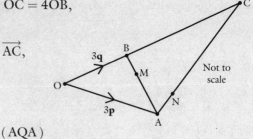

32

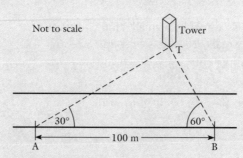

Ceri and Diane want to find how far away a tower, T, is on the other side of a river. To
do this they mark out a base line, AB, 100 metres long as shown on the diagram. Next
they measure the angles at the ends A and B between the base line and the lines of sight
of the tower. These angles are 30° and 60°.

a Use ruler and compasses only to make a scale drawing of the situation.
Use a scale of 1 cm to represent 10 m.
Show clearly all your construction lines.

b Find the shortest distance of the tower, T, from the base line AB.   (AQA)

33 The diagram shows the positions of three airports:
E (East Midlands), M (Manchester) and L (Leeds).

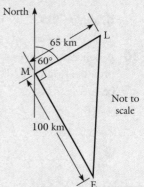

The distance from M to L is 65 km on a bearing of 060°.
Angle LME = 90° and ME = 100 km.

a Calculate, correct to three significant figures, the distance LE.

b Calculate, to the nearest degree, the bearing of E from L.

c An aircraft leaves M at 10.45 a.m. and flies direct to E, arriving at 11.03 a.m.
Calculate the average speed of the aircraft in kilometres per hour. Give your answer
correct to the appropriate number of significant figures.   (OCR)

34 a

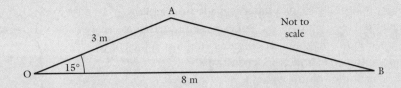

In the triangle OAB, OA = 3 m, OB = 8 m and angle AOB = 15°.

i Calculate, correct to 2 decimal places, the area of triangle OAB.

ii Calculate the length of the side AB.

b

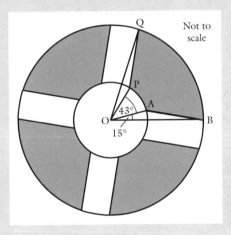

The plan of an ornamental garden shows two circles, centre O, with radii 3 m and 8 m.

Paths of equal width cut symmetrically across the circles.

The shaded areas represent flower beds.

BQ and AP are arcs of the circles.

Triangle OAB is the same triangle shown in part **a** above.

Given that angle POA = 43°, calculate the area of the flower bed PABQ.

Give your answer to an appropriate number of significant figures. (OCR)

35

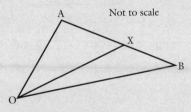

In the diagram X is the point on AB such that AX = 3XB.

Given that $\overrightarrow{OA} = 8\mathbf{a}$ and $\overrightarrow{OB} = 4\mathbf{b}$, express in terms of $\mathbf{a}$ and/or $\mathbf{b}$

a i $\overrightarrow{AB}$

   ii $\overrightarrow{AX}$

b $\overrightarrow{OX}$ (OCR)

36 a A symmetrically shaped timer is made from hollow
hemispheres, cylinders and cones joined together as
shown in the diagram. It contains sand just sufficient to
fill the top cone and cylinder sections.
Calculate the volume of sand.

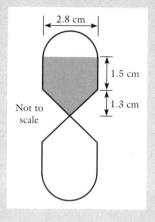

Not to
scale

b When all the sand has run through, it collects as shown in
the diagram. Calculate the height, h cm, of sand in the
cylindrical part of the timer.   (OCR)

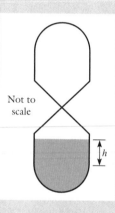

Not to
scale

37

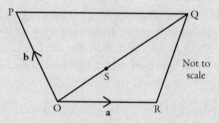

Not to
scale

The diagram shows the position of points O, P, Q and R with vectors **a** and **b** acting
along OR and OP, respectively.

$$\overrightarrow{OR} = \mathbf{a}$$

$$\overrightarrow{OP} = \mathbf{b}$$

$$\overrightarrow{OS} = \tfrac{1}{3}\overrightarrow{OQ}$$

$$\overrightarrow{PQ} = 2\overrightarrow{OR}$$

By expressing $\overrightarrow{PS}$ and $\overrightarrow{RS}$ in terms of the vectors **a** and **b** find the ratio PS : SR and
explain the relationship between the points P, S and R.   (AQA)

38 a The model of the cross-section of a roof is illustrated below.
   BC = 6 cm
   CD = 9 cm
   Angle CDE = 19.5°.

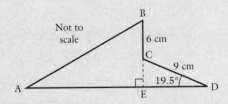

   i Calculate the length of CE.
   ii Triangles ABE and DCE are similar triangles with angle BAE = angle CDE.
     Calculate the length of AB.

b When the roof is constructed, the actual length of BC is 4.5 m. Calculate the area of the cross-section of the actual roof space.   (AQA)

39 a The parallelogram ABCD has vertices at (6, 3), (9, 3) (12, 9) and (9, 9) respectively.

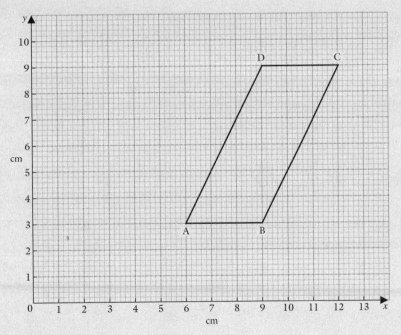

   An enlargement scale factor $\frac{2}{3}$ and centre (0, 0) transforms parallelogram ABCD onto parallelogram $A_1B_1C_1D_1$.
   Draw the parallelogram $A_1B_1C_1D_1$.

b The parallelogram $A_1B_1C_1D_1$ can be transformed back onto the parallelogram ABCD by a single transformation.
   Describe fully this transformation.   (AQA)

40 PQRS and PSTU are parallelograms.

$\overrightarrow{PQ}$ is **a**, $\overrightarrow{PS}$ is **b**, $\overrightarrow{ST}$ is **c**.

Find, in terms of **a**, **b** and **c** expressions in their simplest forms for

a $\overrightarrow{PT}$

b $\overrightarrow{US}$

c $\overrightarrow{PX}$ where X is the mid-point of QT.

d $\frac{1}{2}(\overrightarrow{PQ} + \overrightarrow{PT})$

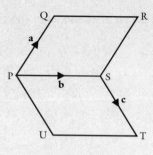

(Edexcel)

41

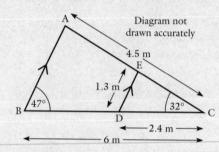

In the diagram BA is parallel to DE

AEC and BDC are straight lines
AC = 4.5 m, DE = 1.3 m, CD = 2.4 m, BC = 6 m
Angle ABC = 47°, Angle BCA = 32°.

a i Calculate the size of angle DEC.

ii Explain how you obtained your answer.

b Calculate the length of EC.

c Calculate the length of AB.

(Edexcel)

42 Describe fully the single transformation which maps the triangle A onto the triangle C.

(Edexcel)

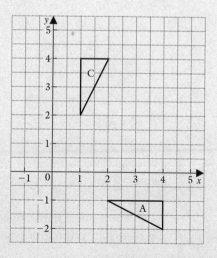

43

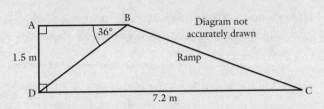

In the diagram calculate the length of BC.   (Edexcel)

44

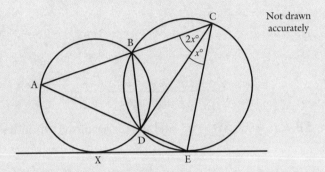

ABC and ADE are straight lines. CE is a diameter.
Angle DCE $= x°$ and angle BCD $= 2x°$.

a Find, in terms of x, the sizes of the angles

   i ABD   **ii** DBE   **iii** BAD.

b Explain why BE $×$ AC $=$ AE $×$ CD.   (Edexcel)

45 A rubbish skip is a prism. The cross-section of the prism is an isosceles trapezium.
This diagram shows the inside measurements of the cross-section.
AB is 4 m long. AD is 1.5 m long. Angle DAB is 70°.

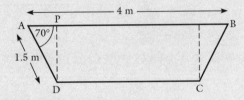

a Calculate the depth of the skip (PD).

b Calculate the length AP.

c Calculate the area of cross-section of the skip.

d The internal width of the skip is 2 m. It is filled with earth. The earth is level with the top of the skip. What is the volume of the earth in the skip? Give your answer to an appropriate degree of accuracy.

e The density of the earth in the skip is 700 kg/m^3. What is the mass of the earth in the skip?   (WJEC)

46 A stained glass window contains triangular pieces of coloured glass, some of which are red. The lengths of the sides of each of the red pieces of glass are 7 cm, 10 cm and 15 cm. Calculate the area of each piece of red glass.   (WJEC)

47 In the rhombus ABCD, E and F are the mid-points of BC and CD respectively.

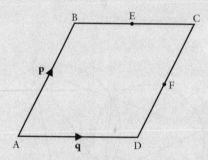

Given that **AB = p** and **AD = q**, find the following vectors in terms of **p** and **q**.

a BD

b AE

c AF

d Show that EF is parallel to BD and is half its length.   (WJEC)

48 A ladder is 16 feet long.
Starting from the position shown, the ladder slips outward from the wall with its ends in contact with the wall and the ground.
Draw five possible positions of the ladder. Hence draw the path of the centre of the ladder.   (AQA)

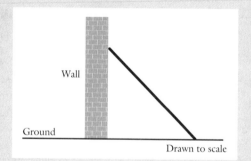

49 In the diagram, X and Y are the mid-points of AO and BO.

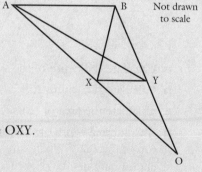

Not drawn to scale

i Write down a triangle which is similar to triangle OXY.

ii Explain why $\dfrac{XY}{AB} = \dfrac{1}{2}$

iii Calculate $\dfrac{\text{area triangle OXY}}{\text{area triangle OAB}}$ (AQA)

50 A man can row a boat at 3 mph in still water.
He wishes to cross a river from A to B at right angles to the river bank. The current is flowing at 2 mph.

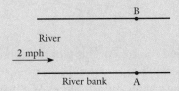

a **i** Draw a vector diagram to help you calculate the angle between the river bank and the direction in which he must row.
ii Calculate the size of this angle.

b The river is 50 yards wide.
How many seconds will it take the man to row across the river?
(1 mile = 1760 yards) (AQA)

51 P, Q, R are the mid-points of the sides of triangle ABC.

Explain why the triangle QPC is congruent
to the triangle ARQ. (OCR)

52 Brian (B) is sitting in his garden.
He notices an aircraft appear above the tree ST. Its bearing from B is 065° and the elevation 30°.
The aircraft disappears from Brian's view $1\frac{1}{2}$ minutes later over the house HG.
Its bearing is now 300° and the elevation 50°.

Assume that the aircraft flew in a straight line at a constant height of 8000 m.

a What was the speed of the aircraft?

b What was the bearing of the aircraft's course?

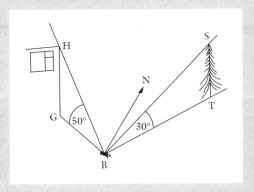

(OCR)

53

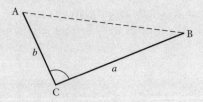

A surveyor is asked to find the area of this plot of ground, the triangle ACB. As she has no accurate instruments with her, she decides to estimate the area. She measures the lengths of AC and BC by pacing and the angle ACB with a simple compass. These are the results and errors:

$$AC = b = 34 \text{ metres} \quad (\pm 2\%)$$
$$BC = a = 87 \text{ metres} \quad (\pm 2\%)$$
$$\text{Angle } ACB = C = 130° \quad (\pm 5\%)$$

a Use the formula Area $= \frac{1}{2} ab \sin C$ to find the largest the area could be.

b Use the cosine rule to find the longest possible length for AB. (OCR)

54 Draw a line AB that is 8 cm long.

a Draw the locus of points, P, which lie above the line AB such that the area of triangle ABP is 12 cm².

b On the same diagram construct the locus of points, Q, which lie above the line AB such that angle AQB is 90°.

c Hence draw all triangles ABC which have C above AB, an area of 12 cm² and an angle of 90°. (OCR)

55 The diagram shows a square inscribed in a circle, centre O, radius 1 cm.

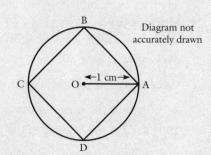

a Calculate the perimeter of the square. Give your answer correct to 2 decimal places.

In the second diagram, points E, F, G and H are added to form a regular octagon.

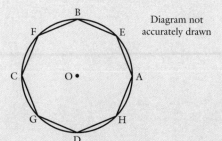

b Calculate the perimeter of the octagon. Give your answer correct to 2 decimal places. (Edexcel)

56 A sweet is in the shape of a sphere.
It consists of a spherical toffee centre with a chocolate coating of uniform thickness.

a The radius of the spherical toffee centre is 0.7 cm.
Calculate the volume of toffee in the sweet.

b The uniform chocolate coating has the same volume as the volume of the toffee centre. Calculate the thickness of the chocolate coating. State the units of your answer.

c 'Fun-size' sweets are spheres with half the diameter of the full-size sweet. Calculate the number of fun-size sweets that can be made using the volume of materials that is used in one full-size sweet. (OCR)

57 An aeroplane can fly at 200 km/h in still air. The aeroplane sets out to fly due north in a south west wind of 30 km/h.
The diagram below shows the air-speed (200 km/h), the wind speed (30 km/h) and the wind direction (from the south west).

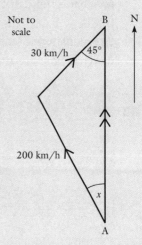

a i Calculate the angle x.
  ii Calculate the ground-speed (represented by AB).

The pilot now flies back due south. The wind speed and direction are unchanged. The air-speed remains at 200 km/h.

b Sketch a vector diagram which shows the speed and direction the plane is heading, the wind speed and direction, and the ground-speed and direction. You do not need to draw the diagram to scale. (OCR)

58 Right-angled triangles can have sides with lengths which are a rational or irrational number of units.
Give an example of a right-angled triangle to fit each description below.

  i All sides are rational.
 ii The hypotenuse is rational and the other two sides are irrational.
iii The hypotenuse is irrational and the other two sides are rational.
 iv The hypotenuse and one of the other sides are rational and the remaining side is irrational. (Edexcel)

59 A scale drawing of a car park is shown.
In the diagram BC = 4 cm, CD = 7.2 cm and angle BCD = 70°.
In the car park the actual length of the side represented by BC is 48 metres.

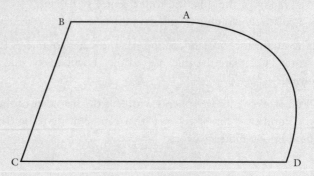

a What is the actual length of the side represented by CD?

b The actual length of the boundary represented by AD is 78 metres.
Calculate the length of AD on the scale diagram. You must show your working.

c What would be the actual size of the angle represented by angle BCD in the car park?

(OCR)

60

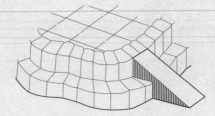

Two steps lead up to the patio in Cathy's garden. She wants to build a wooden ramp so
that she can wheel heavy loads up on to the patio. She decides to make it as shown below.

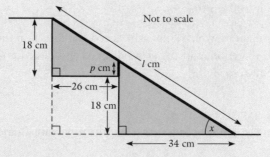

a Calculate the sloping length of the ramp, *l* cm.

b Calculate the area of wood required to make one side of the ramp (the area shown
shaded in the diagram).

c Calculate the angle of inclination of the ramp, *x*.

d For extra strength, Cathy decides to fix a support to fit on the edge of the step.
Calculate the height of this support, shown as *p* cm on the diagram.   (OCR)

61 The diagram represents a right pyramid.
The base is a square of side 2*x*.

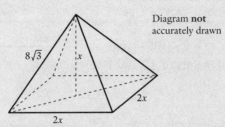

Diagram **not**
accurately drawn

The length of each of the slant edges is $8\sqrt{3}$ cm.

The height of the pyramid is *x* cm.

Calculate the value of *x*.

62

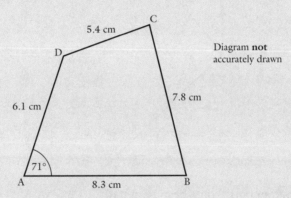

Diagram **not**
accurately drawn

The diagram shows a quadrilateral ABCD.
AB = 8.3 cm, BC = 7.8 cm, CD = 5.4 cm and AD = 6.1 cm. Angle BAD = 71°.

a Calculate the area of triangle ABD.
Give your answer correct to 3 significant figures.

b Calculate the size of angle BCD.
Give your answer correct to 1 decimal place. (Edexcel)

63 A, B, C and D are points on a circle.
AB is parallel to CD.
Lines AD and BC intersect at E.
Angle EDC = 35°.

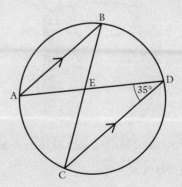

a Write down the size of angle ABE.
Give a reason for your answer.

b i Find the size of angle AEC.

Show all your working clearly.

ii What does this tell you about point E?
Give a reason for your answer. (OCR)

64 a In the diagram XQ = 3PX.

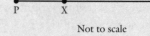

Given that **PX** = **p**, find in terms of **p**

Not to scale

i XQ   **ii** QP

b In the diagram ABCD is a parallelogram.
M and N are the midpoints of AB and DC.
AB = **a** and **AD** = **b**.

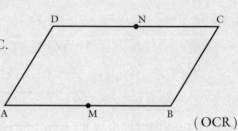

Use a vector method to prove that
AMCN is also a parallelogram.
Show all your working clearly.

(OCR)

65

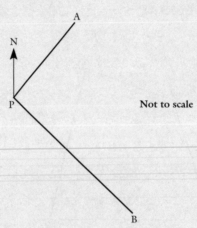

Not to scale

Ship A is 3.4 km from port P on a bearing of 040°.
Ship B is 15 km from P on a bearing of 155°.

a Calculate the distance between the two ships.

b Calculate the bearing of ship A from ship B.

(OCR)

66 A cylinder has radius x and height $4x$.   A sphere has radius r.

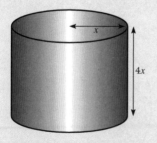

The **total** surface area of the cylinder is equal to the surface area of the sphere.

a Show that $r^2 = \dfrac{5}{2}x^2$

b When $r = 10$, find the value of x. Give your answer in the form $a\sqrt{b}$.   (AQA)

67 OABC is a quadrilateral.

M, N, P and Q are the midpoints of OA, OB, AC and BC respectively.

$\overrightarrow{OA} = \mathbf{a}$, $\overrightarrow{OB} = \mathbf{b}$, $\overrightarrow{OC} = \mathbf{c}$.

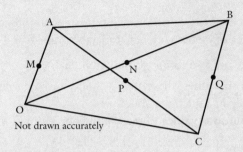

Not drawn accurately

a Find, in terms of **a**, **b** and **c**, expressions for

  i $\overrightarrow{BC}$ **ii** $\overrightarrow{NQ}$ **iii** $\overrightarrow{MP}$

b What can you deduce about the quadrilateral MNQP?
  Give a reason for your answer.   (AQA)

68 ABC are three points on a circle centre O.
  AC = 12 cm, BC = 9 cm. Angle ACB = 25°

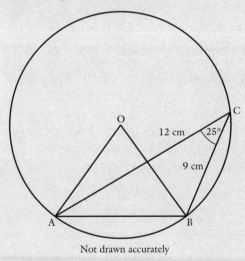

Not drawn accurately

a State the size of angle AOB.

b Find the area of the circle.   (AQA)

4 Handling Data

COLLECTING DATA

Data used for statistical analysis may be in categories such as eye colour, shades of opinion, etc. or it may be numerical.

There are two forms of numerical data:

discrete values; these are exact and distinct, such as the number of people in a queue or shoe sizes.
continuous values; these can only be given in a range on a continuous scale, e.g. the length of a piece of wood.

Numerical data is usually collected by measuring or counting. For example, to collect data on heights of five-year-old boys, we would need to measure and record heights.

To collect non-numerical data, such as eye colour, we first need to decide on the categories to be used to describe eye colour, for example, brown, hazel, blue.

A **questionnaire** is useful when gathering information about opinions. The questions should be short, clear and not phrased so that they lead to a particular answer. For example, when seeking opinions about the desirability of banning cars from a town centre, the question 'Do you agree that, as traffic pollution causes asthma, cars should be banned from the town centre?' is very leading because it ties the basic question with an undesirable aspect of the use of cars. It is also unsatisfactory because there are only two answers: yes or no. Questions should ideally be worded so that a range of opinion is allowed for in predetermined categories,

e.g. agree strongly, agree, disagree, disagree strongly.

SAMPLES

A **sample** is only part of all the relevant information for a statistical investigation.

When a sample does not represent the make-up of the total pool of information from which it is drawn, it is called **unrepresentative**.

If, for example, a sample of four sweets is selected from a bag containing a mixture of coloured sweets, the four sweets could all be green; this sample is not representative of the mixture of colours in the bag. This could be by chance but it might be by design.

A sample that is unrepresentative for reasons other than chance is **biased**.

RANDOM SAMPLES

A random sample is one where the items for a sample are chosen by a method which ensures that every possible item that could be part of the sample has an equal chance of being chosen.

A **simple random sample** is chosen by first allocating a number to each item that could be chosen. The items forming the sample can then be selected by any method that produces numbers at random.

Random numbers can be selected from Random Number Tables or by using the random number function on a scientific calculator.

A **representative sample** has numbers of each distinct group in the same proportion as their numbers in the complete set.

A **stratified sample** produces a representative sample by first dividing the set into groups. Then random numbers are used to select members from each group in numbers proportional to their occurrence in the whole set.

EXERCISE 4.1

1 Harry has to conduct a survey for some Geography coursework. He wants to find out how often people visit his local shopping centre and how far they travel to get there.
Suggest two questions, including options for responses, that Harry could write in order to gather the information he needs.

2 The table shows the distribution of the weights of apples gathered from one orchard.

Weight, w grams	$0 \leqslant w < 20$	$20 \leqslant w < 40$	$40 \leqslant w < 60$	$60 \leqslant w < 80$	$80 \leqslant w < 100$
Frequency	20	406	518	312	45

Write a table showing the number of apples in each weight group needed to give a 10% sample.

3 Rajiv is conducting a survey among the employees in AbCo Ltd. The table shows a breakdown of the employees.

	Full time	Part time
Men	45	8
Women	21	56

Rajiv wants a sample of 20% of the employees.

a Explain why choosing equal numbers of men and women would not give a representative sample.

b How many men who work full time should be chosen?

4 Jane wanted to estimate the proportion of pupils in her school who borrowed books from the school library last week. She asked the first 50 pupils arriving at school one morning 'Did you borrow any books from the school library last week?' and 14 of these said that they had. Jane concluded that 28% of the pupils in her school had borrowed a book from the school library last week.

a Explain why this estimate is likely to be unreliable.

b Suggest another method of choosing the 50 pupils to question so that she would get a better estimate.

c Suggest a question, with four responses, that would give a clearer picture of the use of the school library.

5 The members of a local education authority are conducting a survey of the pupils in their schools to find out the opinions of the pupils on the provision of school lunches.

a Give two factors, with reasons, that should be taken into consideration in order to obtain a stratified sample of pupils.

b The sample is, in fact, chosen by selecting the first ten pupils on each school's register. Give three reasons why this might not produce a representative sample.

6 There are five job centres within Blackstone Unitary Authority. The table shows the number of people interviewed at each job centre.

Dainton	Horley	Jenford	Monkley	Purstock
83	163	282	132	197

The Authority needs a stratified sample of 50 interviewees.
How many should be selected from each job centre?

7 The table shows the number of pupils in Years 7 to 11 in a school.

	Year 7	Year 8	Year 9	Year 10	Year 11
Boys	88	69	75	85	61
Girls	63	70	62	74	80

The governors need a 20% stratified sample of pupils for a survey.

a How many pupils will take part in the survey?

b How many boys will be in the sample?

c How many pupils from Year 9 will be in the sample?

8 David has been asked to do a survey on how pupils at his school travel there and what problems they have with transport.

a Give one reason why taking all first year pupils as his sample would not give reliable results.

b Explain how you would choose a sample to be representative of the whole school.

ORGANISING AND ILLUSTRATING DATA

When information is presented in an unordered form we have **raw data**.

A **frequency table** shows the number of times that each distinct value (or category) occurs.

This frequency table shows the scores when a dice is rolled 20 times.

Score	Frequency
1	4
2	1
3	4
4	3
5	3
6	5

A **grouped frequency table** shows the frequencies of groups of values. This table, for example, shows the distribution of the heights of 55 tomato plants in four groups each of width 10 centimetres.

Height, h cm	Frequency
$20 \leqslant h < 30$	5
$30 \leqslant h < 40$	15
$40 \leqslant h < 50$	25
$50 \leqslant h < 60$	10

Bar charts

A frequency table can be illustrated by drawing a bar for each group (or each value for ungrouped data) whose height represents the number of items in the group. This bar chart illustrates the grouped frequencies of the heights of tomato plants in the table above.

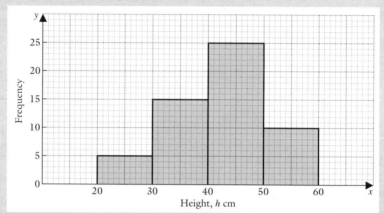

Notice that there are no gaps between the bars; heights are continuous measurements so there is no value on the horizontal scale that could not be a height.

When values are discrete, or in categories, there may be gaps between the bars.

A **frequency polygon** is drawn by plotting the frequency of each group against the mid-point of that group and joining the points with straight lines.

Pie charts

Pie charts are used to show what fraction of the whole list each group or category represents.

The size of a 'slice' is given by the angle at the centre of the circle. This angle is found by first expressing the number of values in the group as a fraction of all the values, and then finding this fraction of 360°.

This pie chart shows what fraction the number of plants in each group is of the total number of plants given.

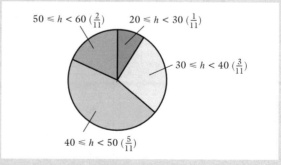

Histograms

A histogram is a bar chart drawn so that the area of each bar is proportional to the frequency of items in the group represented by that bar.

Histograms are used to illustrate grouped data and are particularly useful when the widths of the groups are not equal.

To construct a histogram we make the width of a bar the same as that of the group it represents,

so the height of a bar is equal to $\dfrac{\text{frequency}}{\text{width of group}}$; this fraction is called the **frequency density**.

Frequency density represents the frequency per unit of area and the units involved must be stated on a histogram.

This histogram represents the number of hours worked per month by part-time employees in a company running a chain of restaurants.

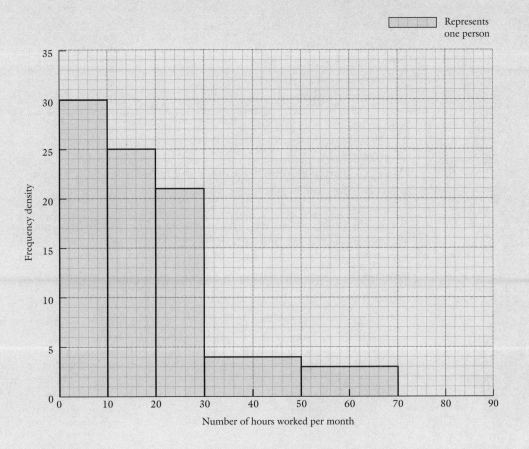

Cumulative frequency
Cumulative frequency is the sum of the frequencies of all values up to and including a particular value or group of values.

This table shows the cumulative frequencies of the distribution of the heights of tomato plants given on page 307.

Height, h cm	Cumulative frequency
$20 \leqslant h < 30$	5
$20 \leqslant h < 40$	20
$20 \leqslant h < 50$	45
$20 \leqslant h < 60$	55

A **cumulative frequency polygon** is drawn by plotting the cumulative frequencies against the upper ends of the groups and joining the points with straight lines. A **cumulative frequency curve** is obtained by drawing a smooth curve through the points.

This cumulative frequency curve is drawn from the table above.

Notice that the vertical axis must start at zero but the horizontal axis can start at any convenient point.

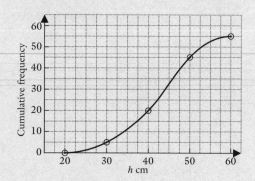

Scatter graphs
We get a **scatter graph** when we plot values of one quantity against corresponding values of another quantity.

When the points are scattered about a straight line, we can draw that line by eye; it is called the **line of best fit**.

We use the words **correlation** to describe the amount of scatter about this line.

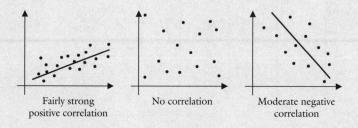

Fairly strong   No correlation   Moderate negative
positive correlation   correlation

A **line graph** gives the value of a variable at specific times: it says nothing about the value of that variable between those times. The graphs are sometimes called **time series**.

This line graph shows Tim's temperature while he was in hospital. It was taken at 4-hourly intervals. The graph shows his temperature at 4 a.m., 8 a.m., etc., but not at 5 a.m. or 6.45 a.m.

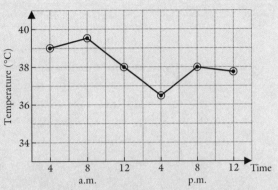

A **moving average** is a way of smoothing values that fluctuate a great deal from one time to the next; seasonal sales for example. The moving average also gives a better indication of trend in these circumstances.

A three-point moving average is calculated by finding the mean of three values at a time, starting with the first three values then dropping the first and adding the fourth, and so on.

A four-point moving average is found in a similar way using four values at a time.

This table shows the three-point moving average for the temperatures in the chart above.

Time	Temperature	Three-point moving average
4 a.m.	39	
8 a.m.	39.5	$(39 + 39.5 + 38) \div 3 = 38.8$
noon	38	$(39.5 + 38 + 36.5) \div 3 = 38$
4 p.m.	36.5	$(38 + 36.5 + 38) \div 3 = 37.5$
8 p.m.	38	$(36.5 + 38 + 37.8) \div 3 = 37.4$
midnight	37.8	

EXERCISE 4.2

1 Candy rolled a dice 50 times and these are the scores that she obtained.

2	3	1	2	4	1	6	4	6	1
5	6	3	5	3	4	2	1	6	3
2	2	3	5	6	1	6	6	6	3
1	1	5	1	2	6	3	1	5	5
4	4	2	3	1	2	4	6	5	6

a Organise these scores into a frequency table.

b Illustrate the distribution of scores with a bar chart.

2 Andy made a survey into the distances travelled by shoppers to a new shopping mall. He collected his information between 1 p.m. and 2 p.m. on a Monday. The table shows his results.

Distance, m miles	$0 < m \leqslant 5$	$5 < m \leqslant 10$	$10 < m \leqslant 15$	$15 < m \leqslant 20$
Frequency	35	18	5	2

a Illustrate this information with
   i a bar chart   **ii** a frequency polygon.
b Draw a pie chart to illustrate this information.
c Which of the three illustrations of this data is the most useful and why?

Petra also made a survey into distances travelled by shoppers to the new shopping mall. She collected her information one Saturday morning. The table shows her results.

Distance, m miles	$0 < m \leqslant 1$	$1 < m \leqslant 2$	$2 < m \leqslant 5$	$5 < m \leqslant 10$	$10 < m \leqslant 20$
Frequency	17	24	13	10	5

d Illustrate this information with a histogram.

Andy and Petra wish to compare their distributions.

e i Suggest what Petra can do with her data to make this easier.
   ii What other method could Petra use to illustrate her data so that a visual comparison can be made?

3 This histogram illustrates the distribution of the times taken by some 5-year-old children to complete a task.

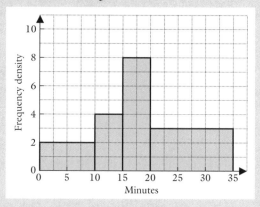

Given that 40 children took from 15 to 20 minutes to complete the task,

a find   **i** how many children took less than 10 minutes
           ii how many children were timed.

b estimate the number of children who took less than $12\frac{1}{2}$ minutes.

4 The heights of 50 trees are shown in the table.

Height (metres)	0–2	2–4	4–6	6–8	8–10
Frequency	6	8	10	15	11

   a Make a cumulative frequency table for these heights.

   b Illustrate the data with a cumulative frequency curve.

   c Use your curve to estimate the number of trees more than 5 metres tall.

5 The table shows the height and age of each of 15 boys.

Age (years)	6	9	7	10	5	6	8	6	9	11	7	8	9	6	10
Height (metres)	1.25	1.38	1.39	1.55	1.20	1.30	1.42	1.33	1.52	1.60	1.48	1.52	1.57	1.41	1.63

   a Draw a scatter diagram to illustrate this information. Start the horizontal axis at 4 and use a scale of 1 cm for 1 year. Start the vertical axis at 1 and use a scale of 1 cm for 0.1 metres.

   b Draw a line of best fit by inspection and use it to estimate the age of a boy who is 1.50 metres tall.

   c Explain whether your diagram can be used to estimate the height of a 20-year-old man.

6 This cumulative frequency graph illustrates the results of a survey in which the speeds of vehicles travelling along the M1 were measured as they passed Newport Pagnell Services on a Sunday evening.

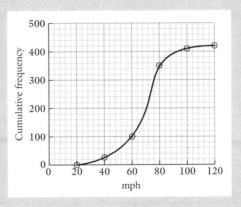

   a How many cars had their speeds recorded?

   b How many of these cars were travelling at less than 60 mph?

   c The speed limit is 70 mph. Estimate what percentage of these cars exceeded the speed limit.

7 The employees of a small office were asked how far they lived from the office and how long the journey into work took them. The results are in the table.

Distance (km)	1	2	3	4	5	6	10
Time (minutes)	10	12	35	30	40	30	45

a Draw a scatter diagram to show these results.

b Draw the line of best fit.

c Estimate how long it will take a new employee who lives 7 km from the office to travel to work.

8 a This histogram shows the distribution of the weights of the luggage of 100 passengers on a morning flight from Heathrow to Tenerife.

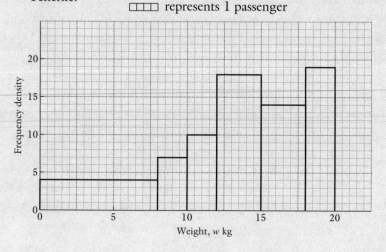

⬜⬜⬜ represents 1 passenger

Use the histogram to find the number of passengers checking in luggage with a weight

i between 15 kg and 18 kg

ii less than 12 kg.

b The table shows the distribution of the weights of the luggage of 100 passengers on a flight from Heathrow to Paris on the same morning.

Weight, w kg	$0 \leqslant w < 8$	$8 \leqslant w < 10$	$10 \leqslant w < 12$	$12 \leqslant w < 15$	$15 \leqslant w < 18$	$18 \leqslant w < 20$
Frequency	60	10	9	9	6	6

i Draw a histogram to show this information.

ii Study the two histograms and make three comparisons between the two distributions.

9 This table shows the monthly sales of a magazine over a six-month period.

Month	January	February	March	April	May	June
Sales	8460	6900	7500	8350	8540	8400

a Calculate the three-point moving averages for these sales.

b Illustrate the figures in the table and the moving average with line graphs.

c Comment on the trend in the sales figures.

SUMMARISING DATA

Measures of central tendency

For a list of values,

the **mean** is the sum of all the values divided by the number of values,

the **median** is the middle value when they have been arranged in order of size, (when the middle of the list is half-way between two values, the median is the average of these two values),

the **mode** is the value that occurs most frequently (there can be more than one mode).

For a grouped frequency distribution,

the **modal group** is the group with the largest number of items in it.

The **median** of a grouped distribution of n values is the $\frac{n}{2}$th value in order of size and is denoted by Q_2. The median can be estimated from a cumulative frequency curve.

The **mean** value of a frequency distribution is given by $\frac{\sum fx}{\sum f}$

where x is the value of an item and f is its frequency.

In the case of grouped data, x is the mid-class value and the mean obtained is an estimate.

Measures of spread

For a list of values,
the **range** is the difference between the largest value and the smallest value.

For a grouped distribution, the **range** is estimated as

the highest end of the last group – the lowest end of the first group.

The **lower quartile** is the value that is $\frac{1}{4}$ of the way through a set of values arranged in order of size and is denoted by Q_1.

The **upper quartile** is the value that is $\frac{3}{4}$ of the way through a set of values arranged in order of size and is denoted by Q_3.

For a grouped distribution, Q_1 is the $\frac{n}{4}$th value, and Q_3 is the $\frac{3n}{4}$th value. Q_1 and Q_3 can be estimated from a cumulative frequency curve.

The **interquartile range** is the difference between the upper and lower quartiles, that is $Q_3 - Q_1$. This gives a measure of the spread of the middle half of a distribution of values. It has an advantage over the range as a measure of spread because it ignores any unusually extreme values.

The diagram shows the median, the upper and lower quartiles and the interquartile range of the distribution of heights of tomato plants described on p. 307.

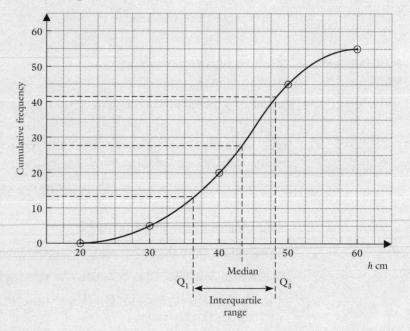

Box plots

A box-and-whisker plot (to give its full name) consists of a box drawn between the upper quartile and lower quartile with a line drawn across the box at the median value. Then lines (the whiskers) are drawn from the edges of the box to the lower and upper ends of the range.

This box plot illustrates a summary of the distribution of the heights of tomato plants described on page 307.

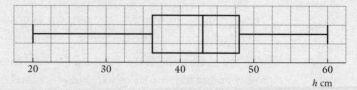

The box represents the middle 50% of the distribution and the whiskers represent the bottom 25% and the top 25% of the distribution.

A box plot does not give any information about the numbers of values in the distribution.

EXERCISE 4.3

1 The frequency table shows the distribution of scores when a dice is rolled 20 times.

Score	Frequency
1	4
2	1
3	4
4	3
5	3
6	5

Find the mean and median score.

2 These are the times, in minutes, taken by each of 10 pupils to complete a mathematics test.

$$10, \ 12, \ 13, \ 15, \ 15, \ 15, \ 16, \ 16, \ 17, \ 18$$

a Find **i** the median **ii** the mode.

b Calculate the mean time.

3 **a** This cumulative frequency graph (first seen in Exercise 4.2) illustrates the results of a survey where the speeds of vehicles travelling along the M1 were measured as they passed Newport Pagnell Services on a Sunday evening.

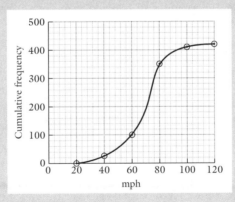

Use this graph to estimate
 i the median speed and the interquartile range
 ii the percentage of motorists that exceeded the speed limit of 70 mph.

b The following Sunday, another survey was made of speeds at the same place. This time there were police cars patrolling on that section of the motorway. The median speed was found to be 64 mph, the interquartile range was found to be 6 mph and the range was from 40 mph to 80 mph. Draw box plots to illustrate the two distributions and compare the two distributions of speeds.

4 The table illustrates the value of houses sold in the last 12 months by an independent estate agent.

Value of houses sold (£1000)	10–49	50–99	100–149	150–199	200–400
Frequency	2	8	6	4	2

a What is the modal value of houses sold?

b Calculate an estimate of the mean value of the houses sold.

The independent agent wants to sell the business.

c Which of the values found in parts **a** and **b** will be of more use to a prospective buyer and why?

5 Andy made a survey into distances travelled by shoppers to a new shopping mall. He collected his information between 1 p.m. and 2 p.m. on a Monday. The table shows his results.

Distance, m miles	$0 < m \leqslant 5$	$5 < m \leqslant 10$	$10 < m \leqslant 15$	$15 < m \leqslant 20$
Frequency	35	18	5	2

a Calculate an estimate for the mean of these distances.

Petra also made a survey into distances travelled by shoppers to the new shopping mall. She collected her information one Saturday morning. The table shows her results.

Distance, m miles	$0 < m \leqslant 1$	$1 < m \leqslant 2$	$2 < m \leqslant 5$	$5 < m \leqslant 10$	$10 < m \leqslant 20$
Frequency	17	24	13	10	5

b Find an estimate for the mean of Petra's distances.

c Compare these two distributions.

6 Zaid weighed each of 50 potatoes on an electronic weighing machine. He recorded each weight and then calculated
the mean weight as 52 grams,
the median weight as 49 grams,
and the interquartile range as 17 grams.
After doing all his calculations, Zaid discovered that the scales had not been properly zeroed and had given each weight as 5 grams greater than the true value.
Give the correct values for the mean, median and interquartile range for the distribution of weights.

7 Some pupils from each of several counties took the same general knowledge test.
These box plots summarise the test scores by pupils in three of those counties.

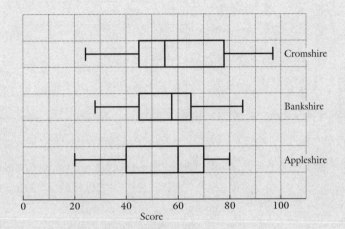

a In which county did 50% of the pupils score 60 or more?

b in which county did 25% of pupils score 40 or less?

c Compare the scores in Bankshire with the scores in Cromshire.

8 Air Wight uses 100-seater aircraft for all its flights.
The table shows the number of empty seats on its flights during May.

Number of empty seats	Frequency, f	Mid-interval value, x	fx
0–4	14	2	
5–9	28		
10–14	21		
15–19	13		
20–24	5		
25–29	6		
30–35	3		

a How many flights did Air Wight operate in May?

b In which interval does the median lie?

c Copy and complete the table. Hence find an approximate value for the mean number of empty seats.

d Give an approximate value for the mean number of passengers on these flights.

9 The times taken by 34 pupils to come to school are recorded in the following cumulative frequency table.

Time, t minutes	Cumulative frequency
$t \leqslant 5$	4
$t \leqslant 10$	10
$t \leqslant 15$	24
$t \leqslant 20$	30
$t \leqslant 25$	33
$t \leqslant 30$	34

a Represent this information on a cumulative frequency curve.
Use $2\,\text{cm} \equiv 5$ units on both axes.

b Use your graph to estimate
i the median journey time
ii the interquartile range
iii the number of pupils who took less than 14 minutes to get to school
iv the percentage of pupils who took more than 18 minutes to get to school
v the probability that one of these pupils, selected at random, took longer than 22 minutes to get to school.

10 The table shows the results of a traffic survey into the speeds of vehicles on the road past the entrance to a school.

Speed, v mph	Number of vehicles
$0 \leqslant v < 10$	3
$10 \leqslant v < 20$	5
$20 \leqslant v < 25$	15
$25 \leqslant v < 30$	24
$30 \leqslant v < 35$	13
$35 \leqslant v < 40$	8
$40 \leqslant v < 60$	2

a Calculate an estimate of the mean speed.

b Draw a histogram to illustrate this information.

c Draw a line on your histogram to show the **median** speed.

PROBABILITY

The probability that an event A happens is $P(A)$ where

$$P(A) = \frac{\text{the number of ways in which } A \text{ can occur}}{\text{the total number of equally likely outcomes}}$$

Since the numerator can never be greater than the denominator,

$$0 \leqslant P(A) \leqslant 1.$$

The probability that an event A does not happen is given by subtracting the probability that it does happen from 1,

i.e. $\qquad P(A \text{ does not happen}) = 1 - P(A \text{ happens}).$

When we perform experiments to find out how often an event occurs, the **relative frequency** of the event is given by

$$\frac{\text{the number of times the event occurs}}{\text{the number of times the experiment is performed}}$$

Relative frequency gives an approximate value for probability.

If p is the probability that an event happens on one occasion, then we expect it to happen np times on n occasions, for example, if we toss an unbiased coin 50 times, we expect $\frac{1}{2} \times 50$, i.e. 25 heads.

To find all the possible combinations of events when two events occur we can make a **possibility table**. This table lists all the possible outcomes when a dice is thrown and a coin is tossed.

		Dice					
		1	2	3	4	5	6
Coin	H	H,1	H,2	H,3	H,4	H,5	H,6
	T	T,1	T,2	T,3	T,4	T,5	T,6

Adding probabilities

Events such that only one of them can happen on any one occasion are called **mutually exclusive**.

We add probabilities when we want the probability that one or other of two events will happen, provided that the events are mutually exclusive. For example, when one dice is rolled,

$P(\text{scoring 5 or 6}) = P(\text{scoring 5}) + P(\text{scoring 6})$

Multiplying probabilities

Events such that each has no influence on whether the others occur are called **independent events**.

We multiply probabilities when we want the probability that two or more events all happen, provided that each event is independent of the others.

For example, when two dice A and B are rolled,

$P(\text{scoring 6 on both}) = P(\text{scoring 6 on A}) \times P(\text{scoring 6 on B})$

EXERCISE 4.4

1 A bag contains 7 red marbles and 9 green marbles. What is the probability that a marble taken from the bag at random is red?

2 The heights of 50 trees are shown in the table.

Height (metres)	0–2	2–4	4–6	6–8	8–10
Frequency	6	8	10	15	11

What is the probability that a random selection of one of these trees will produce a tree that is 4 to 6 metres tall?

3 A trial for a new drug that it is hoped will cure headaches, produced these results when tested on some patients.

No effect	Some relief from pain	Full relief from pain
125	298	77

Find

a the probability that one of the patients, selected at random, found that the drug had no effect

b the probability that one of the patients, selected at random, found that the drug gave some or full relief.

4 Joe rolls an ordinary fair dice three times and obtains a six each time.

a What is the probability that Joe does not obtain a six on his fourth roll?

b Joe rolls the dice 300 times. Approximately how many sixes is he likely to obtain?

5 Two three-sided spinners are such that each of them has its faces numbered 1, 2, 3. They are spun and the scores are added. Find the probability of scoring a total of 5.

6 A car approaching the roundabout from London Road may turn left, carry straight on, turn right or go all the way round the roundabout and return along London Road in the opposite direction. Long-term traffic surveys indicate that the probability that a car will turn left is 0.2 and the probability that it will turn right is 0.3.

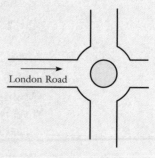

London Road

a What is the probability that the next car will turn left or right?

b What is the probability that the next two cars will both turn left?

c If 5% of the cars return along London Road, what is the probability that the next car will go straight on?

7 The table shows the results of a Mathematics test given to students applying for a place on a Business Studies course.

Mark	0	1	2	3	4	5	6	7	8	9	10
Frequency	0	0	1	2	4	10	7	8	4	4	0

a What is the probability that one of these students, chosen at random, scored more than 5 on this test?

b Two of these students are selected at random. The first student scored more than 5 on this test. What is the probability that the second student also scored more than 5?

8 A geranium plant grown from a packet of seeds will have flowers that are red, white, pink or variegated. No other colours are possible.
The table lists some of the probabilities.

Colour	Red	White	Pink	Variegated
Probability	$\frac{1}{4}$	$\frac{1}{4}$	$\frac{1}{3}$	

Fifty plants are raised from these seeds.

a One plant is chosen at random.
  i What is the probability that the flowers will be variegated in colour?
  ii What is the probability that the flowers will be either white or pink?

b Two plants are chosen at random. What is the probability that they both have red flowers?

9 The table shows how many boys and girls in Class 11N come to school by bus.

	Come by bus	Do not come by bus	Total
Boys	9	4	
Girls	11	6	
Total			

a Copy the table and complete it. How many pupils are there in Class 11N?

b A boy is chosen at random. What is the probability that he will come to school by bus?

c A pupil is chosen at random from the pupils who do not come to school by bus. What is the probability that the pupil is a girl?

d Two pupils are chosen at random. What is the probability that they are both girls?

e Two pupils are chosen at random. Find the probability that they are both boys.

10 The probability of winning a prize with one ticket on a tombola is 0.2. Sally bought *n* tickets.
The probability that every one of Sally's tickets wins a prize is less than one chance in a thousand.
Which is the least number of tickets that Sally bought?

TREE DIAGRAMS Tree diagrams can be used to illustrate the outcomes when two or more events occur.
This tree, for example, shows the possible outcomes when two coins are tossed.

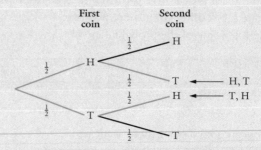

We **multiply** the probabilities when we follow a path along the branches and **add** the results of following several paths,

e.g. the probability of getting a head and a tail is $(\frac{1}{2} \times \frac{1}{2}) + (\frac{1}{2} \times \frac{1}{2})$.

Dependent events (Conditional Probability)
When the number of ways in which an event can happen depends on what has already happened, the probability that the second event occurs is conditional on what happened first.

For example, if two discs are removed from a bag containing 2 red and 2 yellow discs then the probability that the second disc to be removed is red depends on the colour of the first disc removed. If the first disc is red, $P(\text{2nd red}) = \frac{1}{3}$, but if the first disc is yellow, $P(\text{2nd red}) = \frac{2}{3}$. This tree diagram shows all the possibilities.

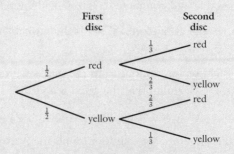

EXERCISE 4.5

1 Jinx Airlines run one daily flight from London to Glasgow. The probability that the flight is delayed is 0.15.
What is the probability that one or both of the next two flights are delayed?

2 Two boxes each contain a mixture of green and red balloons.

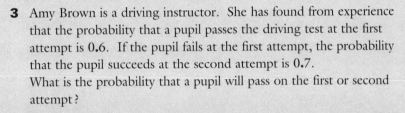

In the small box, there are 3 green and 2 red balloons.
In the big box, there are 10 green and 12 red balloons. One balloon is to be taken at random from each box. Find the probability that both balloons will be red.

3 Amy Brown is a driving instructor. She has found from experience that the probability that a pupil passes the driving test at the first attempt is 0.6. If the pupil fails at the first attempt, the probability that the pupil succeeds at the second attempt is 0.7.
What is the probability that a pupil will pass on the first or second attempt?

4 A bag contains 5 red and 7 blue discs. Two randomly selected discs are removed from the bag. What is the probability that both discs removed are the same colour if

a the first disc is placed back in the bag before the second disc is removed

b the first disc is not placed back in the bag before the second disc is removed?

5 Winston leaves home at 8 a.m. to catch a bus to get to school.
The probability that he has to wait less than 5 minutes for the bus is 0.7.
The probability that he has to wait longer than 15 minutes is 0.03.

a What is the probability that Winston has to wait between 5 and 10 minutes for the bus?

The probability that he is late for school depends on how long he has to wait for a bus:

for a wait of less than 5 minutes, the probability is 0.02,
for a wait from 5 to 15 minutes, the probability is 0.04,
for a wait of more than 15 minutes, the probability is 0.5.

b What is the probability that Winston is late for school?

6 A bag of raffle tickets contains 40 white tickets, 60 pink tickets and 50 yellow tickets. Ben removes one ticket at a time from the bag, picked at random from those left in the bag.

a Find the probability that the first three tickets are white.

b Find the probability that the third ticket that Ben removes is yellow.

7 'Bizzie Lizzie' plants grown from seed will have red, pink or white flowers. The probability that the flowers will be red is 0.3 and the probability that the flowers will be pink is 0.4.
Judy buys two plants, grown from these seeds, that have not flowered yet.

a What is the probability that they will both have red flowers?

b What is the probability that one plant will have red flowers and the other will have pink flowers?

c Find the probability that the two plants have flowers that are of different colours.

8 Four CDs have been won by a group of six people, four of whom are men and two are women. They choose who should have a CD by drawing straws.
What is the probability that the four chosen are all men?

MIXED EXERCISE 4.6

1 A bag contains 7 toffees and 5 mints.

a What is the probability that a sweet taken from the bag at random will be a toffee?

b Another bag contains 4 fruit drops and 6 mints. James takes one sweet from each bag without looking. Complete this tree diagram to show the possible outcomes and their probabilities.

c What is the probability that James takes

i two mints   **ii** exactly one mint?
(WJEC)

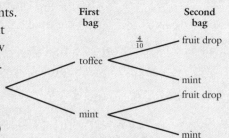

2 Fruit and vegetables which have been grown without the use of chemicals or fertilisers and pest control are called organic. Some of the fruit and vegetables sold in a supermarket are organic, but they are usually more expensive. The management of the supermarket are conducting a survey to find out the opinions of their customers on organically grown fruit and vegetables.

a Suggest **two** factors which should be considered in order to obtain a stratified sample of the customer population for this survey, explaining clearly why you consider these factors to be important.

b The survey is actually conducted by questioning every twentieth customer entering the supermarket on a Monday morning. Give **two** reasons why this method of sampling might produce misleading results.
(WJEC)

3 In a traffic survey, a policeman and a policewoman stand on opposite sides of a road and measure the speeds of approaching vehicles. The policeman measures the speeds of vehicles travelling in an easterly direction just as they enter a 30 mph speed limit. His results are summarised in the group frequency distribution below.

Speed, x mph	Number of cars, f	Frequency density
$10 \leqslant x < 20$	8	0.8
$20 \leqslant x < 25$	5	
$25 \leqslant x < 30$	9	
$30 \leqslant x < 40$	16	
$40 \leqslant x < 50$	9	
$50 \leqslant x < 70$	3	

a Complete the frequency density column in the table.

b Draw a histogram of the data in the table.

c The policewoman standing on the opposite side of the road records the speeds of the vehicles travelling in a westerly direction, just as they are leaving the 30 mph speed limit. Her results are shown in the histogram below.

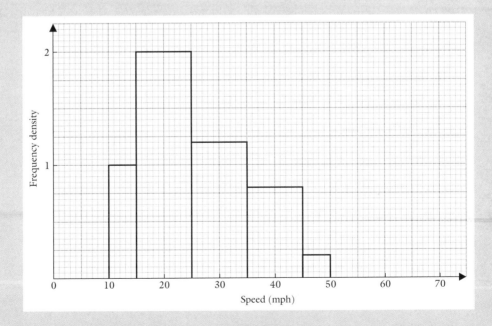

Estimate the number of vehicles whose speeds are shown in the policewoman's histogram that are exceeding the speed limit of 30 mph. (WJEC)

4 At a jam bottling factory, bottles of jam pass through two quality control tests. In the first test, bottles are inspected for flaws such as bubbles and cracks. In the second the labels are inspected for damage. Records show that the probability of passing the first test is 0.95 and, independently, the probability of passing the second test is 0.90. Calculate the probability that a bottle of jam selected at random passes exactly one of the two tests. (WJEC)

5 The table below shows the distribution of the weights of 150 apples.

Weight, w grams	Number of apples	Mid-interval value	
$50 < w \leqslant 60$	23		
$60 < w \leqslant 70$	42		
$70 < w \leqslant 80$	50		
$80 < w \leqslant 90$	20		
$90 < w \leqslant 100$	15		

Calculate an estimate of the mean weight of an apple. (OCR)

6 A doctor's patients are divided by age into groups as shown in the table below.

Age, x, in years	$0 \leqslant x < 5$	$5 \leqslant x < 15$	$15 \leqslant x < 25$	$25 \leqslant x < 45$	$45 \leqslant x < 75$
Number of patients	14	41	59	70	16

a On a copy of the grid below complete the histogram to represent this distribution.

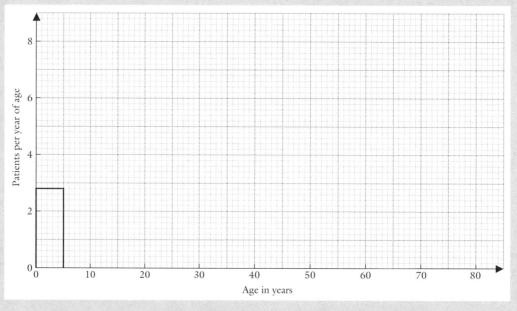

b The doctor wishes to choose a stratified random sample of 40 patients.
Explain, with any appropriate calculations, how this can be done. (OCR)

7 The cumulative frequency graph gives information about the heights of 120 students.

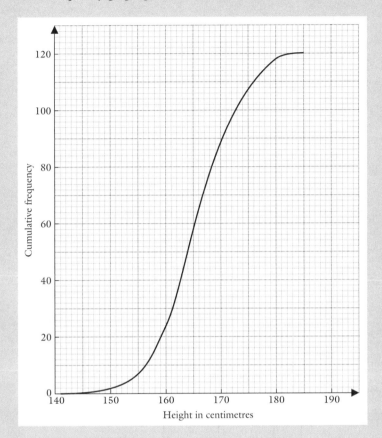

In each of the following parts draw lines on a copy of the graph to show how you obtained your answer.

a Estimate the median height.

b Estimate the interquartile range.

c Estimate the number of students whose height was over 172 cm.   (OCR)

8 In the game of tennis a player has two serves.
If the first serve is successful the game continues.
If the first serve is not successful the player serves again. If this second service is successful the game continues.
If both serves are unsuccessful the player has served a 'double fault' and loses the point.
Gabriella plays tennis. She is successful with 60% of her first serves and 95% of her second serves.

a Calculate the probability that Gabriella serves a double fault.

If Gabriella is successful with her first serve she has a probability of 0.75 of winning the point.
If she is successful with her second serve she has a probability of 0.5 of winning the point.

b Calculate the probability that Gabriella wins the point.   (OCR)

9 The table shows the ages, in completed years, of 100 members of a swimming club.

Age in completed years	Number of members
0 to 9	16
10 to 14	21
15 to 19	15
20 to 29	28
30 to 39	x
40 to 69	y

The histogram below illustrates this information.

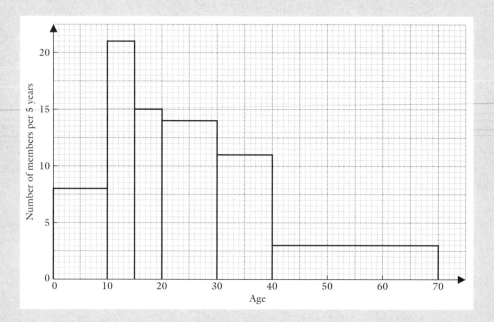

a Explain why the bars have their right-hand ends at 10, 15, 20, 30, 40 and 70 years.

b Use the histogram to find the values of x and y in the table. (OCR)

10 A group of schoolchildren took a Mathematics test and a Physics test.
The results for 12 children were plotted on the scatter diagram opposite.

a Add a line of best fit, by inspection, to a copy of the scatter diagram.

b One pupil scored 7 marks for Mathematics but missed the Physics test.
Use the line of best fit to estimate the mark she might have scored for Physics.

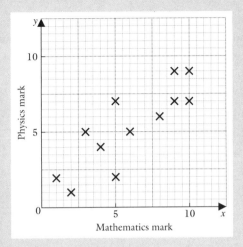

(OCR)

11

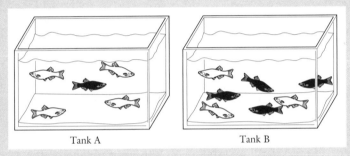

Tank A   Tank B

There are two fish tanks in a pet shop.

In tank A there are four white fish and one black fish.

In tank B there are three white fish and four black fish.

One fish is to be taken out of each tank at random.

Using a tree diagram or otherwise, calculate the probability that

a the two fish will both be white,

b the two fish will be of different colours.   (Edexcel)

12 a Lorraine is writing a questionnaire for a survey about her local Superstore.
She thinks that local people visit the store more often than people from further away.
She also thinks that local people spend less money per visit.
Write **two** questions which would help her to test these ideas.
Each question should include at least three options for a response. People are asked to choose one of these options.

    b For another survey, Lorraine investigates the spending habits of students at her school. The number of students in each year group is shown.

Year group	9	10	11	12	13
Number of students	208	193	197	122	80

Explain, with calculations, how Lorraine obtains a stratified random sample of 100 students for her survey.   (AQA)

13 Sam was making a survey of pupils in his school.
He wanted to find out their opinions on noise pollution by motor bikes.
The size of each year group is shown below.

Year group	Boys	Girls	Total
8	85	65	150
9	72	75	147
10	74	78	152
11	77	72	149
6th Form	93	107	200
			798

Sam took a sample of 80 pupils.

a Explain whether or not he should have sampled equal numbers of boys and girls in year 8.

b Calculate the number of pupils he should have sampled in year 8. (Edexcel)

14 The speeds, in miles per hour (mph), of 200 cars travelling on the A320 road were measured.
The results are shown in the table.

Speed (mph)	Cumulative frequency
not exceeding 20	1
not exceeding 25	5
not exceeding 30	14
not exceeding 35	28
not exceeding 40	66
not exceeding 45	113
not exceeding 50	164
not exceeding 55	196
not exceeding 60	200
TOTAL	200

a On a grid draw a cumulative frequency graph to show these figures.

b Use your graph to find an estimate for

  i the median speed (in mph)
  ii the interquartile range (in mph)
  iii the percentage of cars travelling at less than 48 miles per hour. (Edexcel)

15 In a survey on October 1st, pupils present at Bank School were asked how long they had taken from home to school that morning. Each pupil present ticked one and only one of the following responses:

0 up to but not including 5 minutes	
5 minutes up to but not including 15 minutes	
15 minutes up to but not including 25 minutes	
25 minutes up to but not including 40 minutes	
40 minutes up to but not including 70 minutes	
70 minutes and over	

Exactly 96 pupils ticked the '5 minutes up to but not including 15 minutes' response. No pupil ticked the '70 minutes and over' response.

Exactly 180 pupils ticked the '40 minutes up to but not including 70 minutes' response.

a On a copy of this diagram complete the histogram to show the results of the survey.

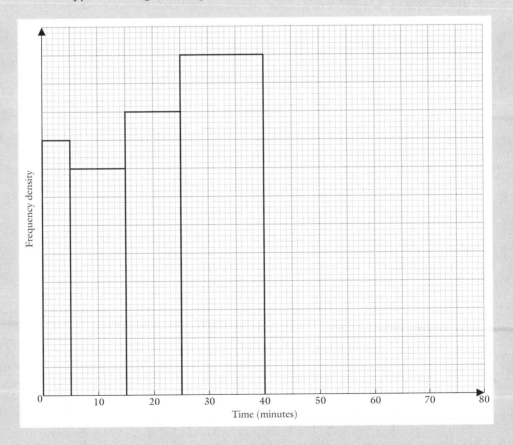

b Calculate the number of pupils who were absent on October 1st given that the total number of pupils at the school is 726. (Edexcel)

16 A bag contains 4 RED and 6 BLUE balls. One ball is chosen and its colour noted.
It is **not** put back into the bag.
A second ball is chosen and its colour noted.

Draw a tree diagram to represent this situation.

  i Find the probability of obtaining **two** RED balls.

 ii Find the probability of obtaining one ball of each colour.   (AQA)

17 The age of each person in a coach party is recorded.
The table shows the number of people in each age category.

Age group (years)	10–	20–	30–	45–	50–	70–100
Frequency	2	0	6	4	22	12

Draw a histogram to represent the data.   (AQA)

18 George passes three sets of traffic lights on his way to work.
The lights work independently of each other.
The probability that he has to stop at any set of traffic lights is 0.35.
What is the probability that George stops at two or three sets of traffic lights?   (AQA)

19 The table below illustrates the age distribution in a village of 360 people.

Age	0–	10–	20–	30–	40–	50–	60–	70–	80–100
Frequency	44	51	59	68	50	35	31	18	4

 a By completing the cumulative frequency table, draw a cumulative frequency diagram to illustrate these data.

Age (less than)	0	10	20	30	40	50	60	70	80	100
Cumulative frequency										

 b Use your graph to estimate

  i the median

 ii the interquartile range.

 c The age distribution of another village has a median age of 45 years and interquartile range equal to 20 years.
Compare the distributions of the age of people in these two villages.   (AQA)

20 A bus company attempted to estimate the number of people who travel on local buses in a certain town. They telephoned 100 people in the town one evening and asked 'Have you travelled by bus in the last week?'
Nineteen people said 'Yes'. The bus company concluded that 19% of the town's population travel on local buses.
Give 3 criticisms of this method of estimation.   (OCR)

21 Of 50 Driving School pupils, 30 pass their car driving test at the first attempt. From past performance it is known that, if a pupil fails at the first attempt, the probability of passing at the second attempt is 0.7.

Calculate the probability that a pupil, chosen at random from the 50 pupils, will pass the car driving test at either the first or second attempt. (OCR)

22 The first one hundred people to enter a cricket ground were asked their ages, which varied between 5 years and 60 years. It was later calculated that the Lower and Upper Quartiles of these ages were 25 years and 38 years respectively.

  a Draw a possible cumulative frequency curve.

  b Use your curve to find the median value. Indicate clearly on your diagram the method you have used.

  c Use your curve to estimate how many of the one hundred people were between 20 and 30 years of age.

  d What is the probability that a person, chosen at random from the one hundred people, will be at least 18 years of age? (OCR)

23 For a school project you have been asked to do a presentation of the eating habits of the pupils in your school.

You decide to interview a sample of pupils.

How will you choose the pupils you wish to interview if you want your results to be reliable? Give **three** reasons for your decisions. (AQA)

24 A spinner, with its edges numbered one to four, is biased.

For one spin, the probability of scoring 1 is 0.2,

          the probability of scoring 3 is 0.3

    and the probability of scoring 4 is 0.15.

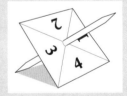

  a Calculate the probability of scoring 2 with one spin.

  b The spinner is used in a board game called 'Steeplechase'. In the game, a player's counter is moved forwards at each turn by the score shown on the spinner.

If the player's counter lands on one of the two squares numbered 27 and 28 (labelled 'WATER JUMP'), the player is out of the game.

| 23 | 24 | 25 | 26 | WATER JUMP 27 | 28 | 29 | 30 | 31 |

    i Ann's counter is on square 26.

      Find the probability that she will **not** be out of the game after one more turn.

    ii Eileen's counter is on square 24.

      Find the probability that her counter will be on square 29 after one more turn.

    iii Peter's counter is on square 25.

      Find the probability that, after two more turns, his counter will be on square 29.

(OCR)

25 Mrs Wilson wants to sell her herd of dairy cows.
A buyer will need to know the herd's average daily yield of milk.
The daily milk yield, p litres, is monitored over 5 weeks.
The table below shows the results of this survey.

Milk yield, p litres	Frequency, f
$140 \leqslant p < 145$	3
$145 \leqslant p < 150$	5
$150 \leqslant p < 155$	9
$155 \leqslant p < 160$	6
$160 \leqslant p < 165$	8
$165 \leqslant p < 170$	4
Total	35

a Mrs Wilson finds the modal class for the daily yield. What is this value?

b Calculate the estimated mean daily milk yield.

c Which is the more suitable average for Mrs Wilson to use?
Give a reason for your answer. (AQA)

26 On each day that I go to work, the probability that I leave home before 0800 is 0.2.
The probability that I leave home after 0810 is 0.05.

a What is the probability that I leave home between 0800 and 0810 inclusive?

b The probability that I am late for work depends upon the time I leave home.
The probabilities are given in the table below.

Time of leaving home	Probability of being late
before 0800	0.01
between 0800 and 0810	0.1
after 0810	0.2

I work 230 days each year. Estimate how many times I would expect to be late for work in a year. (AQA)

27 Sam designs sweatshirts. She needs to know how long to make the sleeves, so she measures the arm lengths of 100 fifteen-year-olds.
Her results are represented by this histogram.

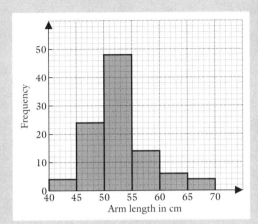

a Use this histogram to complete a copy of the table below.

	Frequency	Cumulative frequency
40 cm < arm length ≤ 45 cm	4	4
45 cm < arm length ≤ 50 cm	24	
50 cm < arm length ≤ 55 cm	48	
55 cm < arm length ≤ 60 cm	14	
60 cm < arm length ≤ 65 cm		
65 cm < arm length ≤ 70 cm		

b Draw a cumulative frequency diagram from these data.

c Use your graph to estimate the median arm length.
Show on your graph how you did this.

d Sam makes sweatshirts to fit arms from 47 cm to 57 cm in length. Estimate the percentage of her sample that her sweatshirts will fit. (AQA)

28 A bag contains 12 sweets, of which 3 are chocolates, 4 are mints and 5 are toffees.
Sweets are taken from the bag at random and are not replaced.
Find the probability that

a the first two sweets taken are both mints

b the second sweet taken is a mint. (OCR)

29 a i This table shows the distribution of spending money for sixty pupils in one week. Calculate the mean amount of spending money.
You may find it useful to complete the table.

Spending money	Frequency	Mid-value	Frequency × Mid-value
£0.00–£0.99	0	0.495	0
£1.00–£1.49	7	1.245	8.715
£1.50–£1.99	12	1.745	20.94
£2.00–£2.49	16	2.245	
£2.50–£2.99	10	2.745	
£3.00–£3.49	6	3.245	
£3.50–£3.99	4	3.745	
£4.00–£4.49	3	4.245	
£4.50–£4.99	2	4.745	9.49
Totals	60		

ii The *actual* mean amount of spending money is £2.59, correct to the nearest penny. Explain why your calculation did not give you this exact amount.

b i Complete the cumulative frequency table for spending money.

Spending money	Cumulative frequency
less than £1.00	0
less than £1.50	7
less than £2.00	19
less than £2.50	35
less than £3.00	
less than £3.50	
less than £4.00	
less than £4.50	
less than £5.00	

ii Draw the cumulative frequency curve for spending money.
iii Use your curve to estimate the median amount of spending money.

c Use your curve to estimate the interquartile range of spending money.

d State **one** advantage of using the mean rather than the median to represent the average spending money.

(AQA)

30 Tom runs a fish farm. He wants to estimate how many fish are in one of the lakes. To do this he takes a sample of 50 fish and marks all of them before returning them to the lake. He takes a further sample of 100 fish and finds 8 of them are marked.
He uses this to estimate the number of fish in the lake.

a How many fish, does he estimate, are in the lake?

b Comment on the reliability of his estimate.

(AQA)

31 Sangita is playing 'heads or tails' with her friend.
She spins a fair coin four times and gets four heads.

a What is the probability that she gets a tail with her next spin of the coin?

b If Sangita spins the coin 600 times in succession, approximately how many times
should she expect to get a tail? (AQA)

32 Fifty competitors took part in this year's Schools' Javelin Championship.
The best distance thrown by each competitor is shown in the frequency distribution below:

Distance thrown, m	Mid-point	Number of competitors
$20 < m \leqslant 30$	25	5
$30 < m \leqslant 40$	35	12
$40 < m \leqslant 50$	45	17
$50 < m \leqslant 60$	55	9
$60 < m \leqslant 70$	65	7

a The diagram below shows a frequency polygon for the best distances in **last year's**
competition.

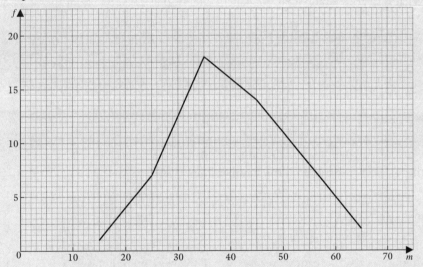

i On a copy of the axes draw a frequency polygon for this year's competition.

ii Use the frequency polygons to make a comparison between last year's and this
year's results.

b **i** Complete the following cumulative frequency table for these results.

Distance thrown, m	20–30	–40	–50	–60	–70
Cumulative frequency	5	17	34		

ii Draw a cumulative frequency curve for this year's results.

iii Use your cumulative frequency curve to estimate the interquartile range.

iv The qualifying distance for the District Sports is 55 metres.
Estimate how many of the competitors would qualify for the District Sports.

(AQA)

33

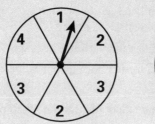 

The diagrams show two fair spinners. Both spinners are spun and the scores are added together.

What is the probability that the sum of the scores is at least 5? (AQA)

34 In an Open University course there are four themes (W, X, Y, Z) and eight topics (a, b, c, d, e, f, g, h).

The examination paper contains seven questions, each involving one theme and one topic. Theme W is twice as likely to occur as any other. All topics are equally likely and each can occur only once.

a What is the probability that the paper contains one question about theme W and topic a?

b What is the probability that the paper contains one question about theme X and topic b?

c What is the probability that the paper contains both these questions?
State any assumption that you make. (AQA)

35 The probability that it will rain tomorrow is r.
The probability that it will be windy tomorrow is w.
The probability that it will rain and be windy tomorrow is b.
Both r and w are marked on the number line below.

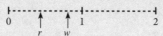

Mark a possible position for b. (AQA)

36 In English football, a team scores 3 points if it wins, 1 point if it draws, 0 points if it loses. The probability that my team will win their next match is 0.4. The probability that it will lose its next match is 0.25.

What is the probability that it will score at least 1 point in its next match? (AQA)

37 The table below gives information about the expected lifetimes, in hours, of 200 light bulbs.

Lifetime, t	$0 < t \leqslant 400$	$400 < t \leqslant 800$	$800 < t \leqslant 1200$	$1200 < t \leqslant 1600$	$1600 < t \leqslant 2000$
Frequency	32	56	90	16	6

a Mr Jones buys one of the light bulbs. What is the probability that it will last at least 800 hours but not more than 1600 hours?

b Draw a frequency polygon to illustrate the information in the table. (OCR)

38 a 50 pupils take an English exam and a Maths exam. The distribution of the marks they obtained is shown in the table below.

	Mark	21–30	31–40	41–50	51–60	61–70	71–80	81–90	91–100
Number of pupils	English exam	0	1	4	20	14	8	2	1
	Maths exam	2	3	6	10	12	10	4	3

The graph below shows the cumulative frequency for the English marks.

i On a copy of the graph, show the cumulative frequency for the Maths marks.

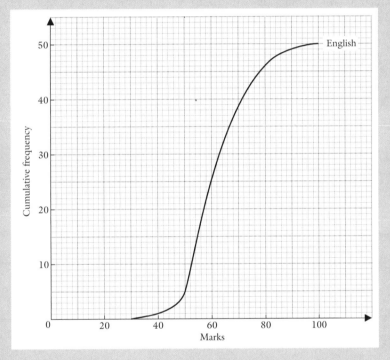

ii Complete the table below

	English	Maths
Median	60	
Interquartile range	14	

iii Use the information in the table to comment on the differences between the two distributions of marks.

b Of the 50 pupils, 30 pass the Maths exam at the first attempt. From past performance it is known that, if a pupil fails at the first attempt, the probability of passing at the second attempt is 0.7. Calculate the probability that a pupil, chosen at random from the 50 pupils, will pass the Maths exam at either the first or second attempt.

(OCR)

39 The diagram shows two boxes A and B.
Box A contains 5 white beads and 3 black beads.
Box B contains 4 white beads and 4 black beads.
A bead is to be taken at random from box A and
placed in box B.
A bead is then to be taken at random from box B
and placed in box A.

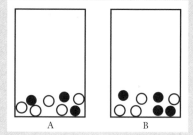

a Calculate the probability that both beads taken will be white.

b Calculate the probability that after both beads have been taken, there will be exactly
5 white beads in box A.
(Edexcel)

40 Information about oil was recorded each year for 12 years.
The scatter graph shows the amount of oil produced (in billions of barrels) and the
average price of oil (in £ per barrel).

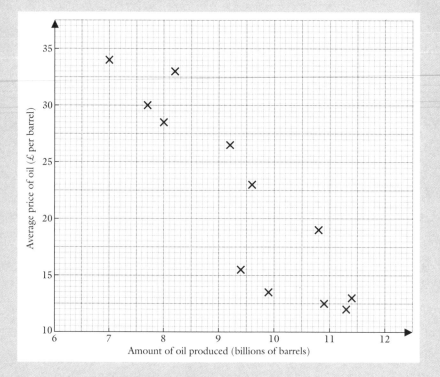

a Draw a line of best fit on the scatter graph.

In another year the amount of oil produced was 10.4 billion barrels.

b Use your line of best fit to estimate the average price of oil per barrel in that
year.
(Edexcel)

41 Kim sowed some seeds in her greenhouse.
10 weeks later she measured the heights of the plants.
Some of the results are shown in the table and the histogram.

Height, h, in cm	Number of plants
$0 < h \leqslant 5$	0
$5 < h \leqslant 20$	30
$20 < h \leqslant 30$	120
$30 < h \leqslant 35$	
$35 < h \leqslant 40$	
$40 < h \leqslant 50$	96
Over 50	0

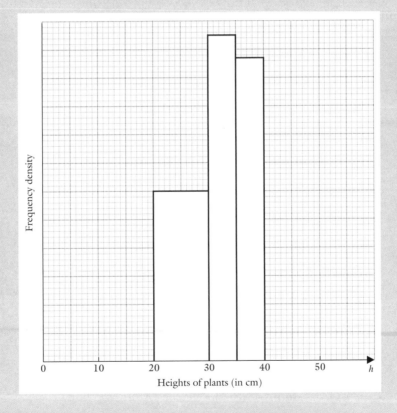

Heights of plants (in cm)

a Use the information to complete the table and the histogram.

Kim had sown 500 seeds.

b Calculate the number of seeds that had not produced plants. (Edexcel)

42 There are 30 students in a class.

The table below gives information about how many boys and girls in the class wear glasses.

	Wear glasses	Do not wear glasses	Total
Boys	4	9	13
Girls	7	10	17
Total	11	19	30

a One of the girls is chosen at random. What is the probability that she wears glasses?

b One of the students is chosen at random from those who do not wear glasses.
What is the probability that the student is a boy?

c Two of the 30 students are chosen at random.
Find the probability that they are both boys.

d Two students are chosen at random from those who wear glasses. Find the probability that they are both girls. (OCR)

43 A survey was carried out to see how much the sixth formers in a school earn each week from part-time employment.

The frequency polygon below shows the distribution of earnings of the male sixth formers.

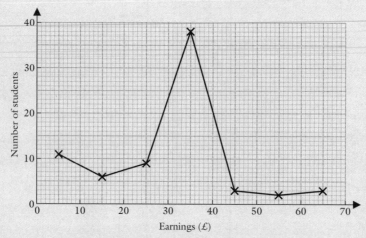

The table below shows the distribution of earnings of the female sixth formers.

Earnings (£x)	Number of females
$0 < x \leqslant 10$	15
$10 < x \leqslant 20$	27
$20 < x \leqslant 30$	14
$30 < x \leqslant 40$	4
$40 < x \leqslant 50$	1

a On a copy of the grid, draw the frequency polygon for the earnings of the females.

b Make two different comparisons between the earnings of the males and the females. (OCR)

44 In the film Teutonic, the Captain and five passengers remain on board a sinking ship.
There are three lifejackets remaining.
The Captain knows that three of the passengers cannot swim.
In his panic he hands out the lifejackets randomly to three of the five passengers.
Calculate the probability that he gives the lifejackets to the three non-swimmers.

(OCR)

45 There are 12 boys and 15 girls in a class.
In a test the mean mark for the boys was n.
In the same test the mean mark for the girls was m.
Work out an expression for the mean mark of the whole class of 27 students. (Edexcel)

46 Leon recorded the lengths, in minutes, of the films shown on television in one week.
His results are shown in the histogram.

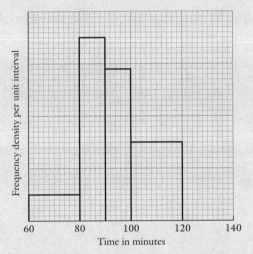

20 films had length from 60 minutes up to, but not including, 80 minutes.

a Use the information in the histogram
to complete the table.

Length in minutes (m)	Frequency
$60 \leqslant m < 80$	20
$80 \leqslant m < 90$	
$90 \leqslant m < 100$	
$100 \leqslant m < 120$	

Leon also recorded the lengths, in
minutes, of all the films shown on
television the following week. His
results are given in the table opposite.

Length in minutes (m)	Frequency	Height
$60 \leqslant m < 100$	72	36
$100 \leqslant m < 160$	x	

b Complete the table giving your answers in terms of x. (Edexcel)

47 The table shows the number of boys and the number of girls in Year 10 and Year 11 of a school.

	Year 10 Group	Year 11 Group
Boys	75	50
Girls	60	30

The headteacher wants to find out what pupils think about a new Year 11 common room. A stratified sample of size 50 is to be taken from Year 10 and Year 11.

a Calculate the number of pupils to be sampled from Year 10.

Two pupils are to be chosen at random to speak to the headteacher.
One pupil is to be chosen from Year 10.
One pupil is to be chosen from Year 11.

b Calculate the probability that both pupils will be boys.

However, the headteacher decides to choose one boy at random from all the boys in Years 10 and 11 together and one girl at random from all the girls in Years 10 and 11 together.

c Calculate the probability that exactly one of the pupils will be from Year 10.   (Edexcel)

48 The table shows the amounts on 10 of Meera's gas bills.

Date of bill	Mar 1999	Jun 1999	Sept 1999	Dec 1999	Mar 2000	Jun 2000	Sept 2000	Dec 2000	Mar 2001	Jun 2001
Amount (£)	17	21	34	40	25	29	40	42	30	32

The time series graph shows Meera's original data.

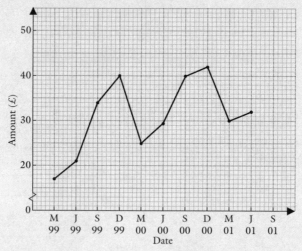

a Calculate the first value of the four-point moving average for the data.

The remaining values of the four-point moving average for the data are

$$30, \quad 32, \quad 33.5, \quad 34, \quad 35.25, \quad 36.$$

b Plot all the values of the moving average on the graph.

c Use your graph to estimate Meera's gas bill for September 2001.   (AQA)

49

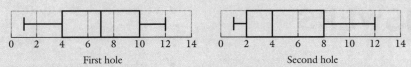

First hole   Second hole

a The box plots summarise the number of shots taken at each of two holes by some golfers in a competition.

  i Find the interquartile range for the number of shots taken at the first hole.

  ii Describe **two differences** between the two distributions.

b William is about to play each of the two holes. The tree diagram shows the probabilities of William getting a 'hole in one', or not, at each hole.

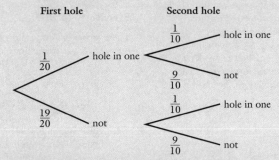

Calculate the probability that William gets a 'hole in one' at the first hole but not at the second hole. (AQA)

50 Jill has a bag containing some black and some white balls.

The probability of taking a black ball from the bag is $\frac{2}{5}$.

a Write down a possible pair of values for the numbers of black and white balls in the bag.

b She puts some more black balls in the bag.

The probability of taking a black ball from the bag is now $\frac{2}{3}$.

Using your pair of values from part **a** find the number of black balls she must have added. (AQA)

51 Pupils in Year 7 are all given the same test.

The test is marked out of 100.

The results for the boys are:
    Lowest score   7 marks
    Highest score   98 marks
    Lower quartile   36 marks
    Median   53 marks
    Upper quartile   66 marks.

a The diagram shows the box plot for the girl's results.

Copy the diagram. Draw a box plot on it to show the information for the boys on the same diagram.

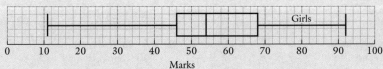

Marks

b Comment on the differences between the boys' and girls' results. (AQA)

ANSWERS

Answers are supplied to questions asking for estimates but there is no 'correct' estimate; we have given a likely value.

Allow a reasonable margin of error for answers read from graphs.

Possible answers are given to questions asking for opinions or reasons or interpretation; any reasonable alternative is also valid.

Diagrams given are intended only to give an indication of shape; no scales are given and axes are not labelled.

Numerical answers are given corrected to three significant figures unless specified otherwise.

The Examination Boards accept no responsibility whatsoever for the accuracy, method or working in the answers given to past examination questions. These answers are the sole responsibility of the authors.

Chapter 1

Exercise 1A (p. 2)

These answers are starting points for discussion; some of these criticisms are in the category of 'bad practice' and the perception of bad practice varies from one individual to another.

1. $2\frac{1}{2} + 1\frac{1}{4} \neq 3\frac{3}{4} - 2\frac{1}{3}$
2. $3x + 11 \neq 3x + 5$ & $11 \neq 3x$ & $6 \neq x$
3. $0.5 \neq 30°$
4. $2\pi r$ is not a formula, it is an expression.
5. Third angle $\neq 70° + 70°$
6. Buns are buns and xp is a sum of money so they cannot be equal, etc.

These are suggestions and can be used as starting points for discussion.

7. Not possible to give more than all of himself (i.e. 100% of himself).
8. This is meaningless – five times less than what?
9. Probably means 'find $4 + 3$, then $5 + 5$ etc.', but reads as 'find $4 + 3 + 5 + 5 + 6 + 2$'.
10. Abbeyfield's score is 203% of what?
11. There is no connection between Dominco's statement and the fact that service engineers are paid commission when they sell a new machine.

Exercise 1B (p. 5)

There are several ways these solutions can be written out. These are suggestions.

1. $x^2 - 3x - 4 = 0$
   i.e. $(x - 4)(x + 1) = 0$
   $\Rightarrow$ $x = 4$ or $x = -1$

2. When $b = 2$ and $c = -1$,
   $a = b + c$ gives
   $a = 2 + (-1)$
   i.e. $a = 2 - 1 = 1$

3. $\widehat{A} + \widehat{B} = 90° + 30° = 120°$
   $\therefore$ $120° + \widehat{C} = 180°$
   $\Rightarrow$ $\widehat{C} = 180° - 120° = 60°$

4. $\sin \widehat{A} = \frac{2}{3}$
   $\Rightarrow$ $\widehat{A} = 41.81\ldots°$

5. Three pounds of these apples costs three times 59 p, that is £1.77.

Exercise 1C (p. 8)

There are several ways these solutions can be written out. These are suggestions.

1. $2x + 3y = 7$ [1]
                      $8x - 5y = 11$ [2]
   $[1] \times 4 \Rightarrow$ $8x + 12y = 28$ [3]
   $[3] - [2] \Rightarrow$ $17y = 17$,   i.e. $y = 1$
   $y = 1$ in $[1] \Rightarrow$ $2x + 3 = 7$,   i.e. $x = 2$

2. If the 4th angle is $x°$
   then $300 + x = 360$ (angle sum of quadrilateral)
   $\Rightarrow$ $x = 60$
   so the size of the 4th angle is $60°$.

3. $AC^2 = AB^2 + BC^2$ (Pythagoras' thm)
   $\therefore$ $AC^2 = 16 + 9 = 25$
   $\therefore$ $AC = 5$ cm

4.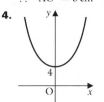
   The graph of $y = x^2 + 4$ does not cross the x-axis so there are no values of x for which $y = 0$, i.e. for which $x^2 + 4 = 0$.

5. If the price excluding VAT is £x
   then $x \times 1.175 = 12.8$
   so $x = 12.8 \div 1.175 = 10.893\ldots$
   The price excluding VAT is £10.89 (nearest penny)

6. F because it could be a rhombus
7. F because it could be a rectangle
8. T
9. T
10. F because it is incomplete.
11. e.g. AB equal *and* parallel to DC and $\widehat{ABC} = 90°$
12. $b^2 - 4ac = 0$

Exercise 1D (p. 10)

1. b $(x + 1)(x + 8) - (x)(x + 9) = 8$
   c $5; (x + 1)(x + 5) - x(x + 6) = 5$
2. $(x + 2)(x + 14) - (x)(x + 16) = 28$
3. a i 38   ii 10
   iii $2S_{10} = 10 \times 38 \Rightarrow S_{10} = 190$
   b 780
   c i $4n - 3$ ii $\frac{1}{2}n(4n - 2)$

4. a $\times 3$ **b i** 3^7 **ii** 3^9

c i $\dfrac{3^8 - 1}{2}$ **ii** $\dfrac{3^{10} - 1}{2}$

d i 3280 **ii** 29 524

e i 3^{n-1} **ii** $\dfrac{3^n - 1}{2}$

5. a e.g. i $(3 + \sqrt{2}) - \sqrt{2} = 3$
     ii $(\tfrac{1}{2})^2 + (\tfrac{1}{4})^2 < \tfrac{1}{2} + \tfrac{1}{4}$

b i $n = 2m + 1 \Rightarrow n^2 = 4m^2 + 4m + 1 \Rightarrow$ odd
  ii $(n) + (n+1) + (n+2) + (n+3) + (n+4)$
    $= 5n + 10 = 5(n+2)$

6. a F: when $n = 2$, $n^2 + 1 = 5$

b T: $n = 2m \Rightarrow n^2 = 4m^2$

c T: n odd $\Rightarrow n^3$ odd and difference between
    two odd numbers is even
    n even $\Rightarrow n^3$ even and difference between
    two even numbers is even

d F: e.g. $\dfrac{\sqrt{6}}{\sqrt{3}} = \dfrac{\sqrt{3} \times \sqrt{2}}{\sqrt{3}} = \sqrt{2}$ which is not
rational

Chapter 2

Exercise 2A (p. 14)

1. $x^2 + y^2 = 4$ **2.** $x^2 + y^2 = 64$

3.

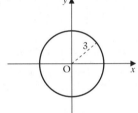

4.

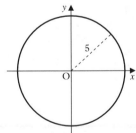

5.

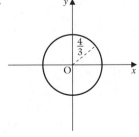

6.

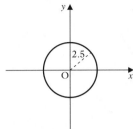

7.

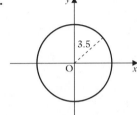

8.

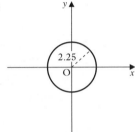

Exercise 2B (p. 14)

1. a $x^2 + y^2 = 49$ **b** $y = 2x - 4$
  c $(4.6, 5.3), (-1.4, -6.9)$
  d $x^2 + y^2 = 49, y = 2x - 4$

2. a $x^2 + y^2 = 36$ **b** $x + y = 4$
  c $(5.7, -1.7), (-1.7, 5.7)$
  d $x^2 + y^2 = 36, x + y = 4$

3. a $x^2 + y^2 = 64$ **b** $4y + 5x + 8 = 0$
  c $(4.0, -7.0), (-6.0, 5.4)$
  d $x^2 + y^2 = 64, 3x + 2y + 4 = 0$

4. a $x^2 + y^2 = 18$ **b** $x - 4y + 4 = 0$
  c $(3.8, 2.0), (-4.3, -0.1)$
  d $x^2 + y^2 = 18x - 4y + 4 = 0$

5. a, b

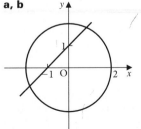

  c $(0.8, 1.8), (-1.8, -0.8)$
  d $x^2 + y^2 = 4, y = x + 1$

6. a, b

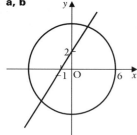

c $(1.9, 5.7), (-3.5, -4.9)$
d $x^2 + y^2 = 36, y = 2x + 2$

7. a, b

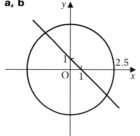

c $(2.2, -1.2), (-1.2, 2.2)$
d $4x^2 + 4y^2 = 25, y = 1 - x$

Exercise 2C (p. 17)

1. $x = 2, y = 4; x = -4, y = -2$
2. $x = 3, y = 4; x = -4, y = -3$
3. $x = 2, y = 3; x = 3, y = 2$
4. $x = 1, y = 2; x = 2, y = 1$
5. $x = 1, y = -3; x = 3, y = -1$
6. $x = 3, y = 5; x = -5, y = -3$
7. $x = 4, y = 3; x = -3, y = -4$
8. $x = 2, y = 5; x = 5, y = 2$
9. $x = 0, y = 3; x = 3, y = 0$
10. $x = 0, y = -4; x = -4, y = 0$
11. $x = 1, y = 3; x = -\frac{9}{5}, y = \frac{13}{5}$
12. $x = 2, y = 5; x = -\frac{7}{5}, y = -\frac{26}{5}$
13. $x = 3, y = -1; x = -\frac{13}{5}, y = \frac{9}{5}$
14. $x = 1, y = 1; x = \frac{17}{13}, y = -\frac{7}{13}$
15. $x = 2, y = 0; x = -\frac{10}{13}, y = \frac{24}{13}$
16. $x = 1, y = -3; x = 3, y = -1$
17. $x = 2, y = 2; x = \frac{3}{2}, y = \frac{5}{2}$
18. $x = 4, y = -2; x = -\frac{8}{3}, y = \frac{14}{3}$
19. $x = 0, y = 3; x = 3, y = 0$
20. $x = 3, y = 4; x = 4, y = 3$
21. $x = 1, y = 1; x = 5, y = -5$
22. $x = 3, y = 2; x = -\frac{3}{2}, y = -4$
23. $x = 3, y = 1; x = -1, y = 5$
24. $x = 3, y = -2; x = 1, y = 2$
25. $x = -1, y = 2; x = \frac{15}{2}, y = -\frac{7}{5}$
26. $x = 1, y = -2; x = -\frac{1}{3}, y = -\frac{26}{9}$

Chapter 3

Exercise 3A (p. 22)

1. $y = 3x$
2. $q = p^2$
3. $V = x^3$
4. $r = \sqrt{A}$
5. $y = \dfrac{24}{x}$
6. $s = \dfrac{r}{10}$
7. $y = 4x^2$
8. $q = -\dfrac{36}{p}$

9.

2	4	5	6	8	10
2	8	12.5	18	32	50

$A = \frac{1}{2}L^2$

10.

3	6	9	12
3	12	27	48

$A = \frac{1}{3}b^2$

11.

2	3	4	5
8	27	64	125

$y = x^3$

Exercise 3B (p. 25)

1. $y = 10x$;

x	2	4	7	8	9.5
y	20	40	70	80	95

2. $C = 6r$;

r	1	3	5	6	8
C	6	18	30	36	48

3. $C = 6n$;

n	100	120	142	260	312	460
C	600	720	852	1560	1872	2760

4. $Y = 10X$;

X	2	4	7	9	11	15
Y	20	40	70	90	110	150

5. **a** 9 **b** 16
6. **a** $\frac{3}{2}$ **b** 20
7. **a** 21 **b** 7
8. **a** 21 **b** 40
9. **a** 24 **b** 15
10. **a** 15 **b** 8
11. **a** 6 **b** 3

Exercise 3C (p. 28)

1.

0	2	3	4	5	8
0	12	27	48	75	192

$y = 3x^2$

2.

2	4	5	6	10
20	80	125	180	500

$s = 5t^2$

3.

−3	−1	0	2	4	7
36	4	0	16	64	196

$y = 4x^2$

4. a 32   **b** ±1
5. a $\frac{3}{4}$ **b** $\pm\frac{1}{3}$
6. a 108   **b** 8

7.

2	4	6	8	10
2	16	54	128	250

$V = \frac{1}{4}H^3$

8.

3	6	9	12	15
9	72	243	576	1125

$y = \frac{1}{3}x^3$

9. a 24   **b** 6
10. a 216   **b** 1
11 a 108   **b** 2

12.

0	1	4	9	25
0	4	8	12	20

$V = 4\sqrt{R}$

13. a 2   **b** 900
14. no; yes, $y = \sqrt{x}$
15 no; $y = \frac{1}{2}x^3$

Exercise 3D (p. 32)

1.

25	50	100	125
20	10	5	4

$N = \dfrac{500}{C}$

2.

12	9	8	6
60	80	90	120

$C = \dfrac{720}{N}$

3.

4	5	6	8	12
30	24	20	15	10

$V = \dfrac{120}{P}$

4. $y = \dfrac{72}{x}$ **5.** $y = \dfrac{2.16}{x}$

6. a $y = \dfrac{1}{x}$

b

$\frac{1}{x}$	0.1	0.2	1	2	4
y	0.1	0.2	1	2	4

c 1; constant of variation

7. a $y = \dfrac{7.2}{x}$

b

$\frac{1}{x}$	0.42	0.56	0.83	1.11	1.25
y	3	4	6	8	9

c 7.2; constant of variation

Exercise 3E (p. 34)

1.

2	4	6	9	12	18
18	9	6	4	3	2

$y = \dfrac{36}{x}$

2.

0.5	1	2	3	6	10
144	36	9	4	1	0.36

$y = \dfrac{36}{x^2}$

3.

0.25	1	4	9	16	25
120	60	30	20	15	12

$q = \dfrac{60}{\sqrt{p}}$

4. a 4   **b** 20   **c** −10
5. a $\frac{4}{3}$ **b** 16
6. a 10   **b** 40
7. a 6   **b** 0
8. a 25   **b** ±10
9. a 4   **b** 12
10. a 56   **b** 2
11. a 2   **c** 3   **e** $\frac{1}{2}$ **g** −1
    b 2   **d** −1   **f** 1

Exercise 3F (p. 36)

1. a 1   **b** 1   **c** 6.25
2. a 21   **b** 6
3. a $y = \frac{3}{4}x^3$ **b** 6   **c** 2
4. a 14   **b** 3
5. a 8   **b** ±3
6. a i 2 **ii** 1.25   **b i** 16 **ii** 49

7.

0	1	2	4	8
0	0.25	1	4	16

8.

0	4	9	16	64
0	0.5	0.75	1	2

9. a i doubled   **ii** halved
    b i multiplied by 9   **ii** divided by 9
10. a i divided by 4   **ii** multiplied by 4
    b i divided by 3   **ii** multiplied by 3
11. a 2   **b** 3   **c** 1   **d** −1
12. a 4 kg   **b** 25 cm
13. 64 m
14. a 1.6   **b** 56.25
15. a 25 cm   **b** 4.8 newtons
16. a £320   **b** 4.5 m
17. a multiplied by 2   **c** increased by 20%
    b multiplied by 5
18. a 0.216 litres   **b** 20 cm
19. a 3   **b** 120
20. a 40 mph   **b** 12 mph
21. a i 250   **ii** 40   **iii** 444

    b $x = \sqrt{\dfrac{k}{n}}$

    c 2.23 cm, round down because rounding up gives less than 800 squares
22. a i 1.26 units **ii** 2.45 units
    b i 1.6 m **ii** 1.2 m
23. D
24. a doubled   **c** quadrupled
    b halved   **d** multiplied by 8

25. a increase by 25%   **b** decrease by 20%
c increase by 56.25%
26. a approximately equal to speed
b approximately equal to $\frac{1}{20}$ of square of speed
c i 146 ft   **ii** 600 ft
d 65.5 mph
e e.g. 3 m (there is considerable variation in car lengths)
f No. The conversion factor used is 1 ft = 30 cm which gives a slight underestimate; speed is only measured in mph in UK but distance can be either in feet or metres.

g

Speed (km/h)	Thinking distance (m)	Breaking distance (m)	Stopping distance (m)
20	4	2	6
50	9	15	24
80	15	38	53
110	21	71	92
140	27	115	142

Chapter 4

Exercise 4A (p. 43)

1. a double   **c** reciprocal
b add 2   **d** cube
2. a −1   **b** 7   **c** −3   **d** −11
3. a 3   **b** −1   **c** 7   **d** 5
4. a 0   **b** 1   **c** 4   **d** 9
5. a 11   **b** 3   **c** 6   **d** 6
6. a 1   **b** −3   **c** 3   **d** −1
7. a 2   **c** 2
b 7   **d** −15.5; $\frac{4}{0}$ has no meaning
8. a 69   **b** −1   **c** 1   **d** −16
9. 1.2   **12.** −5   **15.** 3 or −1
10. 0.2   **13.** ±3   **16.** 1
11. 7   **14.** ±1   **17.** 2 or −1
18. a −4   **b** −6   **c** −3, 2
19. a −15   **b** −7   **c i** 5, −3   **ii** 6, −4
20. a −8   **c** 2
b −7   **d** $\sqrt[3]{9} = 2.08$ to 3 s.f.
21. a $\frac{4}{3}$ **b** 6
22. a i 13   **ii** −3   **b** 1
23. a 8, 6, 2   **c** 6
b decreasing   **d** $x > 6$
24. a 2, 6, 18, 102
b increasing
c 2, 3, 11, 102
d 2 because f(x) increases as x increases from zero and as x decreases from zero

Exercise 4B (p. 46)

1.   **a** $x < 0$
b all x

2.   **a** $x < -1$
b all x

3.   **a** none
b $x > 0$

4.   **a** $x < 0$
b none

5.   **a** $x < 0$
b all x

6.   **a** $x > 1$
b none

7.   **a** $x < -1$ and $x > 1$
b $x < 0$

8.   **a** none
b $x < 0$

9.   **a** $1 < x < 2$
b $x > 1.5$

10.

    a $x < \sqrt[3]{-2}$
    b all x

11. translation by 1 unit upwards

12. translation by 2 units downwards

13. translation 1 unit down

14. reflection in x-axis

15. reflection in x-axis

16. reflection in x-axis

17.

18. translate $y = x^2$ by a units to the right

Exercise 4C (p. 50)

1. a   **c**

    b   **d**

2. a   **c**

    b   **d**

3. a   **c**

    b   **d**

4. a   **c**

    b   **d**

5. a $(x+2)^2 + 2$
b i **ii** 2

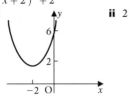

6. a **c** **e**

b **d** **f**

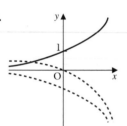

7.

8. a

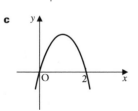

b

c

9. $(x-3)^2 - 6$, 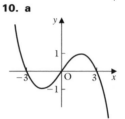 , -6

10. a **d**

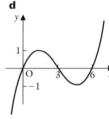

b **e**

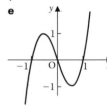

c **f**

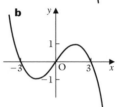 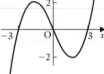

11. a B **b** D **c** C **d** A
12. a translation by 3 units down y-axis, $y = x^2 - 3$
b reflection in x-axis then translation by 4 units up y-axis, $y = 4 - x^2$
c translation by 3 units to right, $y = (x-3)^2$
d translation by 1 unit to left then by 3 units down, $y = (x+1)^2 - 3$
13. b 0
c $y = -(1-x)(x-2)(x-3)$
$= (x-1)(x-2)(x-3)$
14. a **b**

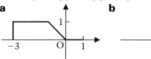

15.

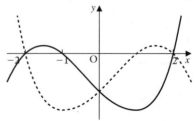

c 3; the graphs of $f(x)$ and $f(-x)$ cross between $x = -1$ and $x = 1$; $-2, 2$

Exercise 4E (p. 55)

1.

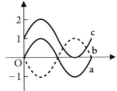

2.

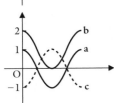

b

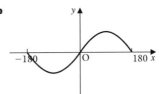

3. b 0, 180, 360   **4 b** 90, 270
5. a $-1 \leqslant \sin x° \leqslant 1$ **b** $-1 \leqslant \cos x° \leqslant 1$
7. $y = \cos x°$ is a translation of $y = \sin x°$ by 90 units to the left.

8.

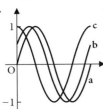

9.

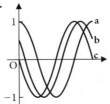

10. b reduces width by factor of 2
   c reduces width by factor of 3
   d stretches width by factor of 2
11. b $-1 \leqslant \sin x° \leqslant 1$ **12. b i** 1 **ii** -1

4. 2; 62, 208   **5.** 2; 38, 142
6. A(0, 2), B(90, 1), C(180, 0), D(270, 1), E(360, 2)
7. a reflection in x-axis; $y = -\sin x°$
   b reflection in x-axis followed by translation of 1 unit up; $y = 1 - \sin x°$
   c reduction of width by factor 2; $y = \sin 2x°$
8. a translation of 1 unit down; $y = \cos x° - 1$
   b reduction of width by factor 3; $y = \cos 3x°$
   c reflection in x-axis; $y = -\cos x°$
9. a from 4 p.m. to 8 p.m.
   b 4 hours 48 minutes
10. a 0, 180, 360   **b** 90, 270
11. a 5   **b** 7
12. a 135   **b** 45
15. 1, the curves only intersect once.
16. a $y = \cos x°$, $y = \dfrac{(x - 20)}{50}$
   b e.g. $y = 30 \sin x°$, $y = 20 - 2x$
   c $(x - 1)^2 = 14 \tan x°$
17. a $-66, -294, 66, 294$
   b $-226, -134, 134, 226$
   c $-264, -96, 96, 264$
18. a $-348, -192, 12, 168$
   b $-168, -12, 192, 348$
   c $-53, -127, 233, 307$
19. a $1.4, -1.4$
   b stretch by a factor of 1.4 parallel to y-axis and a translation by 45 units to the left;
   $y = 1.4 \sin (x° + 45°)$

Exercise 4F (p. 58)

1. a, b

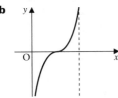

   c same as $y = \tan x$
2. a

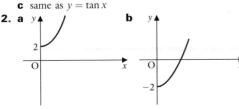

   b

3. a

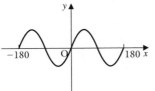

Exercise 4G (p. 62)

1. b

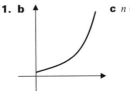

   c $n = 2^t$ **d** $n = 10 \times 2^t$

2. a $P = 1000 \times 1.1^n$ **d** $P = 5000 \times 1.1^n$
   c $\simeq 7.5$, $\simeq 7.25$ **e** £54 173.53

Exercise 4H (p. 63)

1. a

P	9000	6000	4000	2670	1780	1190	790

   b

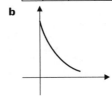

   c 1.7 years, 1.7 years; no, the curve never crosses the horizontal axis.

2. a

m	5	2.5	1.25	0.625	0.313	0.156	0.078

Exercise 4I (p. 65)

1.

−3	−2	−1	0	1	2	3	4	5	6
0.125	0.25	0.5	1	2	4	8	16	32	64

2.

−3	−2	−1	0	1	2	3
8	4	2	1	0.5	0.25	0.125

3. One is the reflection of the other in the y-axis; yes because if $f(x) = 2^x$ then $f(-x) = 2^{-x}$.

4. a $(0, 1)$
  b No. There is no value of x for which $3^x = 0$.
  c

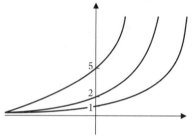

5. a $1.5^0 = 1 = 3^0$; $(0, 1)$, $a^0 = 1$ for all values of a except $a = 0$
  c $(0, 500)$
  d i $(0, 10)$ **ii** $(0, 10)$
6. a 3 **b** 3.6; 1.2 **8.** 4, 1.6
7. 2, 1.8; $y = 2 \times 1.8^{-x}$ **9.** 1000, 1.3
10. a translation by one unit upwards; $y = 2^x + 1$
  b reflection in x-axis; $y = -2^x$
  c translation by one unit downwards; $y = 2^x - 1$
11. b $A = P \times \left(1 + \dfrac{r}{100}\right)^T$ **e** £6573.37
  d 3000×1.04^{20} **f** 7.18%
12. a 79.6 kg **c** 12.8 kg
  b $80 \times (0.995)^n$ **d** 139 days
13. a i 80 mg **ii** Thursday **14.** 6

Exercise 4J (p. 69)

1. a −4, 0, 21
2. a

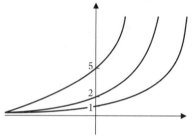

3. 30, 150 **4.** 120, 240
5. a $y = \sin x° + 1$ **b** $y = 2 \sin 2x°$
6. 9, 1.2
7. a C **b** B **c** D **d** A

Chapter 5

Exercise 5A (p. 76)

1. 19, 23 **2.** 48, −96 **3.** 36, 49
4. $1 + 3 + 5 + 7 + 9$, $1 + 3 + 5 + 7 + 9 + 11$
5. $\frac{1}{81}$, $-\frac{1}{243}$ **6.** 42, 68
7. a 3, 5, 7, 9 **c** 0, 3, 8, 15 **e** 2, 9, 20, 35
  b 2, 4, 8, 16 **d** 0, 3, 8, 15 **f** $\frac{1}{2}, \frac{2}{3}, \frac{3}{4}, \frac{4}{5}$
8. $2n + 3$ **9.** $3(n - 1)$ **10.** 2^{n+1}
11. $n(n + 2)$ **12.** n^3 **13.** $n^2 + 1$
14. a $(5, 26)$ **b** $(10, 101)$ **c** $(n, n^2 + 1)$
15. a 6, 11, 16, 21, 26 **c** 8, 12
  b i $5n + 1$ **ii** 151
16. a 1 6 15 20 15 6 1
    1 7 21 35 35 21 7 1
    1 8 28 56 70 56 28 8 1
  b 1, 2, 4, 8, 16, 32, 64, 128
  c 2^{n-1}
17. a $(1st)^2 + (2nd)^2 = (3rd)^2$
  b e.g. $(9, 40, 41)$
18. $\frac{1}{2}, \frac{1}{4}, \frac{1}{8}, \frac{1}{16}, \frac{1}{32}$, for ever
19. a 1, 3, 6, 10, 15 **d** n
  b 210 **e** 2, 3, 4, 5
  c $\frac{1}{2}n(n - 1)$ **f** 21
20. a $1, \frac{5}{2}, \frac{14}{3}, \frac{15}{2}, 11$ **d** $\frac{1}{6}(4n + 1)$
  b $\frac{287}{2}$ **e** $\frac{9}{6}, \frac{13}{6}, \frac{17}{6}, \frac{85}{6}$
  c $\frac{1}{6}n(2n - 1)$ **f** $\frac{81}{6} = 13\frac{1}{2}$
21. 49, 70 **24.** 26, 37
22. 47, 73 **25.** 325, 540
23. 437, 683 **26.** 7, 29
27. a e.g. 3^n, 81, 243, 729 or $6n^2 - 12n + 9$; 57, 99, 153
  b e.g. $3 \times 2^{n-1}$; 24, 48, 96 or $\frac{3}{2}(n^2 - n + 2)$; 21, 33, 48
  c e.g. $(n - 1)^2 + 1$; 10, 17, 26 or multiply previous term by 3 and subtract 1; 14, 41, 122
28. 121 to 122
29. a 146 **b** 99 **c** 335
30. a 22, 27; $u_n = 5n - 3$
  b 14, 84, 204, 374, 594
  c 864, 1184
  d 9, 19, 29, 39, 49
  e $(5n - 3)(5n + 2)$
  f i 2, 9, 21, 38, 60 **iii** $\frac{1}{2}(5n - 1)$
    ii 2, $4\frac{1}{2}$, 7, $9\frac{1}{2}$, 12 **iv** $\frac{1}{2}n(5n - 1)$
31. a 1, 3, 4, 7, 11, 18 **c** 3, 4, 7, 11, 18, 29
  b 2, 3, 5, 8, 13, 21
32. same sequence, but starting 1, 3, 4, ...; yes
33. $n^2 + n + 1$, 111 **36.** $4n^2 - n - 3$, 387
34. $3n^2 + 2n$, 320 **37.** $n^2 - 3n$, 70
35. $12n - 5$, 115 **38.** $3n^2 + 5$, 305

39. $-241, -1265, -5361$
40. $\frac{5}{6}, \frac{6}{7}, \frac{7}{8}$
41. $75, 101, 131$
42. $126, 217, 344$
47. $124, 165, 212$
48. $6, 14, 24$
49. **a** $1, 1, 2, 3, 5, 8, 13, 21$
   b $1, \frac{1}{2}, \frac{2}{3}, \frac{3}{5}, \frac{5}{8}, \frac{8}{13}, \frac{13}{21}, \frac{21}{34}, \frac{34}{55}, \frac{55}{89}$
   c $1, 0.5, 0.6667, 0.6, 0.625, 0.6154, 0.6190,$
       $0.6176, 0.6182, 0.6180$
   d appear to converge to 0.618 to 3 d.p.
50. **a** odd values of N only, uses $\frac{1}{2}(N+1)$ colours
       and N^2 tiles in total, $2(N-1)$ tiles of the last
       new colour, $8x-4$ tiles for Cx except C1
       which needs 5 and the last colour which needs
       $2(N-1)$ tiles, colour $\frac{1}{2}(N-1)$ uses most tiles
       and the first colour uses the least number of tiles
   b even values of N only, uses $\frac{1}{2}N$ colours; $2N$
       tiles of the last colour and $8x$ for all other Cx;
       $C\frac{1}{2}(N-2)$ uses most tiles

Chapter 6

Exercise 6A (p. 85)

1. b and **c**
3. a $\sqrt{3}$ **c** $\sqrt{3}$ **e** $2\sqrt{5}$ **g** $2\sqrt{5}$
   b $\sqrt{3}$ **d** $\sqrt{3}$ **f** $\frac{3}{2}\sqrt{5}$ **h** $3\sqrt{5}$
4. a a: 3   c: 5   e: 5   g: 5
       b: 5   d: 6   f: 6
   b b and c
   c b and g or c and g

Exercise 6C (p. 91)

1. a $\overrightarrow{PR}$ **b** $\overrightarrow{PQ}$ **c** $\overrightarrow{RP}$ **d** $\overrightarrow{QR}$
2. a $\overrightarrow{AC} + \overrightarrow{CB}$
   b $\overrightarrow{AC} + \overrightarrow{CB}$ (or $\overrightarrow{AD} + \overrightarrow{DB}$)
   c e.g. $\overrightarrow{AC} + \overrightarrow{CB}$
3. a $\overrightarrow{AC}$ **c** $\overrightarrow{AD}$ **e** $\overrightarrow{DC}$
   b $\overrightarrow{BD}$ **d** $\overrightarrow{AD}$ **f** $\overrightarrow{AC}$
4. a $\overrightarrow{AD}$ **b** $\overrightarrow{BA}$ **c** $\overrightarrow{AD}$ **d** $\overrightarrow{DC}$
5. a $\overrightarrow{AB}$ **b** $\overrightarrow{BC}$ **c** $\overrightarrow{AC}$
6. a $\overrightarrow{DC}$ **b** $\overrightarrow{DB}$ **c** $\overrightarrow{AB}$ **d** $\overrightarrow{AD}$
7. a

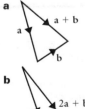

   c

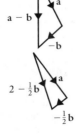

8. a

(diagrams for b, c, d, e with vectors: $a + b$, $-2b$, a, $a - 2b$, $3b$, $2a + 3b$, $-2a$, $\frac{5}{2}b$, $2a$, $-\frac{3}{2}a$, $b - 2a$, $\frac{5}{2}b - \frac{3}{2}a$)

Exercise 6D (p. 94)

1. a Yes; PQ and QR are parallel and have Q in
       common.
   b i $b - a$ **ii** $2a - b$ **iii** $b - 2a$
2. a $q - p$ **c** $\frac{1}{2}(q - p)$
   b $\frac{1}{2}(q - p)$ **d** $\frac{1}{2}(q + p)$
3. a trapezium
   b i $a - b$ **ii** $a - 2b$ **iii** $a + b$
4. a $4b - 6a$ **b** $b - \frac{3}{2}a$ **c** $b + \frac{9}{2}a$
5. a $\frac{1}{2}b$ **d** $\frac{1}{2}(c - b)$
   b $\frac{1}{2}c$ **f** $PQ = \frac{1}{2}BC$; $\frac{1}{2}(c - a) = \frac{1}{2}\overrightarrow{BC}$
   c $c - b$
6. a $\frac{1}{2}a$ **c** $\frac{1}{2}(b - a)$ **e** $\frac{1}{2}b$
   b $b - a$ **d** $\frac{1}{2}b$
   g parallelogram (PQ and SR are // and =)
7. a $b - a$ **c** $\frac{2}{3}(b - a)$ **e** $\frac{1}{3}b + \frac{2}{3}a$
   b $\frac{1}{3}(b - a)$ **d** $\frac{1}{3}b + \frac{2}{3}a$
8. a b **e** $\frac{3}{4}a$
   b $\frac{2}{3}b$ **f** $\frac{1}{2}b + \frac{3}{4}a$
   c $a + \frac{2}{3}b$ **g** $\overrightarrow{RQ} = \frac{3}{4}\overrightarrow{OP}$
   d $\frac{1}{2}b$ **h** $RQ = \frac{3}{4}OP$
9. a They all lie on the same straight line.
   b parallelogram
   c trapezium
   d ECBF
10. a $6a - 2b$ **b** $3a - b$ **d** $1 : 2$
11. a i $b - a$ **iii** $-b$
       ii $-a$ **iv** $a - b$
   b i $b - a$ **iii** $-b$ **v** a
       ii $-a$ **iv** $a - b$ **vi** b
12. a $2x + 2y$ **c** $-2x - y$
   b $3x - y$ **d** $-2x + 3y$
13. a $a - b$ **d** $\frac{1}{2}(b - a)$
   b $b - a$ **e** Yes, both $\frac{1}{2}(b + a)$
   c $\frac{1}{2}(a - b)$

43. $90, 126, 168$
44. $5.7, 6.8, 7.9$
45. $16, -32, 64$
46. $\frac{7}{9}, \frac{4}{5}, \frac{9}{11}$

14. a $b - a$ **e** $\frac{1}{4}(b + 3a)$

    b $a - b$ **f** $\frac{1}{6}(b + 3a)$

    c $\frac{1}{4}(b - a)$ **g** $\frac{1}{12}(b + 3a)$

    d $\frac{3}{4}(a - b)$

15. a $q - p$ **c** $(1 - k)q - (1 - k)p$

    b $k(q - p)$ **d** $kq + (1 - k)p$

Exercise 6E (p. 100)

1. a   **b**

2. b $15\,\text{m/s}$

3. b $59°$

4. b $084°$ **c** $337\,\text{mph}$

5. $20.4\,\text{m/s}, 169°$

6. $6.6\,\text{N}$ at $104°$ to the $8\,\text{N}$ force

7. $65.4\,\text{km/h}$, from $276.5°$

8. $10\,\text{N}$

9. c $112°$

10. a $63°$ to the bank   **c** $1.52\,\text{m/s}$

     b $60°$ to the upstream bank

11. No; $60°$ to the bank upstream, increases speed but lengthens distance when pointed at an angle to the bank upstream and vice-versa; no

Exercise 6F (p. 103)

1. a $\frac{2}{3}ka + \frac{1}{3}kb, \frac{2}{3}ka + \left(\frac{1}{3}k - 1\right)b$

    b 3 **c** $1 : 2$

2. a $ka - \frac{1}{4}b$ **b** $\frac{1}{2}$ **c** $\frac{1}{2}$

3. a hp **e** $kr + (1 - k)p$

    b $r - p$ **f** $h = \dfrac{1}{k}$

    c $(h - 1)p + r$ **g** $k = \frac{1}{2}, h = 2$

    d $k(r - p)$

4. a $q - p$ **d** $\frac{1}{2}p$ **g** $\frac{1}{2}hp + (1 - h)q$

    b $\frac{1}{3}(q - p)$ **e** $\frac{1}{2}p - q$ **h** $\frac{1}{3}kq + \frac{2}{3}kp$

    c $\frac{1}{3}q + \frac{2}{3}p$ **f** $\frac{1}{2}hp - hq$ **i** $h = \frac{4}{5}, k = \frac{3}{5}$

5. a $2b - 2a$ **c** $3a$ **f** rhombus

    b $2b - 4a$ **d** $2b - 2a$

6. a $2c, 2c - a$ **b** $b = \frac{1}{2}a + c$ **c** $\frac{3}{2}$

7. $(-1, 10)$

Exercise 6G (p. 106)

1. a i 6 **ii** $\sqrt{41}$ **c** b and d

     b c and h   **d** g

2. a   **b**

c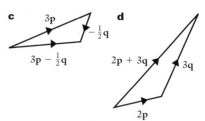

3. a $\overrightarrow{AD}$ **b** $\overrightarrow{AE}$ **c** $\overrightarrow{AB}$ **d** $\overrightarrow{AC}$

4. a $y - x$ **c** $x - y$ **e** $\frac{2}{3}x + \frac{1}{3}y$

    b $\frac{1}{3}(y - x)$ **d** $\frac{2}{3}(x - y)$

5. a $284°$ **b** $25.4\,\text{km/h}$

6. a $\frac{4}{5}p, \frac{1}{5}p, q - p$ **d** trapezium

    b $\frac{1}{5}(q - p), \frac{1}{5}q$ **e** $120\,\text{cm}^2$

    c parallel and $RS = \frac{1}{5}OQ$

Chapter 7

Exercise 7A (p. 110)

These answers are examples of possible responses and can be used as a starting point for discussion.

1. No to all parts.

2. a No.

    b Yes, those who are unhappy, or very happy, with the bank's services.

    c low cost

3. a unlikely to find employed men

    b unlikely to find unemployed men

    c unlikely to find older men

4. a Yes. It isn't possible to catch all grey squirrels.

    b Yes. It isn't possible to measure all the water in a river.

    c Yes. Number keeps changing.

    d Yes. Otherwise there would be no bulbs to sell.

    e No.

5. Not everyone has an equal chance of being selected because not every one has a phone, children (who are most likely to use the recreation ground) are least likely to answer the phone; the question is ambiguous as 'been to' is not the same as 'used'.

Exercise 7B (p. 114)

1. reading across

    a $27, 38$ **b** $608, 278$ **c** $843, 2263$

2. Number a list of students and then use random numbers to select 45 numbers between 1 and 450.

3. e.g. $21, 3, 12$ using the first 6 numbers as follows; $0.69 \times 30, 0.07 \times 30, 0.40 \times 30$

5. Not all numbers have an equal chance of being selected, 1 to 23 are twice as likely to be selected as the others since, e.g. the random numbers 01 or 77 both select 1.

Exercise 7C (p. 117)

1. Catering 59, Academic 28, Engineering 77, Sport 6, Technology 52
2. **a** 3 from each of Years 1, 2, 3 and 4, 4 from Year 5 and 2 from Year 6
   b 6 from Year 1, 5 from each of Years 2, 3 and 4, 6 from Year 5 and 3 from Year 6
3. **a** A 357 kg, B 314 kg, C 200 kg, D 129 kg
   b e.g. mix the whole crop and then select 1 tonne – the proportion of damaged potatoes from each field does not matter
4. Stratified sample by dividing into classes and into sexes as both age and gender are likely to affect responses
5. **a** 26 or 27
   b A 10, B 7 or 8, C 9
   c so that the operation of each machine can be monitored as they may behave differently
   d 400; the sample is selected at random so the proportion of cartons with a faulty quantity of juice in the sample may be different to that of the total number of cartons filled in the hour
6. **a** e.g. at intervals throughout the day for each day of one week in the busiest part of the town centre near the main shops
   b e.g. ask a random selection of people in their homes

Chapter 8

Exercise 8A (p. 125)

1. **a** 46
   b 84
   c 32
   d e.g. range of boys marks greater than girls; median boys mark less than girls.
2. **a** £7000
   b £26 000
   c yes, range of top 25% is £9000 to £26 000
   d 125
3. **a** King Edwards
   b 60 g
   c 130 g
   d e.g. the range of masses of King Edwards is greater than that of the Reds; the median mass of King Edwards is greater than that of Reds.
   e Reds, smaller range
4.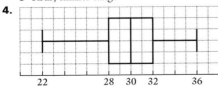

5. **a**

   18   44   59 65   84

   b yes, lowest 25% of marks go up to 44
6. **a**

   Broadsheet

   0   10   20   30

   b e.g. range and median for broadsheet is greater than that for tabloid
7. **a**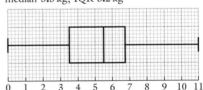

   Outdoors

   3   4   5   6   7   8   9

   b e.g. Greenhouse seedlings are less variable in height and have a greater median than the outdoor ones.
8. **a** median 5.5 kg, IQR 3.2 kg
   b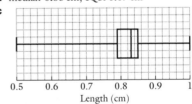

   0 1 2 3 4 5 6 7 8 9 10 11
9. **a** median 0.83 cm; IQR 0.07 cm
   c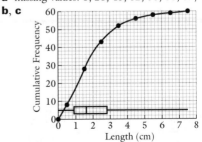

   0.5   0.6   0.7   0.8   0.9   1
   Length (cm)

   d no, the median (i.e. 50%) is less than 0.85 cm
10. **a** missing values: 8, 28, 43, 52, 56, 58, 59, 60
    b, c

    (Cumulative Frequency graph, Length (cm) on x-axis 0–8)

    d the interquartile range because it ignores the few very long screws.

Exercise 8C (p. 134)

1. a i 1   **ii** 0.5   **iii** 1
  b i 10   **ii** 5   **iii** 10

2.

Width	3	3	4	4	6	10
Frequency density	0.67	2.33	2.75	4	3	1.3

3.

Width	30	10	10	10	40
Frequency density	0.3	1.5	2.4	3.6	0.4

4.

Width	30	10	5	10	10	20
Frequency density	0.2	1.1	1.6	1.4	1.4	0.6

5.

Width	5	5	5	3	3	5	4	10
Frequency density	1.2	0.2	3.6	3	4	3.8	2.25	0.4

6.

Width	4	4	4	5	7	4	4	10
Frequency density	8.5	5.5	9.5	6	4	3.3	2.5	1

Exercise 8D (p. 135)

1.

Age	0–3	3–4	4–5
Frequency	15	10	16

2.

Time	0–5	5–15	15–20	20–25
Frequency	50	90	55	30

3.

Height	10–30	30–40	40–50	50–70
Frequency	60	50	70	40

4. a 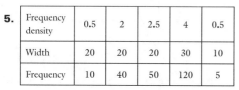 = 5 plants   **b** 30   **c** 80

5.

Frequency density	0.5	2	2.5	4	0.5
Width	20	20	20	30	10
Frequency	10	40	50	120	5

225

6. a

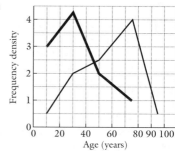

b The mean age is lower and the range of ages is less for the distribution in this question.
c Camberley may be a residential area and Bramberdown may be in a town centre near schools and offices.

7. a

Income (£)	0–2	2–5	5–10	10–20	20–40	40–45
Frequency	36	48	100	120	80	6

b £13.87
8. a 107.5 cm, 9.5 cm

b

Height	90–100	100–105	105–110	110–115	115–120
Frequency	15	15	30	20	10

c

Frequency density	7.5	15	30	20	10

d Equal; the median is the value of the middle item so there are equal numbers of items above and below the median and the area of the bars in a histogram represents the number of items.
e 102.5 cm, 112 cm
9. a 56%
  c 120 mg/100 ml

Exercise 8E (p. 141)

1. Histograms are based on grouped data and show the data spread evenly within each group and this may not be the case.
2. a a key is given
  b 25
  c 22
  d 4 °C
  e 5.2 °C
3. a 56
  b 25
  c 2
4. a 46   **b** 191   **c i** 3   **ii** 6
5. a 65
  b 180
  c Area A has a slightly higher median but larger range than area B.
  d Mean; e.g. better average than the median because it uses all the scores.

6. a The last group does not have an upper end.
   b The median and the quartiles are not in the open ended group; 33 g, 12 g
   c Yes, if we make the last group end at, say, 100 g.

7. a

Frequency density	3	2	4.2	3.2	0.7	0.3

   b In the private school, the mean age is higher and the age range is smaller.

8. The numbers of people in this age range are not given; No, because although the percentages are the same, the actual numbers may have changed; No increase in percentage in any age group and a reduction in percentage in some age groups, the fall is most marked in the 15–29 age range; e.g.

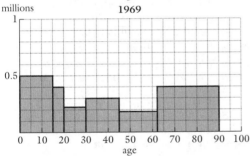

The number of people attending church services in 1989 was lower than in 1979 in all age groups. The largest fall in numbers is in the 15–29 year-olds.

Revision for Higher Tier

1 Number

Exercise 1.1 (p. 151)

1. a $2^3 \times 3^2 \times 7$ **c** 12
   b 36   **d** 126
2. 972
3. a $2\frac{1}{2}$
   b 3, 7, 13
   c 1, $2\frac{1}{2}$, 7, 13
   d 1, 3, 6, 15

4. a 1360   **c** 17   **e** 3   **g** 18   **i** −3
   b 58   **d** 750   **f** 88   **h** 2
5. 604 800
6. 2, 3, 4, 5
7. £73
8. 8; 80 cm
9. a £17   **b** £119
10. a −30   **b** y-values: −30, −5, 4, 3, −2, −5
11. a 60   **b** 1 minute past noon

Exercise 1.2 (p. 154)

1. a 8   **c** $2\frac{5}{12}$ **e** $2\frac{2}{9}$ **g** 2
   b $\frac{1}{12}$ **d** $\frac{11}{12}$ **f** $1\frac{19}{24}$ **h** $1\frac{23}{40}$
2. $1\frac{11}{24}$ inches
3. a £19.50   **c** 1 hour 33 minutes 45 seconds
   b 8.4 m
4. a $\frac{1}{8}$ **b** $\frac{1}{3}$ **c** $\frac{1}{4}$
5. £7.20   **7.** $\frac{1}{20}$
6. $\frac{4}{7}$ **8.** $\frac{67}{162}$

Exercise 1.3 (p. 157)

1. a 2.184   **b** 0.0315   **c** 5.73
2. a 0.63   **c** 0.06   **e** 0.0036
   b 4   **d** 47.2   **f** 420
3. a 7   **c** 3.5   **e** 0.25
   b 16   **d** 0.25   **f** 0.003 125
4. a 9.72   **b** 4.71   **c** 0.0667
5. a $14.5\,\text{m} \leqslant h < 15.5\,\text{m}$
   b $0.1395\,\text{g} \leqslant m < 0.1405\,\text{g}$
6. a 2.75 kg   **b** 6.5 kg
7. 196 cm
8. a 200, >   **c** 4, >   **e** 0.05,   **g** 3,
   b 4, <   **d** 9,   **f** 11,   **h** 3,
9. 12 200 cm^2
10. 12 300 cm^2
11. a 192   **c** 2.58   **e** 0.0499   **g** 2.67
   b 4.31   **d** 9.15   **f** 11.0   **h** 3.01
12. 31 litres
13. a 0.636   **b** 0.294   **c** 5.21   **d** 6.67

Exercise 1.4 (p. 160)

1.

$\frac{2}{5}$	$\frac{3}{20}$	$\frac{9}{25}$	$\frac{3}{100}$	$1\frac{3}{4}$	$2\frac{1}{10}$	$\frac{3}{8}$	$\frac{19}{20}$	$\frac{9}{200}$	$3\frac{3}{25}$
0.4	0.15	0.36	0.03	1.75	2.1	0.375	0.95	0.045	3.12
40%	15%	36%	3%	175%	210%	37.5%	95%	4.5%	312%

2. a $0.\dot{6}$ **b** $0.\dot{1}$ **c** $0.2\dot{7}$ **d** $0.8\dot{3}$
3. a $\frac{7}{9}$ **b** $\frac{7}{90}$ **c** $\frac{142}{9990} = \frac{71}{4995}$
4. a 4%   **b** 6%
5. a £3.75   **b** 160 kg
6. £20   **9.** 4.83%
7. $-\frac{1}{5}$ **10.** £24 324 to nearest £
8. £212.18   **11.** £814.47
12. a Ann Jones, Rudy Williams   **b** 1677

13. £570.58   **15.** 25%

14. £195.56   **16.** 1.57%

17. a 1928 cm^2 **b** 21.6% to 3 s.f.

18. a £12 859.21 **b** 4.73%

19. a 121 000   **b** 6.70%

Exercise 1.5 (p. 164)

1. $(1 + \sqrt{2})(1 - \sqrt{2})$

2. a 2   **d** $1\frac{1}{2}$ **g** x^2y **j** $\frac{1}{8}$

   b $\frac{1}{6}$ **e** $1\frac{1}{8}$ **h** pq **k** 2

   c 4   **f** 4   **i** 3   **l** 25

3. a 1.05×10^8 **b** 2.6×10^{-5}

4. a 36 **b** $\pm\frac{1}{2}$ **c** $\frac{1}{8}$ **d** 1   **e** $\frac{1}{6}$

5. a $5\sqrt{2}$ **c** $3 - 2\sqrt{2}$ **e** $2 + 2\sqrt{2} - \sqrt{3} - \sqrt{6}$

   b $\frac{7\sqrt{2}}{2}$ **d** $\sqrt{2} - 2\sqrt{3}$ **f** 13

6. $-2, \frac{4}{3}, 12, \frac{5}{9}$

7. a 4×10^{-8} **c** 4×10^2

   b 2.5×10^{-1} **d** 7.03×10^2

8. 1.25×10^4 g

9. $\frac{7}{15}$

10. a 1.58   **b** 1.45

11. 6.594×10^{-6}

12. a 3.15×10^{-6} **b** 1.9×10^{30} kg

13. 2.8 m

Exercise 1.6 (p. 167)

1. a $3:1$ **b** $1:25$ **c** $10:1$ **d** $4:1:3$

2. a 62.5 ml   **b** 143 ml

3. 16 kg peat, 4 kg fertiliser

4. 11.5 mm   **6.** 7.5 cm

5. 1.2 km   **7.** 278 sheets to within 1 sheet

8. 1 hour 52.5 minutes (1.875 hours)

9. $5\frac{5}{9}$

10. a $y = \dfrac{15}{\sqrt{x}}$ **b** $1.5625, \left(\frac{25}{16}\right)$

11. Ann £1415.10, Colin £1698.11, Glyn £1886.79

12. a $v = \sqrt{20h}$ **b** 44.7 m/s

13. 1:300

14. a 45.1%   **b** £54 938.27

Mixed Exercise 1.7 (p. 168)

1. a i e.g. 2   **ii** e.g. 12

   b i e.g. $\sqrt{2} + 2\sqrt{2} = 3\sqrt{2}$

     ii e.g. $(\sqrt{2} + 1) + (1 - \sqrt{2}) = 2$

2. 210 g, 200 g

3. a -32.3 **b** $\dfrac{-15 \times 10}{15 - 10} = -30$

4. a 38.5 litres **b** £20.79   **c** 12.9%

5. a £12.42   **b** £15.60

6. a $y = \dfrac{108}{x^2}$ **b i** 3   **ii** ± 2

7. a i $s = 600, u = 10, v = 90$ **ii** 12

   b 12.6 to 3 s.f.

8. £598.03

9. a $h = \dfrac{s^2}{20}$ **b** 20   **c** $9:1$

10. a £343.20   **b** £375

11. a 22.8125 cm^2 **b** 16

12. a $6\sqrt{2}$ **b i** 24   **ii** 2.4 m

13. a 3407   **b** 1999

14. a $\frac{1}{125}$ **b** 25

15. a 0.702 235 540 342 (or as many places as your calculator shows)

   b i $\dfrac{6 \times 10^2}{600 + 200}$ **ii** 0.75

16. 42 mph

17. a 2.3×10^8 km   **c** 7.04×10^{13} miles

   b $1:389$

18. a a number that can be expressed as $\frac{a}{b}$ where a and b are integers

   b $\frac{7}{45}$

19. a 3   **b** 1   **c** $\frac{1}{5}$ **d** $-\frac{1}{3}$

20. Yes by 100 kg if the weights of all the crates are at their upper bounds

21. 4.19×10^6

22. a i rational   **ii** rational

   b i irrational $(= 4\pi)$

     ii rational $\left(= \sqrt{(5 + 4)} = 3\right)$

     iii irrational $\left(= \sqrt{(4 - 1)} = \sqrt{3}\right)$

23. a 5×10^{101} **b** 5×10^{-8} **c** 10^{10}

24. a i e.g. 237

     ii possible values of $\sqrt{n}$ are $\sqrt{236}, \sqrt{237}, \sqrt{238}$ and they are all irrational

   b e.g. 10.24 since $\sqrt{10.24} = 3.2$

25. a $\frac{9}{1}$ **b** $a^{\frac{13}{6}}$ **c** $\frac{5}{2}$

26. a 5.76×10^7 sq miles   **b** 29.2%

27. a 3.65 cm   **b** 15 cm, 14.6 cm

   c i 2

     ii 15 and 14.6 are both 15 correct to 2 s.f.

   d No. Perimeter lies between 29.2 cm and 30 cm and these are not equal when corrected to 2 s.f.

28. e.g. **a i** $\frac{7}{2}$ **ii** $\frac{4}{5}$ **iii** 2

      b i $\sqrt{10}$ **ii** $\sqrt{\frac{4}{5}}$ **iii** $\sqrt{\frac{7}{2}}$

29. a x^{-3} **b** $x^{\frac{3}{2}}$ **c** x^{-2}

30. $\frac{5}{11}$

31. no: diameter of hole can be as small as 60.1 mm and diagonal of cube can be as large as 60.104 mm.

32. 32

33. $3 + 2\sqrt{2}$

34. a i $2^3 \times 3^2, 2^5 \times 3$ **ii** 24 **b** $\frac{5}{11}$

35. 10.20

36. a 2.19×10^{19} **b** 2.3×10^6

37. 7200 million

38. a i $5\sqrt{3}$ **ii** $5\sqrt{2}$ **b** $1 + 2\sqrt{2}$

39. 14

40. a £3573.05   **b** 7.2%

41. a $\frac{1}{9}$ **b** 2^5 **c** $\frac{\sqrt{6}}{2}$

42. 44%

43. **i** true; every third number is divisible by 3
ii true; at least one number is even and one is divisible by 3
iii False; $1 \times 2 \times 3$ is not divisible by 4

44. **a** e.g. $(\frac{1}{2}) > -1$ but $(\frac{1}{2})^2 < (-1)^2$
b e.g. $-1 > -2$ but $(-1)(-1) < (-1)(-2)$
c $(\frac{1}{2})^2 < \frac{1}{2}$

45. **a** false, e.g. $2 + 3 + 5$ is even
b false, e.g. for the set $(19, 23, 29)$, $29 - 19 = 10$

2 Algebra

Exercise 2.1 (p. 178)

1. a $\dfrac{9x}{10}$　　　　**g** $\dfrac{3a - b}{6}$

b $\dfrac{9x + 5}{10}$　　　**h** $\dfrac{2q - 15p}{3pq}$

c $\dfrac{7x - 3}{12}$　　　**i** $\dfrac{3x + 1}{(x - 1)(x + 1)}$

d $\dfrac{7}{2x}$　　　　　**j** $\dfrac{2x - 9}{x(2x - 3)}$

e $\dfrac{2y + 12x}{3xy}$　　**k** $\dfrac{9x^2 - 2x + 28}{(x - 1)(x^2 + 4)}$

f $\dfrac{3x - 1}{x(x - 1)}$　**l** $\dfrac{x^2 - 4}{x(x^2 + 1)}$

2. a $\dfrac{x^2}{3}$　　**e** $\dfrac{3}{2s}$　　**i** $\dfrac{a^3}{3b}$

b $\dfrac{2x}{5}$　　**f** $\dfrac{1}{12x}$　　**j** $\dfrac{3rh}{2}$

c $6x$　　　**g** $\dfrac{2(x + 2)}{2x - 5}$　**k** $\dfrac{2y^3}{x}$

d $\dfrac{1}{2x}$　　**h** $\dfrac{a + 2}{c(2a - 3)}$　**l** $\dfrac{4a^2b^2}{3}$

3. a $2x^2 - 7x - 4$　　**f** $6s^2 + st - 2t^2$
b $28y - 12y^2$　　　**g** $6a^2b - 4ab - 2a^2$
c $3x^2 - 7x - 8$　　**h** $5p^4q - 5p^2q^3$
d $-10a^2 + 41a - 21$　**i** $2\pi r^4 - 6\pi r^3$
e $6p^2q - 4pq^2$　　**j** $3m^3n - 12mn^3$

4. a $(x + 2)(x - 2)$　　**i** $3(x - 3)(x - 1)$
b $(x - 3)^2$　　　　**j** $(3y - 1)(2y + 3)$
c $2t(2s - 3t)$　　　**k** $2A(A^2 - 2)$
d $(x + 4)(x + 2)$　　**l** $(5x - 2)(2x - 3)$
e $(4x + 1)(x + 1)$　　**m** $pq(2p - 3q)$
f $(3x - 2)(2x + 5)$　　**n** $3xy^2(2x^2 - y)$
g $(2t - 5s)(2t + 5s)$　**p** $5\pi ab(3a - 2b)$
h $(3 - 2x)(x + 4)$

5. a 4　　　**b** 25　　**c** $\dfrac{81}{4}$　　**d** $\dfrac{25}{4}$

6. a $(x + 3)^2 - 5$　　**d** $(x - 3)^2 - 13$

b $(x - 4)^2 - 15$　**e** $(x + \frac{3}{2})^2 + \dfrac{27}{4}$

c $(x + 6)^2 - 39$　**f** $(x - \frac{5}{2})^2 - \dfrac{21}{4}$

7. $p = 7, q = -47$
8. $p = 8, q = 2$　　**9.** $a = 6, b = 3$

10. a $\dfrac{2x}{x - 3}$　　**c** $\dfrac{a + b}{b}$

b $\dfrac{x + 3}{x - 3}$　　**d** $\dfrac{(x - 2)}{(2x - 3)}$

Exercise 2.2 (p. 182)

1. a 5　　　**c** 1　　　**e** 8
b 3　　　**d** $\frac{11}{16}$　　**f** $-\frac{3}{10}$

2. a 3　　　**c** $\frac{7}{3}$　　**e** $-5, 4$
b 3　　　**d** $\pm\frac{3}{5}$　　**f** $\dfrac{1 \pm \sqrt{7}}{2}$

3. a $x = 4, y = -1$　　**d** $p = -1, q = 2$
b $x = 2\frac{3}{4}, y = \frac{1}{2}$　**e** $x = 2, y = 3$
c $s = 2, t = 1$　　**f** $x = -1\frac{6}{13}, y = 2\frac{4}{13}$

4. a $s + d = 54$　**b** $s + 2d = 80$　**c** 28

5. a $n > -\frac{1}{2}$
b 0

6. a $3, 4$　**b** $2, 3$
7. a $1, 5$　　　**d** $-1, 1\frac{1}{2}$　　**g** $-\frac{1}{4}, 4$
b $-4, \frac{1}{2}$　　**e** $0, 5$　　　**h** -2
c $-1, 3$　　　**f** $-4, 0, 1$
8. a $x^2 - 6x + 5 = 0$　**b** $5\,\text{cm}$
9. a $x = 1, y = 3; x = -\frac{9}{5}, y = -\frac{13}{5}$
b $x = 2, y = 1; x = -1, y = 2$
c $x = 2, y = 3; x = -\frac{6}{5}, y = -\frac{17}{5}$
d $x = 2, y = 3; x = -\frac{5}{4}, y = -\frac{27}{4}$
e $x = -2, y = -3; x = \frac{3}{2}, y = \frac{5}{3}$
f $x = 2, y = -3; x = -\frac{7}{4}, y = \frac{9}{2}$

10. $-2 < x < 3$
11. a $0.38, 2.62$　　　**c** $-0.73, 2.73$
b $-0.57, 2.91$　　**d** $-1.62, 0.62$
12. 2.6
13. $x = 2.62, y = 0.38; x = 0.38, y = 2.62$
14. $x = 1.18, y = 1.36; x = -0.85, y = -2.69$

Exercise 2.3 (p. 185)

1. a 3　　**b** 24　　**c** 2.3　　**d** $\frac{15}{76}$
2. a $\sqrt{21}$　　**b** $\sqrt{12}$
3. a $c = y - mx$　　　**e** $a = b \pm \sqrt{2s}$

b $x = \dfrac{y - c}{m}$　　**f** $l = \dfrac{T^2g}{\pi^2}$

c $R = \dfrac{100A}{PT}$　　**g** $u = \dfrac{fv}{v - f}$

d $r = \pm\sqrt{\dfrac{C}{\pi}}$　**h** $b = \pm\sqrt{l^2 - a^2 - c^2}$

4. a $P = 4a$　　**b** $A = a^2$　　**c** $A = \dfrac{P^2}{16}$

5. a $t = \dfrac{v - u}{g}$　**b** $v^2 = u^2 + 2gs$

6. a $6, 20, 42, 72, 110$　**b** $6, 24, 60, 120, 210$
7. a $3n + 1$　　　　**c** $\frac{1}{2}n(n + 1)$
b $n^2 - 2$　　　**d** $12 - 2n$
8. a $u_n = 4n + 3$　　**b** $u_n = n^3 - 1$
9. $55, 74$
10. $u_n = (n - 1)^2 - 1; 120$
11. a $500\,\text{m}, 1000\,\text{m}, 1500\,\text{m}, 2000\,\text{m}, 2500\,\text{m}$
b $500n$ metres　　**c** 84
12. a $3, 9, 27, 81$

Exercise 2.4 (p. 190)

1. a i -2 **ii** $\sqrt{20}$ **iii** $(2, 3)$
b i -7 **ii** $\sqrt{50}$ **iii** $(\frac{5}{2}, \frac{9}{2})$
c i $-\frac{8}{5}$ **ii** $\sqrt{89}$ **iii** $(\frac{3}{2}, 2)$
2. a $\frac{1}{2}$ **b** $-\frac{2}{5}$ **c** -4
3. a $2y = x + 6$
b i 6 **ii** 5 **iii** $(5, 5.5)$ **iv** $3\sqrt{5} = 6.71$
4.

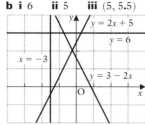

5. $2y = -3x + 6$, $-\frac{3}{2}$
6. a $\frac{1}{3}, -3$ **b** $-\frac{3}{4}, 3$ **c** $2, -4$ **d** $\frac{1}{4}, \frac{3}{4}$
7. a $y = 4x + 1$ **c** $y = -\frac{1}{2}x$
b $3x + 4y = -6$ **d** $y - x = -1$
8. a £20 **b** £44 **c** 75 minutes
d 0.48, cost, in £, per minute spent on repair
9. a $(3, -4)$ is not on the line through the other points
c $x = \frac{4}{5}, y = \frac{7}{5}$
10. $2y = x + 13$

Exercise 2.5 (p. 195)

1. a $-0.5, 1.5$ **b** $-0.62, 1.62$
2. b $y : 0.25, 0.44, 1, 4, 4, 1, 0.44$
c $-1.6, 1.1$
3. a $y : -25, -1, 7, 5, -1, -5, -1, 17$
c max $y = 7.2$ at $x = -0.8$, min $y = -5$ at $x = 2$
d line is $y = 2x - 4$; $-2.4, 1.1, 3.3$
4. a -0.8
b

possible values of k

possible values of k

5. a $y : -21, -5, -1, -3, -5, -1, 15$
b i $y = -4$; $-1.9, 0.35, 1.53$
ii $y = x + 2$; 2.46
6. a, b

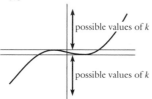

c $x = 5.9, y = 1.1$; $x = 1.1, y = 5.9$

7. a

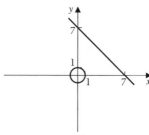

b there are no solutions
8. a A **b** $m = kr^2$
c Yes, because the mass varies as the square of the radius and, since mass $\propto$ volume and volume $= \pi r^2 d$ where d is the thickness, d must be constant.
d $9.55 \, \text{g/cm}^3$
9. a $V(\text{£000s}) : 10, 8, 6.4, 5.12, 4.10, 3.28, 2.62, 2.10, 1.68$
c i £8940 **ii** 3.11 years (3 years 1.3 months)
10. a $n : 1, 2, 4, 8, 16, 32, 64$
c $n : 2.83, 11.3, 45.3$
d i 5.7 **ii** 18
e i 6 hours 30 mins **ii** $10\frac{1}{2}$ hours
11. c i $8 \, °\text{C}$ per minute **ii** $12 \, °\text{C}$ per minute
d No. The alcohol curve is less steep than the water curve for $t > 4$.
12. a C **b** E **c** B **d** A **e** D
13. $20 = ka^0$, $30.2 = ka^1$; $k = 20, a = 1.5$
14. b 0.9
15. a 0, number of houses is not changing
b 32, number of houses is increasing at 32 per year

Exercise 2.6 (p. 201)

1. $x \leqslant 2, y \leqslant 2x + 4, y \geqslant 3 - x$
2. a $y = 4x - 6$
c $\frac{1}{2}$
d

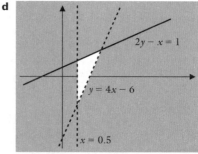

3. b $x \leqslant -0.24, x \geqslant 4.24$
4. $(2, 1), (3, 1), (3, 2)$
5. a $x \geqslant 50, y \geqslant 100, 1.5x + y \leqslant 450$
b 425
6. a $y \geqslant \frac{1}{6}x, y \leqslant x, y \leqslant -\frac{2}{3}x + 5$
b $y \leqslant -x, y \geqslant -\frac{1}{6}x, y \leqslant \frac{2}{3}x + 5$

Exercise 2.7 (p. 205)

1. **a** 90 seconds
   b No. No part of the graph has zero gradient.
   c $0.27 \, \text{m/s}^2$
   d cyclist slowed down to a stop
   e 347.5 m
   f 3.86 m/s
2. **a i** 0   **ii** $1.55 \, \text{m/s}^2$
   b i 0–10 seconds   **ii** 10–20 seconds
3. **a** 4 minutes
   b $8 \, \text{km/h/min} = 480 \, \text{km/h}^2$
4. **b** 6   **c** acceleration (m/s^2)

Exercise 2.8 (p. 210)

1.

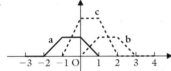

2. **a** $\frac{1}{2}$
   b

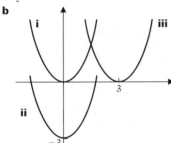

3. **a**   **c**

   b   **d**

4. **a** translation by 3 units along x-axis to the left;
   $y = (x + 3)^2$
   b reflection in x-axis followed by translation of
   3 units up y-axis; $y = -x^2 + 3$
   c translation of 2 units down the y-axis;
   $y = x^2 - 2$
   d translation of 2 units to the right along x-axis;
   $y = (x - 2)^2$
5. **a** 6   **c** reflection in y-axis

6. **a** $(7, -1)$ **c** $(5, -2)$
   b $(5, -3)$ **d** $(\frac{5}{2}, -1)$
7. **a**

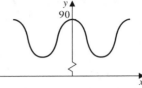

   b

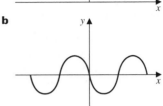

8. **a, b**

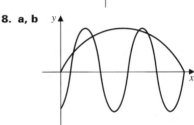

9. **b** 0, 180, 360
   d i

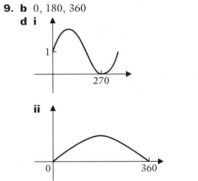

   ii

10. 30   11. **b** 49, 131
12. **c** translation by 3 units up parallel to y-axis and
    1 unit to the left parallel to x-axis
13. **b i** 7.36 a.m. and 4.24 p.m.
    ii 3 m below rung A

Mixed Exercise 2.9 (p. 213)

1. **a** $y = 3 - \frac{3}{4}x$
   b i $2n(7 - 2n)$ **ii** 1, 2, 3
2. **a** 6.375, 7.296
   b 1.5 and 1.6
   c 1.6
3. **a i** $n > 11$ **ii** 12
   b i $14x < 50 + 8x$
   ii $x < 8\frac{1}{3}$; for a hire period of less than 8
   months, Vincent's scheme is cheaper but
   for a period of more than 8 months,
   Rowena's scheme is better value

4. a $25(G + 10)$ **b** $G = \dfrac{C - 121}{14}$

5. a $117 = k + 4m$, $149 = k + 6m$ **b** 16

6. a $x > -3\frac{3}{5}$ **b** $x = 35$, $y = 15$

7. $c = \dfrac{100s}{p + 100}$

8. a 7812 **b** $d = \pm\sqrt{\dfrac{k}{N}}$ **c** 3.95 cm

9. $x + y = 40$, $2x + 3.5y = 92$;
8 at £3.50 and 32 at £2.00

10. $-2, -1, 1, 2$

11. i y $\pm 4\frac{1}{2}$ **ii** $x = 2\frac{3}{4}$

12. $-5.5, 0.5$ **13.** $3x(x - 2)$

14. a $u_n = n^2$ **b** $u_n = n^2 + n$

15. a $C = LF$ **b** 245 m **c** 648 kHz

16. a $1.8x$ km/h

  b i $\dfrac{6660}{x}$ **ii** $\dfrac{3700}{x}$

  c $\dfrac{6660}{x} - \dfrac{3700}{x} = \dfrac{10}{3}$, 888 km/h

17. a $\dfrac{2V}{W(E + R)}$ **c** 5, 10

18. a 2 m **b ii** $a = -5$, $b = 40$ **c** 20.75 m

19. $x = \frac{1}{2}$, $y = 2\frac{1}{2}$

20. a $x \geqslant 3.5$ **b** $-2 \leqslant x \leqslant 2$

21. a $\dfrac{8x - 2}{(x + 2)(x - 4)}$

  c i $(x - 5)^2 - 31$ **ii** $5 \pm \sqrt{31} = -0.57$ or 10.57

22. a i $4x^3y^2$ **ii** $16p^4q^6$

  b i $3a^2b^2(3b + 5a)$ **ii** $(x + 12)(x - 5)$

  c

23. $p = 4$, $q = -3$

24. $-0.64, 3.14$

25. a $\dfrac{x - 1}{x - 2}$ **b** $0, -2$ **c** $-1, 5$

26. a $8a^5$ **b** 6.45, 1.55 **c** $y = \dfrac{3x - 10}{2x - 5}$

  d $x = 1$, $y = 4$; $x = 6\frac{1}{2}$, $y = -1\frac{1}{2}$

27. $3, -1$

28. a $3x^3 + 4x$ **c** $8x^6y^3$

  b $(a - b)(x + y)$ **d** $x = 2\frac{1}{2}$, $y = -6$

29. $x = -\frac{2}{5}$, $y = -7\frac{4}{5}$; $x = 6$, $y = 5$

30. a 2, 3, 4, 5 **c** $n(n - 1)$

  b $x = 5$, $y = -1$ **d** $6a^5b^2$

31. $\frac{1}{4}$

32. a $(x - 3)(x + 3)$ **b** $\dfrac{x - 3}{2x - 1}$

33. a $a = 25$, $b = -5$

  b 7.24, 2.76

  c -5

35. a $\dfrac{17a}{30}$ **b** $\frac{1}{3}$ **c** $-6, 8$

36. b i $(4x - 9)(x - 1)$

  ii $(2y + 3)(2y - 3)(y + 1)(y - 1)$

37. b $0, -2.2$

38. 0.8 m/s^2

39. b $22°$, $158°$

40. P

41. a D **b** e.g.

42. a T : 14, -2, -10, -14, -16

  b, c T

  d 2.7 minutes

43. b (The solid line is the original graph.)

  i
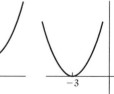

  ii

  iii

  iv

44. a i no acceleration (maximum speed)
  ii deceleration
  b 9 m/s^2

45. a 24 **c** reflection in y-axis **d** 3

46. b $-0.4, 2.4$ **c** $x \leqslant -0.3$, $x \geqslant 3.3$

47. a **b** **c**
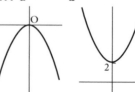

48. $y \geqslant 1$, $y \leqslant x - 2$, $x + y \leqslant 6$

49. a $y = 2x + 8$ **b** $(-4, 0)$

50. a $(6, 5\frac{1}{2})$ **b** 2

51. a ii $120°$, $240°$ **b ii** 4.33 m above

53. a −0.7, 1, 2.7   **b** $k < 0$ or $k > 4$
   c ii $y = x^3 - 3x^2$
54. a i B   **ii** C   **iii** A   **iv** D
   b

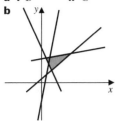

55. b −241.2°, 61.2°, 118.8°
56. a number of bacteria: 20, 60, 180, 540, 1620,
   4860
   b $n = 3^t \times 20$
   d i 1030 (± 150)
     ii rate per hour at which the number of
        bacteria are increasing after $3\frac{1}{2}$ hours
57. a −2, −1, 1   **c** 0.80, −0.55, −2.25
   b 1.52
58. a iv   **b ii**   **c v**
59. b 63.5   **c i** 243.5   **ii** 296.5
60. b −0.3 °C per minute
   c rate at which temperature is falling, in °C per
     minute, 80 minutes after being left to cool
61. b $x = 3.27, y = 3.55$ **d** $-\frac{3}{4}$
62. i (−0.5, 0)
   iii the lines are parallel so there is no point of
     intersection
63. a $n = 2 \times 3^t$
   b i 60
     ii the rate of spread of the rumour (number
        of people/hour hearing the rumour) after
        3 hours
64. b ii −1.4, 0.8, 3.5
65. a 5 mg   **b** 5, 2.89, 1.67, 0.96, 0.56
   d 3.45 hours
   e more rapidly; rate of decrease is faster because
     dividing by a larger power of 3
66. a i $(x - 6)(x + 2)$ **ii** −2, 6
   b A(6, 0), B(−2, 0)   **c**

67. 1.5
68.

69. a $W = BH^2$ **b** 100 kg
   c i 54.4   **ii** mass per metre of height
70. a $s = 2t$ **b** 0, 20, 25
   c i 10 seconds   **ii** 2 ft/s²
71. a 30°, 150°, 390°, 510°
   b 150°, 210°, 510°, 570°

3 Shape, Space and Measure

Exercise 3.1 (p. 235)

1. a 5.24 km   **e** 0.5 kg   **i** 330 mm
   b 24 cm   **f** 0.67 m   **j** 0.42 m
   c 60 g   **g** 139 cm   **k** 5200 mg
   d 7.5 g   **h** 2.45 m   **l** 7 kg
2. a 82.6 mm   **e** 60 000 mg   **i** 24 ft
   b 2400 kg   **f** 3520 miles   **j** 72 in
   c 6 ft   **g** 7 ft 1 in   **k** 18 480 ft
   d 46 yd 2 ft   **h** 67 oz   **l** 180 in
3. a 157 lb   **e** 12 st 8 lb   **i** 4000 m²
   b 54 cwt   **f** 4 lb 6 oz   **j** 25 ha
   c 32 in   **g** 11 ft   **k** 15 acres
   d 335 cm   **h** 80 in   **l** 2662 sq yd
4. a 5600 ml   **e** 10 gl 5 pt   **i** 15 gallons
   b 320 pints   **f** 8.206 litres   **j** 76 litres
   c 420 litres   **g** 12 litres   **k** 0.645 litres
   d 77 ml   **h** 420 cm³   **l** 35 ml
5. a 5.6 cm   **e** 1500 mm²   **i** 10 cm³
   b 72 m   **f** 800 000 mm²   **j** 5300 mm³
   c 3 ft   **g** 8500 cm²   **k** 216 sq in
   d 16 yd   **h** 7000 cm³   **l** 2 sq ft
6. a 4.5 m   **e** 128 in   **i** 23.4 cm
   b 3.5 g   **f** 910 kg   **j** 155.94 g
   c 13 ft   **g** 375.2 cm   **k** 1010 mm²
   d 34 cwt   **h** 4964 lb   **l** 524 sq in
7. a 11 lb   **f** 4.1 kg   **k** 22.5 litres
   b 80 km   **g** 31.25 miles   **l** 4 gallons
   c 3 m   **h** 8.8 oz   **m** 300 acres
   d 80 in   **i** 5 cm   **n** 40 ha
   e 16.7 yd   **j** 4 tons   **p** 16 in
8. a 20 m/s   **c** 0.6 g/cm³   **e** 200 kg/m³
   b 800 km/h   **d** 55 mph   **f** 126 km/h

Exercise 3.2 (p. 240)

1. a i 20.4 cm   **ii** 24.3 cm²
   b i 37.8 cm   **ii** 80.6 cm²
   c i 214 mm   **ii** 3630 mm²
   d i 48 cm   **ii** 100 cm²
   e i 46.8 cm   **ii** 152.5 cm²
   f i 136 mm   **ii** 1032 mm²
2. a i 49.6 cm   **ii** 81.8 cm²
   b i 83.3 cm   **ii** 154 cm²
3. a 7.4 cm   **b** 6.3 cm   **c** 9 cm   **d** 43.0 m
4. a i 226 cm³   **ii** 207 cm²
   b i 1950 mm³   **ii** 1440 mm²
   c i 1020 m³   **ii** 679 m²
   d i 463 cm³   **ii** 290 cm²
5. a 17 900 mm³
   b 154 000 cm³
6. a 45.4 cm   **b** 80.3 cm   **c** 803 cm²   **d** 12.8 cm
7. a i 194.4 cm³   **ii** 0.175 kg
   b 478 ml   **c** 4 cm
   d i 697 cm³   **ii** 380 cm²   **iii** 4.74 kg
   e i 511 cm³   **ii** 10.2 cm
   f 20 cm²

8. a i 3.125 cm **ii** 30.7 cm^2 to 3 s.f.
  b 335 ml
  c 10.9 cm to 3 s.f.
9. 10 cm
10. a 62.8 m
   b 4270 m^2
   c 513 m^3
11. a R   **b** P   **c** U
12. 1
13. a 14 140 cm^3
   b 2830 cm^2 (taking the surface area as that of the metal globe), 283 cm^3 (assuming card is flat)
   c 707 cm^2

Exercise 3.3 (p. 247)

1. a $a = 74$, $b = 58$, $c = 74$, $d = 122$
  b $e = 50$, $f = 130$
  c $g = 52$, $h = 26$, $i = 26$, $j = 128$
  d $k = 36$, $l = 74$, $m = 61$
2. a $p = 42$
  b $x = 65$, $y = 35$, $z = 145$
  c $q = 38$, $r = 72$, $s = 72$, $t = 70$, $u = 65$
  d $a = 60$, $b = 30$, $c = 60$
3. a i 30   **ii** 20
  b i 144°   **ii** 156°
  c i 20   **ii** 24
  d i Yes, 8   **ii** No
  e i Yes, 30   **ii** No
4. a $z = 126$ **c** $x = 72$, $2x = 144$
  b $x = 120$ **d** $a = 36$, $b = 108$, $c = 36$
5. 110°, 150°, 135°, 120°, 65°, 140°
6. $x = 70$, $y = 110$, $z = 30$
7. a $a = 72$, $b = 44$, $c = 92$
  b No, interior angles not all the same
8. a $(6, -3)$, 49 sq units
  b $(5, 3)$, 20 sq units
  c $(5, -2)$, 21 sq units
  d $(-2, 5)$, 50 sq units
9. a C(5, 1), D(5, 5)
  b C(5, −2), D(−2, −1)
  c D(0, 6), E(0, 3), 30 sq units
  d (5, 4), (−1, 2)
10. a rhombus   **b i** 144° **ii** 36°

Exercise 3.4 (p. 251)

1. a D, E and J; F and H **b** E and C; B and G
2. a No
  b Yes, equiangular and sides in same ratio
  c Yes, both rectangles with sides in same ratio
  d Yes, equiangular
3. a △ABC and △JKL, 2 angles and a pair of corresponding sides equal
  b △GHI and △ABC or △JKL, equiangular
4. b i 15 cm   **ii** 8 cm
5. a 3 : 5   **b** 20 cm   **c** 9 : 25
6. a 9 : 25   **b** 24.3 cm^3

7. a 4 : 3   **b** 320 cm^3
8. a i 1 : 1.03   **ii** 1 : 1.05
  b i 90 mm   **ii** 78 mm
  c i 261 mm   **ii** 224 cm^2

Exercise 3.5 (p. 255)

1. a $d = 88$, $e = 136$
  b $a = 47$, $b = 94$, $c = 43$
  c $f = 124$, $g = 62$, $h = 118$, $i = 62$, $j = 28$
  d $a = 55$, $b = 80$, $c = 10$, $d = 30$
2. a $a = 82$, $b = 71$, $c = 27$
  b $d = 60$, $e = 60$, $f = 72$, $g = 48$
  c $3x = 33$, $4x = 44$, $y = 11$
  d $a = 48$, $b = 48$, $c = 42$, $d = 6$, $e = 42$
4. a 57°   **b** 66°   **c** 57°   **d** 63°
5. a $x°$ **b** $90° - x°$ **c** $90° - y°$
6. a 33°   **b** 73°   **c** 72°
7. a 106°   **b** 74°   **c** 53°
8. a 105°   **b** 75°   **c** 30°   **d** 75°
9. a $(90 - 2x)°$ **c** $(90 - 2x)°$ **e** $(90 - 3x)°$
  b $(90 - 5x)°$ **d** $5x°$

Exercise 3.6 (p. 259)

1. b 87 m
2. a i 84°   **ii** 66°
  b i 9.8 cm   **ii** 12 cm
3. a 170 m   **c** 115 m
4. b

5. b 3.9 m   **6. b** e.g.

7. c 5.6 m
8. Arc of circle, centre C, radius BC, from initial position of C to car body

Exercise 3.7 (p. 265)

1. a i 8.4 cm   **ii** 10.5 cm
2. a 13.0 cm   **b** 12.6 cm
3. 1.2 m
4. a 3.36 m
  b $1.7^2 - x^2 = 2.9^2 - (3.36 - x)^2$; 0.859 m, 2.50 m, 1.47 m
5. a i 5.25, 5.15 **ii** 8.35, 8.25
  b 6.573 cm, 6.364 cm to 4 s.f.
  c 6.5 cm to the nearest half centimetre

6. a $> 90°$ **b** $= 90°$ **c** $< 90°$
7. a 10.7 cm **b** 11.9 cm **c** 14.4 cm **d** 9.64 cm
8. b 13.7, −1.7 **c** 13.7 cm, 12.7 cm, 18.7 cm
9. a 41.8° **b** 10.5 cm
10. a 20.7 m **b i** 60.6° **ii** 10.2 m
11. a 143.1° **b** 3.2° **c** 497 m **d** 298 m
12. a 27.39° **b** 24.84°
13. 2.74 km
14. b 216 yd **c** 203 yd, 032.4° **d** 268 yd, 340°
15. b $\frac{1}{4}$ square unit **f** 1.31 units

Exercise 3.8 (p. 270)

1. b i $c = \begin{pmatrix} 0 \\ -4 \end{pmatrix}, d = \begin{pmatrix} -6 \\ -4 \end{pmatrix}$

 ii same size but reflected in a vertical line
2. a i $\overrightarrow{AD}$ **ii** $\overrightarrow{BA}$ **iii** $\overrightarrow{DC}$
 b i $\overrightarrow{AB} + \overrightarrow{BD}$ **ii** $\overrightarrow{CD} + \overrightarrow{DB}$
3. a i Yes. AX is parallel to XC and X is a
  common point.
 ii Yes. DX is parallel to XB and X is a
  common point.
 b i, ii $a − b$ **iii** $a + b$ **iv** $a + b$
 c parallelogram, opposite sides are parallel and equal
 d diagonals bisect each other
4. a i $−p$ **ii** $q − p$ **iii** $2q − 2p$
5. b 52 km/h, 138°
6. a $b − a, \frac{1}{2}(b − a), \frac{1}{2}(b + a), \frac{1}{2}b − a$
7. a i $3r − q$ **ii** $\frac{1}{2}(3r − q)$ **iii** $\frac{1}{2}(3r + q)$
 b $3r$
 c i parallelogram **ii** trapezium
8. a i $q − p$ **ii** $\frac{1}{2}q − \frac{1}{2}p$

Exercise 3.9 (p. 274)

1. a i 90° anticlockwise rotation about $(0, 1)$
 ii reflection in $y = 1$
 b reflection in $y = x + 4$
 c translation by $\begin{pmatrix} -3 \\ 3 \end{pmatrix}$
 d e.g. translation by $\begin{pmatrix} -3 \\ 0 \end{pmatrix}$ followed by reflection
  in $y = 2.5$
2. d reflection in $y = x$
3. a e.g. rotation by 90° clockwise about $(0, 0)$
  followed by translation $\begin{pmatrix} 0 \\ -2 \end{pmatrix}$
 b rotation by 90° clockwise about $\left(-\frac{1}{2}, 0\right)$
4. a $\left(0, 1\frac{1}{2}\right), 2$ **b** $(12, 1), (10, 5), (8, 5)$
5. a $(11, 4), −3$
 b $(3.5, 7), \frac{1}{3}$
 c rotation of 180° about $\left(8\frac{1}{2}, 5\right)$
6. a 10 000 m²
 b i 4 cm **iii** 9 cm
 ii 6.5 cm **iv** 8.875 cm²
7. a $1\frac{3}{5}$ **b** 8 cm **c** 53° **e** $\frac{4}{5}$
8. a $\frac{2}{3}$ **b** $\frac{4}{9}$
9. b $y = x − 2$

10.

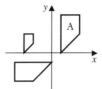

Exercise 3.10 (p. 279)

1. a i 6.40 cm **iii** 10.3 cm
 ii 9.85 cm **iv** 11.0 cm
 b i 66.0° **ii** 54.6°
 c 38.7°
 d i 235 cm³ to 3 s.f. **ii** 38.25 cm²
2. a 0.243 m³
 b 12.5°
 c 2.57 m²
 d 1.38 m
3. a 7.18 cm
 b 63.8°
 c 70.8°
4. a $\sqrt{12}$ cm
 b $\sqrt{\dfrac{32}{3}} = \dfrac{4\sqrt{6}}{3}$ cm
 c $\dfrac{16\sqrt{2}}{3}$ cm³

Mixed Exercise 3.11 (p. 281)

1. a 17.5 m
 b 54.5°
2. a $\widehat{APB} = 106.3°$
 b 18.5 m
3. 19.8 minutes
4. a 175.4 cm²
 b 26.8 cm to 27.0 cm
 c 37.4°
5. a 11.25 cm²
 b 5.50 cm²
 c i 4.5 cm **ii** 12.4 cm²
6. a 12.4 m **b** 72.1°
7. 3.2 cm
8. 492.4 km/h on bearing 050°
9. a 23.3 cm **b** 62.7°
10. a i $\frac{1}{2}c − a$ **ii** $c − 2a$
 b AD : OE = 1 : 2
11. 50.9 m
12. a $x°$ **b** $(90 − x)°$ **c** $x°$ **d** $2x°$
13.

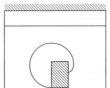

14. a 6.25 cm ⩽ AC < 6.35 cm
 b 4.41 cm, 4.08 cm
 c 4 cm to nearest cm

15. 25.07 cm

16. a 35.3 cm²

b i 9.42 mm² **ii** $\frac{\pi x}{40} - \frac{9}{2}\sin x$

17. a 42 cm³

b 384 cm²

c i 16 **ii** 64

18. a 41.9 cm **b** 6.67 cm **c** 27

19. a 13 N **b** 3 N **c** 9.43 N

20. a 130° **b** 65°

21. a 63°

b enlargement, centre O by scale factor 2.24

22. a 339 cm³ **b i** 50 tubes **ii** 4635 cm³

23. a 8.20 m **b** 15.3°

24. a 90° (angle in a semicircle)

25. 14.5°

26. a $4\pi r^2 + 3\pi rR$

b only formula containing terms with (length)² terms only

27. a 90° **b i** 108° **ii** 72°

28. 618 cans

29. a 80° **b** 74°

30. b 6.8 km

31. b 2p + 4q

c they lie on a straight line

32. b 43 m

33. a 119 km **b** 183° **c** 330 km/h

34. a i 3.11 m² **ii** 5.16 m **b** 31 m²

35. a i 4b − 8a **ii** 3b − 6a **b** 2a + 3b

36. a 11.9 cm³ **b** 1 cm

37. $\overrightarrow{PS} = \frac{2}{3}\mathbf{a} - \frac{2}{3}\mathbf{b}$, $\overrightarrow{SR} = \frac{1}{3}\mathbf{a} - \frac{1}{3}\mathbf{b}$;

PS : SR = 2 : 1; P, S and R lie on a straight line

38. a i 3.00 cm **ii** 27 cm **b** 71.6 m³

39. b enlargement by scale factor $\frac{3}{2}$, centre (0, 0)

40. a b + c **c** $\frac{1}{2}\mathbf{a} + \frac{1}{2}\mathbf{b} + \frac{1}{2}\mathbf{c}$

b b − c **d** $\frac{1}{2}\mathbf{a} + \frac{1}{2}\mathbf{b} + \frac{1}{2}\mathbf{c}$

41. a i 101° **b** 1.8 m **c** 3.25 m

42. rotation 90° anticlockwise about O

43. 5.35 m

44. a i $(90 - x)°$ **ii** $x°$ **iii** $(90 - 2x)°$

b both are expressions for twice the area of △ACE

45. a 1.41 m **c** 4.92 m² **e** 6880 kg

b 0.513 m **d** 9.83 m³

46. 29.4 cm²

47. a q − p **b** $p + \frac{1}{2}q$ **c** $q + \frac{1}{2}p$

48.

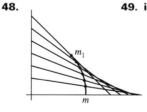

49. i OAB **iii** $\frac{1}{4}$

50. a ii θ = 48.2° **b** 46 seconds

52. a 262 m/s **b** 266°

53. a 1283 m³ **b** 116 m

54.

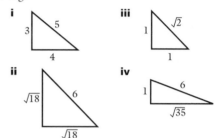

55. a 5.66 cm **b** 6.12 cm

56. a 1.44 cm³ **b** 0.18 cm **c** 8 sweets

57. a i 6.09° **ii** 220 km/h

58. e.g.

i

3, 5, 4

iii

1, √2, 1

ii

√18, 6, √18

iv

1, 6, √35

59. a 86.4 m **b** 6.5 cm **c** 70°

60. a 70.0 cm **c** 31.0°

b 612 cm² **d** 2.4 cm

61. 8 **62. a** 23.9 cm² **b** 78.4°

63. a 35°, angles in the same segment

b i 70° **ii** E is the centre of the circle

64. a i 3p **ii** −4p

b e.g. $\mathbf{AM} = \mathbf{NC} = \frac{1}{2}\mathbf{a}$ so a pair of opposite sides are equal and parallel.

65. a 16.7 km **b** 346° **66. b** $2\sqrt{10}$

67. a i c − b **ii** $\frac{1}{2}$c **iii** $\frac{1}{2}$c

b parallelogram; e.g. $\mathbf{NQ} = \mathbf{MP}$ so a pair of opposite sides are equal and parallel.

68. a 50° **b** 129 cm² (128.56... cm²)

4 Data Handling

Exercise 4.1 (p. 305)

1. e.g. How often do you visit this centre?
Less than once a week?
Once a week?
More than once a week?

How far have you travelled to get here?
Less than 2 miles?
Between 2 and 5 miles?
More than 5 miles?

2. frequency: 2, 41, 52, 31, 4

3. a there are fewer men than women in the company

b 9

4. a e.g. the early arrivals may all come by bus from the same area

b stratified sample using classes as the groups

c e.g. How many books did you borrow from the school library last week?
none? 1? 2? 3 or more?

5. a age because appetites change as children get older and gender as boys and girls may make different choices of food

b not every pupil has an equal chance of being selected; schools are different in size; if a school's register is arranged in order of classes, only pupils from one age group will be selected; if the register is sorted into girls and boys only one gender will be selected.

6. Dainton 5, Horley 10, Jenford 16, Monkley 8, Purstock 11

7. a 145
b 76
c 27

8. a e.g. not representative as more first year pupils than older pupils are likely to use the school bus
b select a sample from each class.

Exercise 4.2 (p. 311)

1.

Score	1	2	3	4	5	6
f	10	8	8	6	7	11

2. c e.g. a pie chart because it shows proportions which are more important than actual numbers
e i e.g. use the same sized groups, so combine the first three groups
ii e.g. pie chart

3. a i 20   **ii** 125   **b** 30

4. a

Height (m)	–2	–4	–6	–8	–10
Cum f	6	14	24	39	50

c 31

5. b 7–8 years old
c No. There is no correlation between age and height after mid-teens.

6. a 420   **b** 100   **c** 55%

7. c about 40 minutes

8. a i 21   **ii** 33
b ii ranges equal, the total weight and the variability of the weight on the Paris flight is lower.

9. a 7620, 7583, 8130, 8430
b

c upwards

Exercise 4.3 (p. 317)

1. 3.75, 4
2. a i 15 minutes   **ii** 15 minutes
b 14.7 minutes
3. a i 72 mph, 16 mph   **ii** 57%
b ii median speed is lower and the IQ range is smaller, i.e. cars travelled at a more uniform speed.
4. a £50 000–£99 000   **b** £123 000
c e.g. the mode because it gives the most common house prices and is not distorted by the two very high prices that may not be repeated
5. a 5.33 miles   **b** 3.48 miles
c the mean distance travelled on Saturday is lower but the distances are slightly more variable
6. 47 g, 44 g, 17 g
7. a Appleshire   **b** Appleshire
c the range in Cromshire is greater than in Bankshire, median higher in Bankshire
8. a 90   **c** 12
b 10–14 empty seats   **d** 88
9. b i $12\frac{1}{2}$ minutes   **ii** 7 minutes
iii 21   **iv** 17.6%   **v** $\frac{3}{43}$
10. a 27.3 mph
b

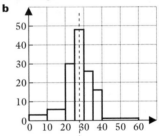

c line that divides the area of the histogram in half (28 mph)

Exercise 4.4 (p. 322)

1. $\frac{7}{16}$ **4. a** $\frac{5}{6}$ **b** 50
2. $\frac{1}{5}$ **5.** $\frac{2}{9}$
3. a 0.25   **b** 0.75
6. a 0.5   **b** 0.04   **c** 0.45
7. a $\frac{23}{40}$ **b** $\frac{22}{39}$
8. a i $\frac{1}{6}$ **ii** $\frac{7}{12}$ **b** $\frac{1}{16}$
9. a 30   **b** $\frac{9}{13}$ **c** $\frac{3}{5}$ **d** $\frac{136}{435}$ **e** $\frac{26}{145}$
10. 5

Exercise 4.5 (p. 325)

1. 0.2775   **4. a** $\frac{37}{72}$ **b** $\frac{31}{66}$
2. $\frac{12}{55}$ **5. a** 0.27   **b** 0.0398
3. 0.88   **6. a** 0.0179   **b** 0.331
7. a 0.09   **b** 0.24   **c** 0.66
8. $\frac{1}{15}$

Mixed Exercise 4.6 (p. 326)

1. a $\frac{7}{12}$

b

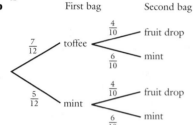

First bag   Second bag

$\frac{7}{12}$ toffee
  $\frac{4}{10}$ fruit drop
  $\frac{6}{10}$ mint

$\frac{5}{12}$ mint
  $\frac{4}{10}$ fruit drop
  $\frac{6}{10}$ mint

c i $\frac{1}{4}$ **ii** $\frac{31}{60}$

2. a e.g. age because attitudes to organic food tend to change with age; income because organic produce is dearer

b e.g. excludes most people with full-time jobs; excludes most people who do one big shop each week;

3. a frequency density: 0.8 1.0 1.8 1.6 0.9 0.15

b

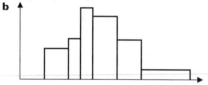

c 15

4. 0.14   **5.** 72.5 g

6. a

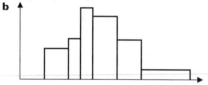

b $\frac{1}{5}$ of each age group; i.e. 3, 8, 12, 14, 3

7. a 165.5   **b** 9 cm   **c** 23

8. a 0.02   **b** 0.64

9. a data given as completed years

b $x = 22$, $y = 18$

10. b 6

11. a $\frac{12}{35}$ **b** $\frac{19}{35}$

12. a e.g.

How far did you travel?   Less than 1 mile
From 1 to 5 miles
More than 5 miles

How much did you spend?   Less than £10
From £10 to £20
More than £20

b $\frac{1}{8}$ of each year group

Year group	9	10	11	12	13
Sample size	26	24	25	15	10

13. a No. There are more boys than girls in that year so he should ask more boys.

b 15 pupils

14. b i 43.5 mph   **ii** 38–49 mph   **iii** 71.5%

15. a

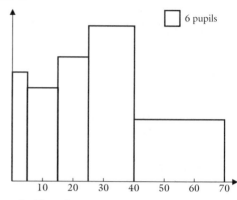

b 60 pupils

16. i $\frac{2}{15}$ **ii** $\frac{8}{15}$

17.

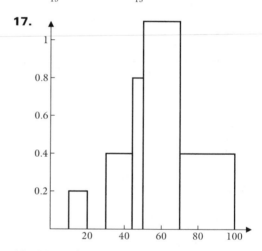

18. 0.281 75

19. a

Age	0	–10	–20	–30	–40	–50	–60	–70	–80	–100
Cum f	0	44	95	154	222	272	307	338	356	360

b i 33 years   **ii** 31 years

c population in the 2nd village is older with a less evenly distributed age range

20. e.g. excludes households without a phone; the person who answers the phone may not be representative of the whole household; may have used a bus in a different town

21. 0.88

22. a

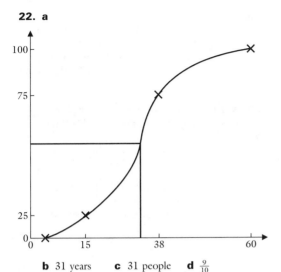

b 31 years   **c** 31 people   **d** $\frac{9}{10}$

23. e.g. choose pupils at random from the register so not biased by who you know/like...
stratify the sample to include the correct proportion of each year group
and stratify the sample to include the correct proportion of boys and girls – age and gender may both affect eating habits

24. a 0.35   **b i** 0.45   **ii** 0   **iii** 0.06

25. a $150 \leqslant p < 155$
  b 156 litres
  c mean – the mode ignores the fact that two groups have almost the same frequency

26. a 0.75   **b** 20 days

27. a

f	4	24	48	14	6	4
Cum f	4	28	76	90	96	100

  c 52 cm   **d** 76%

28. a $\frac{1}{11}$ **b** $\frac{1}{3}$

29. a i mean: £2.50
  ii assumes that the amounts are evenly spread within each group and this cannot be true in this case
  b i Cum f: 0 7 19 35 45 51 55 58 60
  iii £2.30
  c £1.15
  d mean as it includes all the amounts and so reflects the total spending power of the group

30. a 625
  b assumes that fish are evenly mixed at time of second sample and that all fish are equally likely to be caught – both assumptions are not necessarily valid

31. a $\frac{1}{2}$
  b 300

32. a i

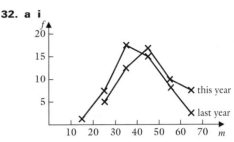

  ii modal group is higher, so is the minimum distance
  b i Cumulative frequency: 5 17 34 43 50
  iii 17 m
  iv 11 competitors

33. $\frac{29}{48}$

34. a $\frac{1}{20}$
  b $\frac{1}{40}$
  c $\frac{1}{800}$ assume topics are independent of themes

35. $b < r$ ($b = rw$) so should be marked to the left of r ($\simeq 1$ cm from O)

36. 0.75

37. a $\frac{53}{100}$
  b

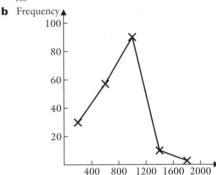

38. a i

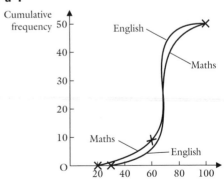

  ii Maths: median 63, IQ range 22
  iii marks for Maths are slightly higher (on average) but more widely spread than the marks for English
  b 0.88

39. a $\frac{25}{72}$ **b** $\frac{5}{9}$

40. b about £16.50 per barrel

41. a Frequency density: 0   2   12   23   21.4   9.6   0

      Frequency: 0   30   120   115   107   96   0

   b 32

42. a $\frac{7}{17}$ **b** $\frac{9}{19}$ **c** $\frac{26}{145}$ **d** $\frac{21}{55}$

43. b e.g. males have a wider range of earnings and a higher average than the females.

44. $\frac{1}{10}$

45. $\dfrac{12n + 15m}{27}$

46. a missing values: 70, 58, 60   **b** $\frac{x}{3}$

47. a 31   **b** $\frac{25}{72}$ **c** $\frac{7}{15}$

48. a 28

   b

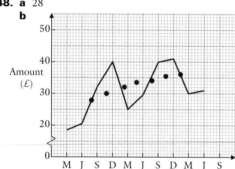

   c £44

49. a i 6

     ii e.g. median for first hole is lower than the median for the second hole, the interquartile range for the first hole is higher up the scale than that for the second hole.

   b $\frac{9}{200}$

50. a 2 black and 3 white (or any multiple of these)

   b 4 (or twice the first number of black balls)

51. a

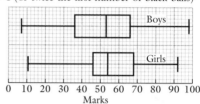

   b boys' marks have a greater range than the girls' and a lower median.

INDEX